Springer Proceedings in Physics 37

Springer Proceedings in Physics

Managing Editor: H. K. V. Lotsch

Volume 30 *Short-Wavelength Lasers and Their Applications*
Editor: C. Yamanaka

Volume 31 *Quantum String Theory*
Editors: N. Kawamoto and T. Kugo

Volume 32 *Universalities in Condensed Matter*
Editors: R. Jullien, L. Peliti, R. Rammal, and N. Boccara

Volume 33 *Computer Simulation Studies in Condensed Matter Physics: Recent Developments*
Editors: D. P. Landau, K. K. Mon, and H.-B. Schüttler

Volume 34 *Amorphous and Crystalline Silicon Carbide*
Editors: G. L. Harris and C. Y.-W. Yang

Volume 35 *Polycrystalline Semiconductors*
Editors: H. J. Möller, H. P. Strunk, and J. H. Werner

Volume 36 *Nonlinear Optics of Organics and Semiconductors*
Editor: T. Kobayashi

Volume 37 *Dynamics of Disordered Materials*
Editors: D. Richter, A. J. Dianoux, W. Petry, and J. Teixeira

Volumes 1 – 29 are listed on the back inside cover

Dynamics of Disordered Materials

Proceedings of the ILL Workshop
Grenoble, France, September 26-28, 1988

Editors: D. Richter, A. J. Dianoux, W. Petry, and J. Teixeira

With 202 Figures

Springer-Verlag Berlin Heidelberg New York
London Paris Tokyo

Dr. Dieter Richter
Dr. Albert J. Dianoux
Dr. Winfried Petry
Institut Laue-Langevin, Centre de Tri 156X, F-38042 Grenoble Cedex, France

Dr. José Teixeira
Laboratoire Léon Brillouin, CEN Saclay, F-91191 Gif sur Yvette Cedex, France

ISBN-13: 978-3-642-93421-6 e-ISBN-13: 978-3-642-93419-3
DOI: 10.1007/978-3-642-93419-3

Softcover reprint of the hardcover 1st edition 1989

Printing: Weihert-Druck GmbH, D-6100 Darmstaat
Binding: J. Schäffer GmbH & Co. KG., D-6718 Gruŋstadt
2154/3150-543210 – Printed on acid-free paper

Preface

This book contains the proceedings of a workshop on the "Dynamics of Disordered Materials" held at the Institut Laue-Langevin in Grenoble in September 1988. Containing 10 review articles and about 40 contributed papers, the book surveys recent developments in both theory and experiments on disordered materials in general and more specifically on glasses.

A large part of the book concerns the recently proposed mode-coupling approach to the behaviour of a viscous liquid around its glass transition. Because of its well-defined predictions for the time-dependent scattering function, this theory is at the origin of a large number of experiments using scattering techniques.

In the neighbourhood of the glass transition the relevant dynamics extend over a broad range in time scales and the application of quite different experimental techniques becomes essential. Many contributions report on experiments using dielectric relaxation, NMR or light scattering techniques. The majority of the papers use, however, neutron scattering techniques – time-of-flight, backscattering and spin echo – because they access a large dynamic range as well as the momentum dependence of the response function.

Most of the experiments presented in this book are in good agreement with the mode-coupling theory and are also analysed with reference to more classical approaches such as free volume theories using the Vogel-Fulcher law or the Kohlrausch rule to describe the structural relaxation near the glass transition. Some points remain to be clarified such as the significance of a critical temperature which is found to lie above the glass transition temperature. Its independent determination is certainly necessary in order to substantiate the fitted exponents. Furthermore, most of the materials undergoing a glass transition are molecular liquids, or polymers having intramolecular degrees of freedom which are not explicitly taken into account in the theory. This problem concerns particularly the so-called β-relaxation, i.e. the description of the glass dynamics when the diffusive motions are frozen. Amorphous metals and possibly colloidal hard-sphere liquids are probably the only exception to this situation. But at least on amorphous metals, experiments cannot be performed near their glass transition temperature.

One signature of disordered materials is the occurrence of an excess vibrational density of states at low frequencies. It has been the goal of considerable experimental and theoretical efforts to reach a microscopic understanding of this phenomenon. Some situations are presented where the density of states can be fully understood by taking into account the peculiarities of intra- and intermolecular motions. In this respect, important progress has been achieved in determining the density of states of a disordered material by low-energy inelastic neutron scattering, in particular on coherent scatterers.

Another approach to the dynamics of disordered materials is the fracton picture which was originally developed to describe the excitations of self-similar "fractal" objects. Several contributions are concerned with the dynamics of such fractals mainly studying the vibrational density of states. Some contributions discuss the application of the fracton concept to materials like glasses without a fractal structure in considering a connectivity length within the material as the main ingredient of the theory.

We believe that the dynamics of disordered materials will remain a challenge for both theoreticians and experimentalists. Because of their nature, such materials are extremely complex: their structure is between those of solids and liquids and their dynamics extends over more than 15 decades in time. In spite of this complexity and the large dynamic range, it is astonishing what similar behaviour is displayed by such diverse glasses as ionic, molecular, polymer or even biological glasses. It is a challenge for physicists and chemists to find the reason for this unique behaviour and certainly much work has to be done before a complete understanding of these materials can be achieved. We hope that this book, in describing the present situation, will provide help and incentive for further work in this exciting domain.

The editors would like to thank the contributors for promptly preparing their manuscripts. We express gratitude to the Institut Laue-Langevin and the Laboratoire Léon Brillouin for financial help. We would like to acknowledge P. Egelstaff for his critical summary and are grateful to B. Maier and his staff who helped to organize the workshop, in particular to Mrs Diana Dijoux as the workshop secretary.

Grenoble,
Paris
October 1988

A.J. Dianoux
W. Petry
D. Richter
J. Teixeira

Contents

Part III Low Frequency Excitations

Part IV Spin Glasses and Orientational Glasses

Part I

Introduction to Neutron Scattering

Inelastic Scattering of Neutrons as a Tool to Study the Glassy State

P.A. Egelstaff

Department of Physics, University of Guelph, Guelph, Ontario N1G2W1, Canada

Cold, thermal and slow neutrons have been used extensively to study the dynamics of condensed matter and it is reasonable to expect an equal or greater impact of these techniques in the case of the dynamics of glassy substances. But there are significant differences in this case compared to crystalline solids or to liquids for example, and the impact of these differences will be explored.

In many investigations models have been proposed to enable the data to be interpreted, and some of the more popular of the models will be reviewed. Both the successes and the limitations of the models will be discussed. This review will be concluded with an attempt to forecast some future directions for research in this field and their possible outcomes.

1. INTRODUCTION

Slow, thermal and cold neutrons have been used for many years to solve problems in condensed matter physics in physical chemistry, biology and engineering. Their usefulness has driven a continuous search for better neutron sources and better techniques for using neutrons. The study of glasses is now one of these driving forces.

Neutrons are a versatile and powerful tool for several reasons. First, neutron wavelengths can be chosen to be about equal to the spacing between atoms, or to the size of a defect or to another interesting feature and the corresponding neutron energy is usually comparable to the energy levels that determine the chemical and physical properties under study. Diffraction effects from neutrons scattered by nearby nuclei will show up structural effects, while the energy exchange between neutrons and the sample will be related to dynamical effects: and simultaneous analysis will provide significant new information. The scattering from an isolated nucleus is isotropic and elastic, so that there are no complications due to form factors, selection rules or competing cross sections.

Secondly, neutron cross sections are distributed virtually at random, both between isotopes of the same element and throughout the periodic table. Thus adjacent elements in the table can be separated, light elements can be studied in the presence of heavy elements, and most importantly the scattering properties of a sample can be altered without changing its chemistry. In addition the scattering by differently oriented, but otherwise identical, nuclei can be different (for example in the case of hydrogen this effect is very large). Because the true incoherent scattering from such random effects contains no structural information, it is possible to study single particle dynamical effects also.

This remarkable range of properties has led to the application of neutrons to many classes of problems, limited only by the suitability of the neutron sources and their associated neutron spectrometers. In many experiments it is possible

Springer Proceedings in Physics, Vol. 37 **Dynamics of Disordered Materials**
Editors: D. Richter · A.J. Dianoux · W. Petry · J. Teixeira

to measure the absolute cross section for an effect, the position and width of a peak and the dependence of scattered intensity on angle of scatter and/or on energy change of the neutrons during the scattering process. For most of these experiments the interpretation has been unique and quantitative in an absolute sense. At this meeting we hope to see how these virtues of neutron scattering can be exploited in the study of glasses.

One of our limitations will be in the instruments available today. Dynamic and structural phenomena in this field cover an enormous range. One illustration of this is seen through listing the significant time scales studied in neutron experiments. These extend from the lifetimes of metastable states (e.g. 45 days for glassy sulphur[1] at room temperature and less at elevated temperatures) and different viscous relaxation times, through the relaxation times near the liquid glass transition point (e.g. $\sim 10^{-9}$ sec for an ionic glass[2]), the low frequency vibrations in glasses, (e.g. $\sim 10^{-12}$ sec for silica[3]), the time relevant to molecular vibrations (e.g. $\sim 10^{-14}$ sec for silica[4]) and the time relevant to driving molecular breakup and chemical reactions ($\sim 10^{-16}$ sec). An equally extensive list could be given for length scales also. To cover these possibilities with adequate resolution (including angular resolution and angular range in addition to energy resolution) requires many different specialized spectrometers. Moreover severe requirements are placed on the sample environments; in order to cover the wide range of temperature, densities and chemical compositions of interest. And finally the scope for isotopic substitution experiments in this field is very large, but this work requires substantial effort and expense[5]. It is therefore not surprising if the application of the several neutron methods to this field is still in its infancy, in spite of about 23 years[6] of development.

In this paper we attempt to describe the current state of work on dynamical phenomena, and so it will be assumed that the cross sections for total scattering and elastic or quasi-elastic scattering are known. However in some cases these data are poorly known and in many cases are only moderately well known. For a program of inelastic work to be successful there should be a parallel program of extensive and accurate diffraction experiments. The general principles will be described in the next section, and this is followed by sections on a variety of neutron methods available today and some possible near term developments. No attempt will be made to review the literature, but selected experimental examples will be taken to illustrate the points under discussion.

2. THE GENERAL PRINCIPLES BEHIND NEUTRON SCATTERING DATA

A schematic diagram of an inelastic scattering experiment on an isotropic sample is given at figure 1. In practice the observed data must be corrected for backgrounds, container scattering, multiple scattering, absorption factors and for resolution effects before the differential cross section for single scattering is obtained. These steps have been explained elsewhere[7], and here we assume that the experiment has been designed[7] to make these corrections straightforward so that good quality data has been obtained. In addition we assume that adequate calibrations have been made to define ω and Q accurately, and to determine the absolute cross section scale accurately. In the limited time available for some experiments, not all of these points will have been covered properly and so in reading some publications the reader will be left to contribute to the interpretation.

The isotropy of the sample may be tested by repeating the experiment in figure 1 for several sample orientations, when the data should be unchanged. Also because each data point represents the accumulation of several hundreds (sometimes, several thousands) of neutron scattering events the data will be an

FIGURE 1

Schematic diagram of a neutron inelastic scattering experiment: θ is angle in the scattering plane and ϕ is an azimuthal angle. Frequently the monochromators and detectors are multiplexed, and for isotropic samples a wide range of ϕ angles are binned together. The various collimators have not been shown.

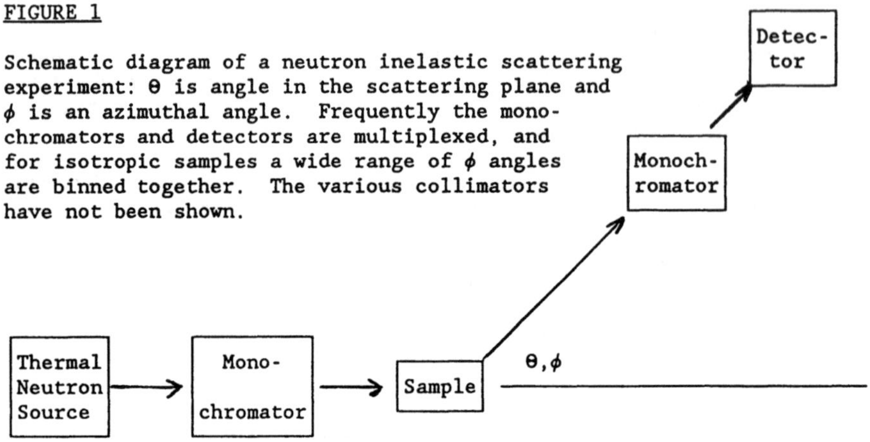

average over all the states of the sample accessible within this time frame, usually about 12 hours or so.

The important variables[8] in such a scattering experiment are the momentum ($\hbar \underline{Q}$) and energy ($\hbar\omega$) transfer, which are given by the conservation laws:-

$$\hbar \underline{Q} = \hbar(\underline{k}_o - \underline{k}) \quad ; \quad \hbar\omega = \frac{\hbar^2}{2m}(k_o^2 - k^2) \tag{1}$$

where $\underline{k}_o$ and $\underline{k}$ are incident and scattered wave vectors respectively, and m is the neutron mass. If the sample is isotropic, the direction of $\underline{Q}$ is unimportant so that the first of these equations may be replaced by:

$$Q^2 = k_o^2 + k^2 - 2kk_o \cos\theta \tag{2}$$

where θ is defined in figure 1. It is evident that the mass of the neutron restricts the range of these variables that is accessible. This restriction is seen most clearly by eliminating k from the second of equations (1) and equation (2); and then plotting ω v.s. Q for given incident energies and angles (θ) of scatter, as is done in many textbooks.

A simplifying assumption is usually made about the sample, namely that the isotopes of each element are randomly arranged allowing elemental scattering amplitudes to be used. Hence if the sample contains n elements, the differential cross section becomes:

$$\frac{d^2\sigma}{d\Omega d\omega} = \frac{k}{k_o} \sum_{i=1}^{n} \sum_{j=1}^{n} b_i^{coh} b_j^{coh} S_{coh}^{ij}(Q,\omega)$$

$$+ \frac{k}{k_o} \sum_{i=1}^{n} (b_i^{inc})^2 S_{inc}^{i}(Q,\omega) \tag{3}$$

where the b_i^{coh}, b_i^{inc} are the coherent and incoherent scattering amplitudes for the ith element respectively, and $S(Q,\omega)$ denotes a van Hove scattering function. Because of the isotropy of the sample these functions are even in Q, but are odd functions of ω. Ideally one varies the isotopic composition of several elements to vary b_i and b_j, so obtaining a set of simultaneous equations which may be solved to obtain all the individual S-functions appearing in equation (3). In current practice this is not done, and normally combinations of these functions are studied as described by Enderby and Gullidge[5]. This means that the final data is a sum of scattering amplitude weighted functions. Frequently however, either the coherent or the incoherent terms will dominate the total in equation (3), and this simplifies the theoretical discussion.

Each term in the sums in equation (3) is the Fourier transform of a space-time correlation function $G_{ij}(r,t)$, denoting the probability of finding atoms of elements i and j at a separation of $\underline{r}$ at time t:

$$S^{ij}_{coh}(Q,\omega) = \frac{1}{2\pi}\int e^{i(\underline{Q}.\underline{r}-\omega t)}\left[G_{ij}(\underline{r},t)-\rho_{ij}\right]d\underline{r}\,dt \tag{4a}$$

where ρ_{ij} is an average density. G(r,t) is sometimes referred to as a density-density correlation function. Thus the complete decomposition of equation (3) would make a wealth of detailed information available. But as indicated above, ingenuity is required to extract an interpretable piece from the total listed at equation (3). The structure factor $S^{ij}(Q)$ is often used in the discussion of data and simply related to the van Hove function by:-

$$S^{ij}(Q) = \int_{Q_{constant}} S^{ij}_{coh}(Q,\omega)d\omega \tag{4b}$$

and its transform is thus related to the space time function at t = 0.

When the sample is in an equilibrium state the asymmetry in ω is accounted for by the detailed balance factor, and for either coherent or incoherent scattering a symmetrical function can be defined by:

$$\tilde{S}^{ij}(Q,\omega) = e^{\hbar\omega/2k_BT}\, S^{ij}(Q,\omega) \tag{5}$$

It should be noted also, that the Fourier transform of an odd function is a complex function, so that in general $G_{ij}(r,t)$ will be complex. Its real and imaginary parts may be related[9] through the condition (5) in the equilibrium case only. In contrast the fact that S(Q,ω) is even in Q, yields real space time functions for cases where the asymmetry in ω is unimportant. The space-time functions, $G_{ij}(r,t)$, may be discussed in classical terms if:

$$\text{the measured time} \sim \frac{2\pi}{\omega} >> \frac{h}{k_BT} \quad ; \quad \text{and} \tag{6a}$$

$$(\text{the measured distances})^2 \sim \left(\frac{2\pi}{Q}\right)^2 >> \frac{h^2}{2M_i k_BT} \tag{6b}$$

where M_i is the mass of an atom of element i. By defining non-dimensional variables, these limits are expressed simply as:-

$$\alpha_i << 1 \quad ; \quad \beta << 1 \tag{7}$$

where $\alpha_i = \hbar^2Q^2 / 2M_i\, k_BT$ and $\beta = \hbar\omega/k_BT$.

The first order quantum correction on the time-scale is a term in $\hbar$, whereas the leading correction on the distance r-scale is $O(\hbar^2)$. Thus it is usual to discuss quantum corrections in terms of the detailed balance factor (equation 5) since this includes the first order correction[9]. However it does not include the second order correction properly.

When the sample is a stationary system (that is it is not evolving with time), a useful properly of Fourier transforms can be exploited. Namely the transform of $\omega^2 S(Q,\omega)$ is a velocity-velocity correlation function (or equally a current-current function), and the transform of $\omega^4 S(Q,\omega)$ is a force-force correlation function. In some cases this allows a simpler physical interpretation than by considering a density-density correlation function.

In the above a number of conditions have been specified, upon which some commonly used expressions depend. However in the study of glasses it is possible that the system may not be isotropic or may not be in thermal equilibrium or may be evolving with time. Although the study of $S(Q,\omega)$ under such conditions may serve as a useful experimental tool, it is worthwhile reexamining some of the above results to determine whether they are still applicable.

3. SPECTRAL DENSITIES

It is convenient to start with a definition of a spectral density which is quite general and so has a meaning for every physical system. The usual choice is to consider the Fourier transform of the velocity correlation function for similar atoms in a (classical) system, and define:

$$\frac{1}{\pi}\int_0^\infty \langle v(0)\, v(t)\rangle \,\mathrm{Cos}\,\omega t\, dt = z(\omega) \qquad (8)$$

where $v(t)$ is the velocity of an atom at time t. The initial mean square velocity $\langle v(0)^2\rangle = k_B T/M$ in a classical system, and hence the area of $z(\omega)$ is known.

This choice is convenient for several reasons. The function behaves well mathematically (in contrast to the single particle density for example), the limit as $\omega \to 0$ is proportional to the diffusion constant and for a system of harmonic oscillators $z(\omega)$ is equal to the usual density of states. Moreover this function is simply related[10] to the self or incoherent part of the van Hove scattering function, by:-

$$\lim_{Q\to 0} \left\{ \frac{\omega^2\, S_{inc}(Q,\omega)}{Q^2} \right\} \longrightarrow z(\omega) \qquad (9)$$

For systems that exhibit small quantum effects, a simple generalization of the above is possible[10]. If we redefine $z(\omega)$ via:-

$$\lim_{Q\to 0} \left\{ \frac{2k_B T}{\hbar}\, \omega \mathrm{Sinh}\left(\frac{\hbar\omega}{2k_B T}\right) \frac{\tilde{S}_{inc}(Q,\omega)}{Q^2} \right\} \longrightarrow z(\omega) \qquad (10)$$

the meaning of $z(\omega)$ is essentially unchanged* and its area can be shown to be $k_B T/M$ from the formula for the first ω moment of $S(Q,\omega)$.

---

* The usual Bose factor: $\dfrac{1}{\omega[(\exp \hbar\omega/k_B T)-1]} = \dfrac{\exp(-\hbar\omega/k_B T)}{2\,\omega\, \mathrm{Sinh}(\hbar\omega/2k_B T)}$,

gives the detailed balance factor and the pre-factor in equation (10).

In cases where the sample contains several elements and the data are not separated into individual $S_{inc}(Q,\omega)$ by isotopic substitution methods, a scattering amplitude weighted function is obtained namely:-

$$\bar{z}(\omega) = \frac{1}{\sum \left(b_i^{inc}\right)^2} \sum_i \left(b_i^{inc}\right)^2 c_i z_i(\omega) \tag{11}$$

where c_i is the concentration of element i.

In practice the incoherent cross section of many samples is a small part of the total. However the coherent term is the sum of two van Hove functions - a self term and an interference term, corresponding to single particle correlations and correlations between pairs of different atoms, respectively. The self term behaves in an identical way to the incoherent term, while the intensity from the interference term oscillates in sign as a function of Q. Thus an "incoherent" approximation is used in some cases, namely to assert that if data at constant ω are plotted as a function of Q, a smooth (usually monotonic) line through the data will be equal to the sum of all single particle effects. This approximation is better at higher Q, since the relative size of the interference term decreases as Q increases. Hence provided it can be shown that the physics of the sample allow the extrapolation to $Q = 0$ (in equation 10) to be used over a wide range of Q, it is possible to use equation (10) for smoothed coherent data starting from reasonably large Q. Also usually this approximation will be better at high ω (because the wavelength difference between the incident and scattered beams reduces interference effects): hence the extrapolation to $Q = 0$ can be used usually to find the high ω part of $z(\omega)$ in the case of coherent scatterers. Egelstaff and Schofield[11] discuss this method and describe a way of improving the extrapolation by an iterative calculation.

Either as a complementary or as an alternative method, theoretical models of the system may be used to calculate $S(Q,\omega)$ in terms of a density of states. In common with earlier authors Carpenter and Pelizzari[12] assumed that a glass could be treated as a harmonically vibrating system of atoms with fixed equilibrium positions, and that the system was in thermal equilibrium. They calculated the one and two phonon contributions to $S(Q,\omega)$ since, if both Q and T are low enough, these terms are the only important ones. A number of authors[13,14] have assumed that in addition to these assumptions, the incoherent approximation is valid and then the inelastic part of $S(Q,\omega)$ becomes:-

$$\omega \sinh\left(\frac{\hbar\omega}{k_B T}\right) \frac{\tilde{S}_{inel}(Q,\omega)}{Q^2} = \frac{\hbar}{2k_B T} \sum \frac{b_{ti}^2}{\sum b_{ti}^2} F_{one}^i(Q,\omega) c_i z_i(\omega) + O(Q^2) \tag{12}$$

where, for this case, $F_{one}^i(Q,\omega) = \exp(-2W_i)$ is a Debye-Waller factor and is independent of ω. Thus equation (11) is recovered at low Q where all $W_i \to 0$. The areas of the $z_i(\omega)$ are normalized to $k_B T/M_i$; some authors normalize to unity and then this factor should be included in F_{one}^i. In this equation $\tilde{S}_{inel}$ is the sum of coherent and incoherent terms and b_{ti} is the total scattering amplitude for element i. The b_{ti} are defined by the equation:-

$$4\pi b_{ti}^2 = \left(\frac{A_i + 1}{A_i}\right)^2 \sigma_{fa}^i \tag{12b}$$

where A_i is the atomic weight of element i, and σ_{fa}^i is its free atom scattering cross section.

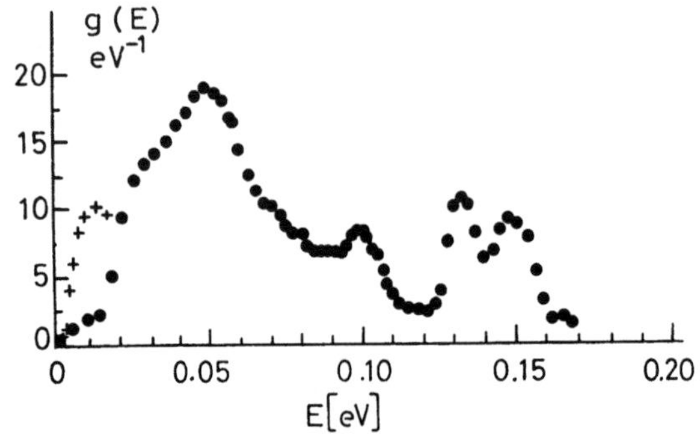

FIGURE 2

The measured[3,14] density of states for α-silica at room temperature. Circles are from ref. 14 and are reliable for hν > 15 meV; the crosses denote the low energy data of ref. 3. (E = hν)

Results for silica obtained by this method have been published by Leadbetter and Stringfellow[13] and by Carpenter and Price[14]. The latter are shown in figure 2, together with the low frequency data of Buchenau et al[3]. It can be seen that data over a relatively wide frequency range has been obtained. The use of equation (12) implies that the derived $z(\omega)$ would agree with the $z(\omega)$ obtained via the extrapolation to $Q = 0$ method. An inspection of the figures of Carpenter and Price[14] shows how this method could be effective over their range. These authors say that their data below 15 meV are only approximate, because of contamination from the wings of their elastic peaks.

At low frequencies a model for the coherent scattering is needed. The simplest approach of this kind is to redefine $F_{one}(Q,\omega)$ in equation (12). In an early model of this kind Egelstaff[15] treated the elastic structure factor as a continuous distribution of Bragg reflections within a generalized one phonon treatment of damped sound waves in liquids. As a simple example of this method we assume that the sample is monatomic and that all modes have the same velocity, and then his result would be:

$$F_{one}(Q,\omega) = \frac{e^{-2W}}{2qQ} \int_{Q-q}^{Q+q} K\, F_{elast}(K)\, dK \tag{13a}$$

where $q = \omega/c$ and c is the common mode velocity. The function $F_{elast}(Q)$ is defined by:

$$S(Q,\ \omega = 0) = F_{elast}(Q)\ \delta(\omega) \tag{13b}$$

and $S\ (Q,\ \omega = 0)$ is the structure factor for elastic scattering. It is evident that this integral involves both Q and ω, but if we assume that F^{i}_{elast} is nearly constant between the limits of integration, then it reduces to:-

$$F_{one}\ (Q,\omega) \simeq e^{-2W}\ F_{elast}(Q) \tag{13c}$$

As we have a monatomic sample, this simplification leads to a factorization of equation (12) into purely ω and purely Q dependent factors. Any approximation which says that the cross section is proportional to the product of a density of states factor and a model dependent structure factor, we shall call an uncoupled approximation.

Buchenau[16] developed the ideas of Carpenter and Pelizzari[12] in deriving more general expressions for $S(Q,\omega)$, and for incoherent scattering he specialized to the case of long wavelength modes and assumed that the amplitude of motion of different kinds of atoms is the same. Consequently the ω dependence of the $z_i(\omega)$ in equation (12) are the same, and the uncoupled approximation can be used. For coherent scatterers Buchenau et al[3] used similar assumptions, and within the framework of equation (12) their result may be written (since $M_i z_i(\omega)$ is independent of i, in this case):

$$\frac{\sum b_{ti}^2 F_{one}^i(Q,\omega)/M_i}{\sum b_{ti}^2} \equiv \left| \sum_j b_j \, e^{-W_j} \, e^{iQ.R_j} \frac{(Q.\epsilon_j)}{Q \, M_j^{1/2}} \right|^2 \left\{ \sum b_j^2 \right\}^{-1} \tag{14}$$

where $\underline{R}_i$ is the equilibrium position vector for the jth atom of the glass and $\underline{\epsilon}_j$ is a polarization vector. When $q \ll Q$ this expression yields a generalization of (13c) for phonon scattering in polyatomic samples. For other models, including rotational modes for example, new model dependent results[3] based on equation (14) will be obtained.

An interesting prediction follows from the use of the uncoupled approximation: namely that the Q dependence of the intensity at all frequencies is the same. This prediction may be tested by measuring the angular distribution as a function of energy transfer, and its shape should be independent of ω varying only by an amplitude factor. This experiment is easily accomplished on a time-of-flight spectrometer, which is essentially a constant ω machine. In the case of silica the data of Buchenau et al[3] verify this prediction at low frequencies, and so $F_{one}(Q)$ is then obtained from experimental data. Consequently $z(\omega)$ is readily evaluated, and was included in figure 2. These data differ from Carpenter and Price[14], but are thought to be more accurate due to the higher energy resolution used in this work. However in none of the experiments has the data been separated experimentally into silicon and oxygen functions, to check whether the methods are valid for each function.

Thus we may summarize the various methods of finding $z(\omega)$, as follows:

(i) general method: $Q \longrightarrow 0$ limit in equation (10) for incoherent scattering
(ii) first incoherent approximation: use of the general method for coherent data, after smoothing interference effects.
(iii) one-phonon, harmonic approximation: use of equation (12) for incoherent scattering
(iv) second incoherent approximation: use of equation (12) on smoothed coherent data
(v) one-phonon, harmonic approximation followed by the uncoupled approximation: use of equation (14) in equation (12) with a model for the modes of motion.

In the case of silica at room temperature, all the methods should give the same $z(\omega)$ within reasonable limits. However it is clear that if either very high precision results or high temperatures were required, these methods are unlikely to be successful except for the general method used on near-perfect data. Some understanding of the physics of the sample is thus required before deciding which methods to use in any given case.

4. LOW FREQUENCY MODES in GLASSES

The spectral density for a glass and the corresponding crystalline material are well known from specific heat data to differ significantly at low frequencies, and this has been confirmed in a number of neutron experiments. There has been speculation that this difference is related to modes that are characteristic of the glassy state, and thus it is concluded that they should be investigated in detail. Both coherent and incoherent neutron scattering experiments have been proposed. However in practice experiments on these modes have been done with coherently scattering material. For this reason it would be convenient if neutron scattering methods had been developed to the point where they overlapped with methods using Brillouin scattering of photons, so that spectral peaks related to single modes could be studied. Unfortunately this situation can be achieved only in special cases, such as dense gases[17]. For normal glasses the best opportunity for Brillouin experiments would be at the highest feasible temperatures where the intensity as $Q \rightarrow 0$ is large (i.e. $\rho k_B T \chi_S$; where χ_S is the adiabatic compressibility). But even this case is not yet practical. Thus at present the experimental study of low frequency modes in glasses has involved the low frequency part of the coherent $S(Q,\omega)$ at relatively high Q. Interference effects are expected to be important in such work, as the wavelength of the scattered neutrons is similar to that for the incident neutrons.

Egelstaff[6] pointed out that for silica near the first maximum the structure factor at 1.55 A^{-1}, the neutron spectra were dominated by an elastic component. No inelasticity was observed in the range $0.5 < Q < 2$ A^{-1} and in order to see significant inelastic intensity it was necessary to work with Q above 2 A^{-1}. The essentially elastic character of the first peak in the diffraction pattern of silica was made very clear in the work of Buchenau et al[3]. In addition they showed that the second peak (at Q = 3.1 $Å^{-1}$) is strongly inelastic (in the range 0.5 to 10 meV). This difference is striking, and should be correlated with the difference in the neutron and X-ray diffraction patterns. This comparison is made in figure 3. The 3.1 $Å^{-1}$ peak is seen by neutrons and not by X-rays, and since X-ray scattering is mainly due to silicon and neutron scattering mainly due to oxygen, these combined results suggest that the motion of the oxygen is the main contributor to the 3.1 $Å^{-1}$ peak, while diffraction by relatively stationary molecular centres is the main contribution to the 1.55 $Å^{-1}$ peak. In this case, the comparison of neutron and X-ray results is similar to the isotopic substitution method with neutrons alone.

In addition Buchenau et al[3] used several models with equation (14), and showed that the 3.1 $Å^{-1}$ peak in silica involved the rotational modes of a five tetrahedron unit which emphasizes the oxygen-oxygen features. These modes were important over the whole range of their neutron data, and they inferred that sound wave modes would contribute only at very low frequencies (~ 0.1 THz). On the other hand in amorphous germanium Buchenau et al[18] showed that the sound wave contribution was relatively more important. Effects due to a sound wave model and a bond bending model were studied: they concluded that the sound wave region extended up to 2 THz, while the bond bending region extended downwards to 0.8 THz with a considerable overlapping region. Additional data on germanium extending to higher Q and higher energy transfers has been reported by Maley et al[19], and therefore a fairly complete picture is available for both silica and germanium at room temperature. The specific heat may be calculated successfully from the neutron density of states, and Buchenau et al[3,18] conclude that it is not necessary to invoke "fractons" to account for the low frequency properties of either of these materials.

If the low frequency part of the spectrum could be measured accurately by a low Q method, it would be possible to determine the ω^2 region of $z(\omega)$ and study

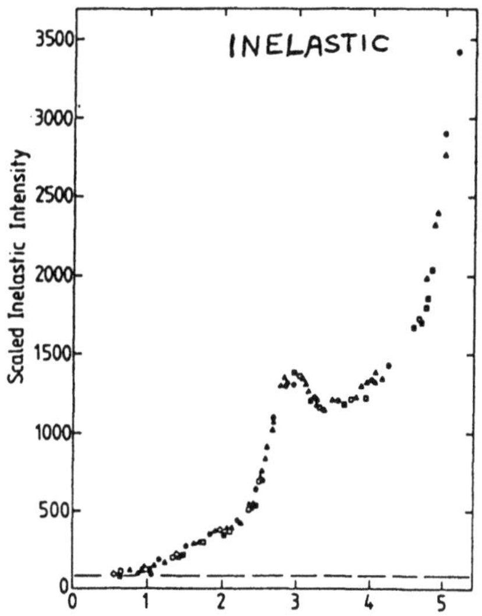

FIGURE 3

A comparison between the neutron diffraction[6] pattern, the X-ray diffraction pattern[6] and the neutron inelastic scattering intensity[3] as a function of Q for ν between 0.3 and 1.8 THz. Note the presence of the 3.1 $Å^{-1}$ peak in both sets of neutron data and its absence in the X-ray data.

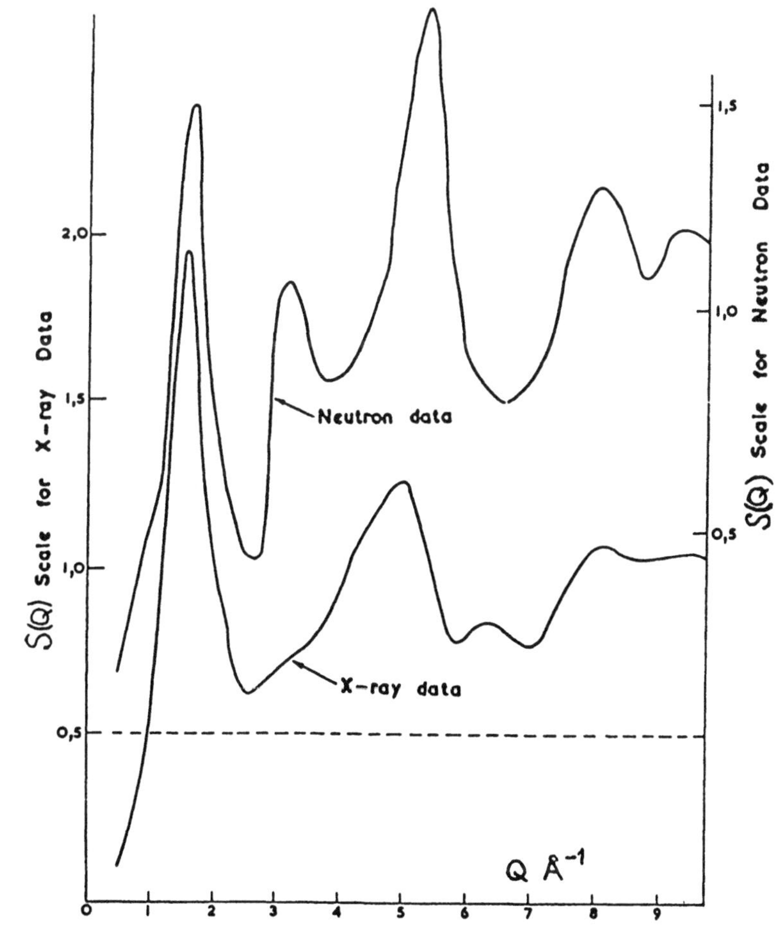

where deviations from this form occur. This should allow the suggestion[20] that deviations occur related to a crossover from sound waves to fractons. Also if low Q neutron Brillouin scattering work were possible, the frequency-wave number relation for the sample could be studied and the fracton picture tested further. Courtens et al[20] have used photon Brillouin scattering to obtain such data on silica aerogels, and have concluded that the crossover to fracton behaviour is at the upper limit of their data. It is possible that the scattered neutron intensity would be larger for aerogels than normal silica and hence be a better case for neutron Brillouin scattering experiments. However it should be remembered that to observe a Brillouin peak the momentum and energy transfer equations (1) require the neutron velocity to be significantly larger than the velocity of the mode.

Thus to conclude this section, we note that experiments at high Q near diffraction maxima have been analyzed in some detail and shown to fit a picture of molecular-type motions at high frequencies and sound modes at lower frequencies (~ 1 THz). This work has not provided evidence for "fractons" in dense glasses. On the other hand it would be worth attempting neutron Brillouin scattering experiments on samples that have exhibited fracton behaviour in other experiments. If successful these experiments would extend such data to higher frequencies and provide a more severe test of theoretical predictions.

5. MEMORY FUNCTIONS and the LIQUID-GLASS TRANSITION

Many theories of the dynamic structure factor for fluids are based on linear response theory and may be expressed in terms of a complex memory function $M(Q,\omega)$. For monatomic classical systems the most common expression is:

$$S(Q,\omega) = \frac{S(Q)}{\pi} \frac{\omega_0^2 \operatorname{Re}\{M(Q,\omega)\}}{|\omega_0^2 - i\omega M(Q,\omega) - \omega^2|^2} \tag{15}$$

and $\omega_0^2 = k_B T Q^2 / MS(Q)$. The real and imaginary parts of $M(Q,\omega)$ are related by a Kramers-Kronig relationship. Although this expression is theoretically 'exact', it is (in practice) a relationship between two functions $S(Q,\omega)$ and $M(Q,\omega)$: however the memory function is believed to behave in a simpler way than $S(Q,\omega)$ and to be more amenable to theoretical prediction. Thus to complete the theoretical description we need either a model or a theory for $M(Q,\omega)$. In many early treatments empirical models were used. For example the theoretical formulae given in previous sections could be re-cast into the form shown at equation (15), and then the particular function $M(Q,\omega)$ required to generate them would be evident. The starting equation (15) is a general equation, and in this section two examples have been chosen, one each at high and low frequencies, to illustrate the experimental features underlying the mode-coupling theory of glasses.

Egelstaff and Gläser[22] point out that, although in general a complex function $M(Q,\omega)$ cannot be fully determined from a limited range of a real function $S(Q,\omega)$, there are a number of special frequencies at which both the real and imaginary parts may be determined from experimental data. Their method has been applied[23] to the metallic glass $Mg_{70}Zn_{30}$, using the data of Suck et al[24,25] with $\omega > 5 \times 10^{12}$ rad/s. The memory function is broken into two parts, one similar to that found in liquids and another part, related to the elastic scattering intensity, having an infinite relaxation time. A relaxation time for the liquid-like part of ~ $1/4\omega_0$ was found over a Q range of 2 to 7 Å^{-1}, and the relative size of this term increased with Q. The ratio of the infinite relaxation time to the infinite limit $S(Q,\omega \to 0)$ is a finite quantity in this model, although it varies with Q. The observed variation was about a factor of 5 increase from 2 to 7 Å^{-1}. These facts about $Mg_{70}Zn_{30}$ may be characteristic ones for glasses, but more analysis of this kind is needed.

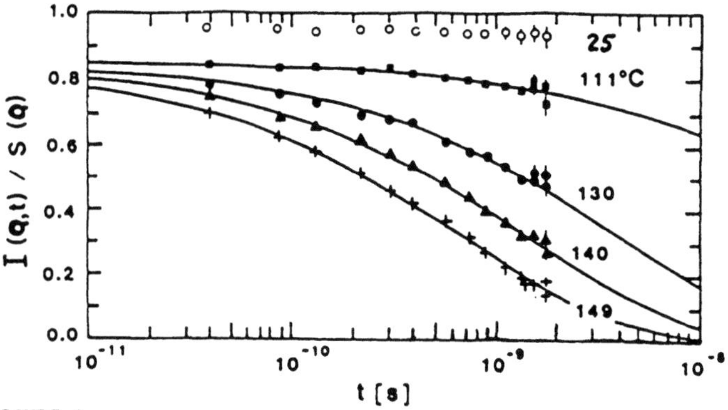

FIGURE 4

The data of Mezei et al[2] on the intermediate scattering function, I(Q,t) for the ionic glass $Ca_{40}K_{60}(NO_3)_{140}$ at Q = 1.7 $Å^{-1}$. The glass transition temperature is about 60 °C, and these data show the development of an 'infinite' relaxation time as the glass transition is approached. The lines are a fit of theoretical formulae to the data.

In the theory of the liquid-glass transition, (Leutheusser[26], Bengtzelius et al[27]) the memory function is divided in a similar way into a normal liquid-like part and a part which includes the very slow structural motions. The later leads to infinite relaxation times at the liquid-glass transition point. To calculate this part a "two-mode coupling approximation" is made. Mathematically[26,27] this approximation relates the slow part of the memory function to the dynamic structure factor through an integral equation. Solution of this equation coupled with equation (15) yields M(Q,ω). However it is difficult to make predictions for a particular glass which maybe compared to experimental data. Nevertheless the overall framework is consistent with the above observations on $Mg_{70}Zn_{30}$, and with very low frequency experiments by Mezei et al[2]. They used the spin-echo technique which measures the Fourier transform of S(Q,ω) with respect to ω, or I(Q,t). Some of their results for the ionic glass $Ca_{40}K_{60}(NO_3)_{140}$ in the liquid phase are shown in figure 4, for Q = 1.7 $Å^{-1}$ which is near the major peak in S(Q). The relaxation time increases dramatically as the transition temperature is approached, and is too long to be measured below T_G. These results will be discussed in later papers.

6. LARGER ENERGY TRANSFER EFFECTS

Interference effects extend far into the inelastic spectrum, and may be observed in a number of ways. For example oscillations in the ω-width of S(Q,ω) plotted as a function of Q (known as de Gennes oscillations) have been studied for many years. These oscillations are not exactly in phase with those in the static structure factor, and the phase difference is related to the atomic motions in the sample, and may be used for testing models. The de Gennes oscillations are not included within the uncoupled approximation for example.

Interference oscillations may be observed in several ways. For example Maley et al[19] plot the ratio of intensities for two modes in amorphous germanium (at 8.4 and 2.4 THz). They observed interference peaks as a function of Q, with a peak to valley ratio of ~ 2 near Q = 4 $Å^{-1}$. These results were used as a test of dynamical models proposed by Weaire and Alben[28]. Because the interference effects in glasses are stronger than in liquids (for example) and extend to higher values of Q, it is expected that such effects will become a useful test of theoretical models when accurate data become available.

The impulse approximation[29] has been used to interpret the scattering of high energy neutrons from matter, where the energy shifts are comparable to those calculated by assuming free atom recoil takes place. In this case the incident neutron energy is comparable to or larger than the molecular binding energies. The energy distribution of the scattered neutrons consists of a series of peaks, one for each nuclear mass present. Thus the elemental $S(Q,\omega)$ functions are separated in frequency and are easily identified. The width in frequency of each function is a measure of the momentum spread for atoms of that element, and therefore the light elements show the greatest shifts and widths and so are the easiest to study. With the advent of accelerator based neutron sources having sufficient intensity to work in this energy range[30], this experimental field may be a fruitful one for the study of glasses.

7. DISCUSSION

The outline given in previous sections has covered the principal ways in which neutron inelastic scattering is being used to advance the study of glasses. Detailed discussion has been avoided since that will be covered by other papers. In many cases however, the current limitations in this field became evident. From the experimental point of view there are two immediate challenges.

First the isotopic substitution method should be exploited to study $S^{ij}(Q,\omega)$ for individual elements and for cross correlation functions between elements. This is difficult partly because of the experimental time involved, but mainly because of the limited range of isotopes available. The data, however, should form a severe test of models. Secondly the gap between the inelastic scattering experiments being done in the neighbourhood of the first peak of $S(Q)$ and the photon Brillouin scattering experiments should be closed. If equipment can be built to cover this gap, we should be able to study the properties of modes in the range between the continuum region and the region dominated by structural effects.

The underlying assumptions behind many glass models should be tested more severely. For example, the assumption of thermal equilibrium may be tested at each frequency through the detailed balance factor. Careful measurements are required at given Q values for equal magnitudes in energy gain and energy loss. More detailed studies of the anisotropy of samples and how structure and dynamics varies with real time may be worthwhile in some cases.

Some of the models described earlier are worth more detailed investigation. For example the uncoupled approximation is valid for small values of the momenta of the modes and careful measurements should show how it fails gradually as ω is increased. A very high energy resolution experiment (e.g. by the spin echo method) of $S(Q,\omega)$ in the neighbourhood of the 1.55 $Å^{-1}$ diffraction peak of silica would reveal how this peak decays for $\omega \neq 0$, and hence test low frequency models severely. Another example of the need for careful measurements is the determination of the mean square amplitude of atoms vibrating about their equilibrium positions. This requires accurate diffraction data at very high Q using high energy incident neutrons, so as to obtain very high resolution in r-space. In principle all the partial structure factors are required. Improvements in all the above data would improve both the determinations of and the understanding of the density of states functions.

The direct measurement of memory functions is a topic in its infancy. This is worthwhile as a guide to future theoretical developments. It requires accurate measurements at a number of specified frequencies, and in addition to its value as a severe test of theoretical models, it would serve as a means of comparing glasses at a basic level.

Sophisticated studies of the liquid-glass transition, of different categories of glasses (inorganic, ionic, metallic, etc.) and of related materials such as gels, porous glasses etc. are all rapidly developing fields due to new ideas developed recently and the application of the techniques discussed here.

References

1. R. Winter, T. Bodensteiner, C. Szornel and P.A. Egelstaff. Proc. 4th Int. Conf. on Non-Cryst. Materials (Los Angeles 1988).
2. F. Mezei, W. Knaak and B. Farago, Phys. Rev. Letts. 58, 571, (1987).
3. U. Buchenau, M. Prager, N. Nücker, A.J. Dianoux, N. Ahmad and W.A. Phillips, Phys. Rev. B34, 5665, (1986).
4. J.M. Carpenter and D.L. Price. Phys. Rev. Letts. 54, 441, (1985).
5. J.E. Enderby and P.M.N. Gullidge. In "Neutron Scattering" Ed. D.L. Price and K. Sköld, Part B, p. 471 (Academic Press 1987).
6. P.A. Egelstaff. In "Physics of Non-Crystalline Solids" p. 127, Ed. J.A. Prins (North-Holland 1965) - also see review on p. 664.
7. P.A. Egelstaff. In "Neutron Scattering" Ed. D.L. Price and K. Sköld, Part B, p. 405 (Academic Press 1987).
8. L. van Hove, Phys. Rev. 95, 249, (1954).
9. P. Schofield, Phys. Rev. Letts. 4, 239, (1960).
10. P.A. Egelstaff, In "Inelastic Scattering of Neutrons", I.A.E.A. (Vienna), p. 38 (1961).
11. P.A. Egelstaff and P. Schofield, Nuc. Sci. Engng., 12, 260, (1962).
12. J.M. Carpenter and C.A. Pelizzari, Phys. Rev. B12, p. 2391 and 2397 (1975).
13a. A.J. Leadbetter and M.W. Stringfellow, In "Neutron Inelastic Scattering", I.A.E.A. (Vienna), 501, (1972).
13b. A.J. Leadbetter and A.C. Wright, J. Non-Cryst. Solids I, 23, (1972).
14. J.M. Carpenter and D.L. Price, Phys. Rev. Letts. 54, 441, (1985) and J. Non. Cryst. Solids 92, 153 (1987).
15. P.A. Egelstaff, AERE Report R4101 (1962).
16. U. Buchenau, Z. Phys. B. 58, 181 (1985).
17. P.A. Egelstaff, G. Kearley, J.-B. Suck and J.P. Youden, to be published.
18. U. Buchenau, M. Prager, W.A. Kamitakahara, H.R. Shanks and N. Nücker, Europhys. Letts. 6, 695 (1988).
19. N. Maley, J.S. Lannin and D. Price, Phys. Rev. Lett., 56, 1720 (1986).
20. S. Alexander and R. Orbach, J. Physique Lett. (Paris), 43, L625, (1982).
21. E. Courtens, J. Pelous, J. Phalippou, R. Vacher and T. Woignier, Phys. Rev. Letts., 58, 128 (1987).
22. P.A. Egelstaff and W. Gläser, Phys. Rev., 31, 3802 (1985).
23. P.A. Egelstaff, Z. Phys. Chem., 156, S311, (1988).
24. J.-B. Suck, H. Rudin, H.J. Güntherodt and H. Beck, J. Phys. C14, 2305 (1981).
25. J.-B. Suck, H. Rudin, H.J. Güntherodt and H. Beck, J. Phys. C, L1045 (1980).
26. E. Leutheusser, Phys. Rev. A29, 2765 (1984).
27. U. Bengtzelius, E. Götze and A. Sjölander, J. Phys. C17, 5915 (1984).
28. D. Weaire and R. Alben, J. Phys. C7, 489 (1974).
29. S.W. Lovesey. In "Theory of Neutron Scattering from Condensed Matter" Vol. 1. Sec.3.4, Oxford University Press (1984).
30. R. Siliver (Ed.) Proc. Workshop on High Energy Excitations in Cond. Matter. LANL Report LA-10227-C (1984).

Part II

Dynamics Around the Glass Transition

Neutron Scattering Anomalies Near the Glass Transition Singularity

L. Sjögren and W. Götze *

Institute of Theoretical Physics, Chalmers University of Technology, S-412 96 Göteborg, Sweden

The mode coupling theory of strongly coupled fluids predicts the existence of a singularity marking the transition from liquid to glass dynamics. Near the singularity the particle motion is governed by two critical time scales and the correlation functions are described by two scaling laws. As a result the neutron scattering cross section is predicted to exhibit glass transition anomalies near some critical temperature T_c located typically 30 to 150° K above the transformation temperature T_g. The most important features of those anomalies are reviewed. Suggestions for the analysis of scattering data are made.

Many aspects of the dynamics of simple classical liquids under normal conditions can be understood by proper extension of Boltzmann's kinetic equation or of Maxwell's viscoelastic theory. There are, however, some phenomena which cannot be treated along those lines like e.g. the long time anomalies or the non Lorentzian velocity spectra. Mode coupling theories have been invented to treat those delicate liquid state features. Quite generally, evaluating single mode and pair mode excitation spectra within one closed approximation scheme, a microscopic treatment of the liquid state dynamics was developed [1]. It was noted four years ago, that a special version of a mode coupling theory [2] and a schematic mathematical model reflecting some essentials of mode coupling equations [2, 3] exhibit a dynamical phase transition with features , which were in some respects quite similar to those discussed since long for liquid glass transitions. The original observations were corroborated by recent research.The theory treats the liquid glass transition as a new paradigm of condensed matter dynamics; its essence is the existence of a subtle singularity causing anomalies on microscopic length scales and mesoscopic time scales. Neutron scattering is the most advanced experimental tool to explore the space time region relevant for the new theory. Indeed, the recently published experimental data [4, 5, 6] demonstrate, that neutron scattering can reveal several new things about the old liquid glass transition problem. Therefore we shall give in this article a summary of those theoretical results, which might be of relevance for the interpretation of neutron scattering data performed for simple undercooled glass forming liquids.

---

* On sabbatical leave from: Physik-Departement der Technischen Universität München, D-8046 GARCHING, FRG and Max-Planck-Institut für Physik und Astrophysik, D-8000 MÜNCHEN, FRG

Springer Proceedings in Physics, Vol. 37 **Dynamics of Disordered Materials**
Editors: D. Richter · A.J. Dianoux · W. Petry · J. Teixeira

The central quantity of the mode coupling theory of liquids is the correlation function or correlator formed with density fluctuations ρ_q for wave vector **q**, the normalized coherent intermediate scattering law

$$\Phi_q(t) = < \rho_q^*(t)\, \rho_q > / S_q \ . \qquad (1a)$$

$S_q = < |\rho_q|^2 >$ denotes the static structure factor. We will need also the Laplace transform for frequency ω: $\Phi q(\omega) = LT[\ \Phi_q(t)\ ](\omega) = \Phi_q'(\omega) + i\Phi_q''(\omega)$. Its imaginary part, the density excitation spectrum, determines the dynamical structure factor

$$S(q\omega) = \Phi_q''(\omega)\, S_q / \pi \ . \qquad (1b)$$

Let us note also the dynamical compressibility $\chi_q(\omega) = \chi_q'(\omega) + i\chi_q''(\omega)$, the ratio of density fluctuations $<\rho_q>$ and small external fields creating this fluctuation; both quantities varying with wave vector q and frequency ω. One gets for temperature T:

$$\chi_q(\omega) = [\ \omega\Phi_q(\omega) + 1\ ]\, S_q / (k_B T) \ . \qquad (1c)$$

The mentioned correlation functions are expressed in terms of the known characteristic oscillator frequency of liquids $\Omega_q = (q^2v^2/S_q)^{1/2}$, v denoting the thermal velocity, and a current relaxation kernel $M_q(\omega)$. The latter has the same analytic properties as the correlator. The expression looks most simple if formulated for the compressibility:

$$\chi_q(\omega) = -(q^2v^2/k_B T) / [\ \omega^2 - \Omega_q^2 + \omega\Omega_q^2 M_q(\omega)\ ] \ . \qquad (2a)$$

Obviously, $M_q(\omega)$ is a generalization of an oscillator damping constant. It can also be viewed as analogue of a phonon self energy or polarization operator as considered in the theory of anharmonic crystals. For $q \to 0$, $\omega \to 0$, (2a) reduces to a Navier-Stokes equations result and therefore $M_q(\omega)$ can be interpreted also as a generalization of a longitudinal sound damping constant. The kernel is written as a sum of two contributions $M_q(\omega) = m_q^o(\omega) + m_q(\omega)$. The first one varies smoothly with frequency and control parameters like temperature T. If we concentrate on frequencies ω smaller than the typical microscopic ones for liquids, it can be treated as a friction constant: $m_q^o(\omega) = i\nu_q$, $\omega \leq 10^{12}\ s^{-1}$. The frequencies Ω_q, ν_q do not show any drastic temperature variation in the regime under discussion. The non trivial point of the theory is the mode coupling expression for the second contribution:

$$m_q(t) = \sum_{kp} V(q;kp)\, \Phi_k(t)\, \Phi_p(t) \ . \tag{2b}$$

The coupling coefficients V are given in terms of static two-and three-particle correlation functions, where the latter are expressed in terms of the former by some factorization approximation. The formulated equations for $\Phi_q(t)$ are closed; they represent the most simple version of the theory [2]. The equations ignore density fluctuations due to activated hopping processes; we will come back to this problem later. It is an essential point of the theory that S_q, the total coherent scattering cross section for neutrons with momentum transfer q, is anticipated to be a smooth function of wave vector q and temperature T. This implies that there occurs neither a structural phase transition nor nucleation. Notice, that these statements should not be considered as a result of the theory but rather as restrictive assumptions about the experimental situation to be considered. Let us add two comments. First, the structure factor S_q exhibits a pronounced peak at some wave vector q_o, reflecting the strong correlations of the particle positions for microscopic distances. This peak implies, that the by far most important contributions to (2b) come from the shell $k \sim p \sim q_o$. In this sense the theory deals primarily with microscopic length scales [2]. In the long wave length limit $q \to 0$ the theory reproduces hydrodynamics. The relaxation kernel tends to a transport coefficient in this case. The $q \sim q_o$ anomalies then lead via (2b) to anomalies of the kernel for $q \to 0$. Hence anomalies of transport coefficients and elastic moduli appear within the theory, but they are corrolaries rather than the essence of the discussion. Second, shear excitations or tagged particle correlators do not enter the formulated basic equations of motion (2). There are further equations for these quantities [2], which yield in particulaar viscosity and diffusivity. However, these equations need $\Phi_q(t)$ as input, and in this way the anomalies for viscosity and the like are consequences to be understood from those exhibited by the density fluctuations.

If the temperature is shifted through some critical value T_c equations (2) bring out a sharp transition from ergodic to non ergodic density motion. Remember, that the structure factor S_q and hence all parameters Ω_q, V(q;kp) vary smoothly for $T \sim T_c$. The transition is not connected with anomalies of the free energy and therefore it can be called a dynamical phase transition. More explicitly, one finds density fluctuations to relax to zero on the high temperature side:

$$\Phi_q(t \to \infty) = 0 \ , \ T > T_c \ . \tag{3a}$$

The corresponding density excitation spectrum is continuous:

$$S(q,\omega) = \text{cont.} \;, \; T > T_c \;. \tag{3b}$$

One expects this result for liquids, of course. But let us point out, that the continuous spectrum is obtained even if the stochastic damping term is ignored $\nu_q = 0$. In this case the continuum is produced exclusively by non linearities. It is produced predominantly by non linear effects under realistic strong coupling conditions $T \sim T_c$, $\nu_q \sim \Omega_q$. On the low temperature side one finds density fluctuations to arrest spontaneously

$$\Phi_q(t \to \infty) = f_q > 0 \;, \; T < T_c \;. \tag{3c}$$

The spectrum exhibits an elastic peak on top of a continuum

$$S(q,\omega) = S_q f_q \delta(\omega) + \text{cont.} \;, \; T < T_c \;. \tag{3d}$$

Quantity f_q, the non ergodicity parameter or glass form factor, is the Debye-Waller factor of the amorphous solid structure obtained for $T \leq T_c$. It is a continuous positive function of wave vector q. If one approaches the critical temperature T_c from the strong coupling side, the form factor approaches a non zero limit [2]:

$$f_q \to f_q{}^c > 0 \;, \; T \to T_c - 0 \;. \tag{3e}$$

The appearance of an ideal glass phase for $T < T_c$ and an ideal transition for $T = T_c$ is a result of our neglecting activated processes. Including these processes the elastic peak is broadened and the transition is smoothened to a continuous cross over [7] (see the later discussion in connection with Fig. 9). We will stick to describing the idealized case for reasons of simplicity. To avoid confusion let us mention already here, that the critical temperature T_c is located within the supercooled liquid region, typically about 30-150°K above the calorimetric glass transition temperature T_g.

The incoherent scattering law $S^s(q,\omega)$ exhibits a similar ergodic to non ergodic transition at T_c [2]. The corresponding form factor f_q^s, the Mössbauer-Lamb factor of the solid, is a bell shaped function of wave vector q. The small q variation determines the mean squared particle displacement or localization length r_0:

$$f^s_{q \to 0} = 1 - (r_0 q)^2 \;. \tag{3f}$$

At the critical point, r_0 increases to a finite value:

$$r_0 \to r_0^c < \infty \ , \ T \to T_c - 0 \quad . \tag{3g}$$

For the hard sphere model [2] and for Lennard-Jones systems [8] r_0^c was found to be the value expected from the classical Lindemann melting criterion.

The characterization of the glass as a non ergodic state was implied in many previous discussions. Our point is twofold. First, we do not postulate some non ergodicity but show how it follows from the microscopic equations of motion for strongly interacting particles. In particular we can predict some general properties of f_q and evaluate it for simple models. Second, we find that part of the spectrum, which is denoted by "cont." in equations (3b,3d), to exhibit a subtle low frequency singularity, which depends sensitively on the separation parameter

$$\varepsilon = (T_c - T)/T_c \ . \tag{4}$$

It is this singularity, which we have proposed to call the glass transition singularity [7]; its characterization is the main issue of the theory.

Let us consider the form factors in detail, since they are quantities, which can be measured as (quasi) elastic peak intensity in a neutron scattering experiment. For a classical harmonic system the form factors would vary with temperature like:

$$\log[f_q] = a_q - b_q T \ , \ \text{harm. system} \ . \tag{5a}$$

A similar relation holds for f_q^s. In deviation from such a regular variation the theory predicts a much more rapid critical decrease of $f_q^{(s)}$ if T increases towards T_c [2]. Near T_c one finds the asymptotic law [9]

$$f_q = f_q^c + \begin{cases} h_q \sqrt{\varepsilon} + O(\varepsilon) \ ; \ T \to T_c - 0 \ , & (5b) \\ O(\varepsilon) \ ; \ T \to T_c + 0 \ . & (5c) \end{cases}$$

More explicitly, for $T > T_c$ the area under the quasielastic line, the α peak, varies smoothly with temperature but for $T < T_c$ it varies critically according to a $\sqrt{\varepsilon}$ law. Other quantities like f_q^s, r_o, elastic or dielectric moduli exhibit similar variations [9], but these differ in the respective values corresponding to f_q and h_q. Measurements of $h_q > 0$ and $h_q^s > 0$ would be very relevant, since these quantities rule the intensity of von Schweidler relaxation and β relaxation, to be discussed below. Experimental verification of the $\sqrt{T_c - T}$ laws, (5b), would be the most convincing way to identify the critical temperature T_c from the low temperature side. The first quantitative result in this sense was achieved for polybutadien [5] by determining r_0 via incoherent scattering for $T < T_c$. The coherent scattering results for Ca-K-NO_3 [4] confirm the predictions (5b,5c) for $T < T_c$ as well as for $T > T_c$ qualitatively.

The first measurements of the critical decrease of f_q were done by CHAMPENEY and SEDGWICK [10], who performed Rayleigh scattering of Mössbauer radiation for a whole series of simple organic glass formers. Since this important piece of work was not acknowledged adequately in the recent discussions we reproduce as representative example their result for di-ethyl-phthalate as Fig 1 and add some comment. Plotted is actually a counting rate N semi logarithmically. Let us assume for simplicity that the α line is Lorentzian of width $1/\tau$. Similarly we assume the resolution curve of the Mössbauer experiment to be Lorentzian of width γ, where γ essentially is twice of the reciprocal nuclear life time. Then one gets

$$N = N_0 \begin{cases} f_q, & \text{if } \gamma\tau \gg 1 \quad (6a) \\ \gamma\tau f_q, & \text{if } \gamma\tau \ll 1 \quad . \quad (6b) \end{cases}$$

The authors [10] interpreted their results on the basis of equation (6b) anticipating f to decrease according to (5a), and attributing the observed decrease of N to a strong decrease of τ. This interpretation is obviously incorrect. An anomalous decrease of f_q is detected in Fig. 1 already for temperatures close to the calorimetric glass transition temperature T_g. For $T \sim T_g$ one knows τ to be of order 10-1000 seconds. Since $\gamma \sim (10^7\ s^{-1})$, one

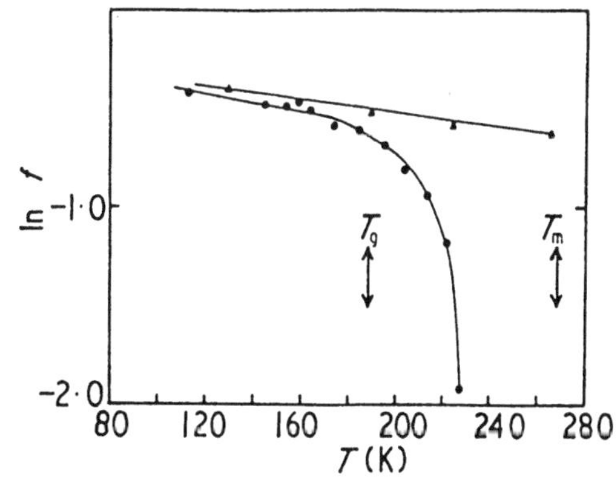

Fig.1 as discussed in text . Taken from ref. [10].

can safely use (6a) for $T \sim T_g$. Hence the remarkable temperature variation of f_q observed for $T < T_g+20°$ say, cannot be attributed to τ variations; rather it signalizes a critical softening of the glass structure. A fit to (5b) would be possible; however for temperatures near the expected T_c (6a) becomes unvalid. For $T \sim T_c$ one finds $\eta \sim 10^3$ poise and hence $\tau \sim 10^{-7} \sim 10^{-9}$ seconds; and this implies $\gamma\tau \sim 10^0 \sim 10^{-2}$. If one crosses over from (6a) to (6b) the predicted cusp in (5b,c) is smeared out, and the constant predicted for N in (5c) is replaced by a function decreasing strongly with increasing T. A quantitative analysis of the data in Fig. 1 is not possible then, since the one experiment cannot fix two unknowns f and τ. Similar remarks apply to the neutron scattering data for f_q published for 3-α-naphtyl-benzene (TαNB) [6]. We do not see a justification to identify N/N_0 with f in the low f region nor do we see one for the power law fitting procedure used in reference [6] (note a further comment below in connection with Fig. 7).

In order to evaluate the wave vector dependence of the form factors one must discuss quantitatively internal structure of molecules, distinguish between one component systems and mixtures, understand the difference between Coulomb and van der Waals forces.This is not possible at present and results are available only for some models: the hard sphere system [2] Fig. 2, Lennard-Jones systems [8] and hard sphere mixtures [11]. The full curves for f_q and $f_q{}^s$ in Fig. 2 refer to the critical point; the corresponding structure factor is shown as curve B. The dashed curves refer to $\varepsilon \approx 0.03$, the corresponding structure factor is shown as curve A. The critical amplitudes h_q and h_q^s are obtained as the difference between dashed and full curves. The compressibility of the system is S_q/k_BT. Hence it is easy to compress the fluid if S_q is large and density fluctuations arrest easily for such wave vectors. Similarly, a small S_q makes it difficult to induce density fluctuations, in particular spontaneous ones. So f_q is large at the maximum position q_o of the structure factor and it oscillates in phase with S_q. For large q coherence gets lost and f_q approaches the one particle distribution $f_q{}^s$. It is more difficult to break the well arrested structure at q_o by temperature fluctuations than at such q, where it is difficult to achieve arrest anyway. So h_q exhibits a minimum at q_o and oscillates opposite in phase to S_q. This obvious explanation [2] makes it evident that Fig. 2 should present also the qualitative features of realistic systems. Since Ca-K-NO_3 [4], polymers [5] or organic liquids [6] are much softer structures than the hard sphere system, one expects the mentioned oscillations to be smaller for those systems than the ones shown in Fig. 2.

Let us now turn to the true dynamics. The low frequency spectra near the critical point are governed by two characteristic frequencies denoted by ω_ε and $\omega_\varepsilon' = 1/\tau$. Both approach zero if T approaches T_c ; they describe slowing down of the dynamics. The ratio $\omega_\varepsilon'/\omega_\varepsilon$ also tends to zero, so that there

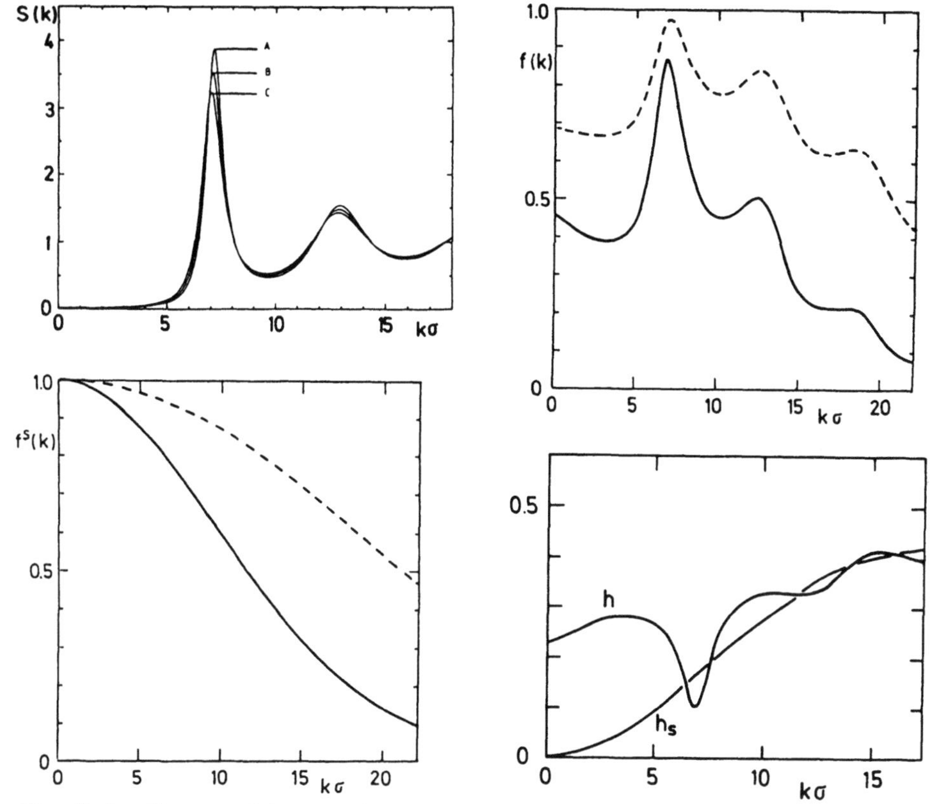

Fig. 2 As discussed in text. Taken from ref.[2].

appear two regimes of low frequency dynamics, which are separated the more the smaller $|T - T_c|$.

$$\omega_\varepsilon \to 0\,,\; 1/\tau \to 0\,,\; \omega_\varepsilon\tau \to \infty \;\text{ if } \varepsilon \to 0\;. \tag{7}$$

On scale ω_ε the dynamics is described by a first scaling law [9]:

$$\Phi_q(t) = f_q^{\,c} + h_q c_\varepsilon f_\pm(t\omega_\varepsilon)\;,\; c_\varepsilon = c_0\sqrt{|\varepsilon|}\;,\; \varepsilon \gtrless 0\;. \tag{8a}$$

Similar formulas hold for other quantities like incoherent scattering law, elastic moduli etc. $f_q^{\,c}$ and h_q are the same as introduced above in (5b). The master functions $f_\pm$ are independent of the separation parameter ε. They are the same for all quantities. In particular they are independent of wave-vector q and agree for coherent and incoherent correlations [9]. Let us explain the asymptotic character of (8a) more explicitly for the compressibility spectrum $\chi_q''(\omega) \propto$

$\omega S(q,\omega)$. It depends sensitively on frequency ω and temperature T. But if it is considered in units of c_ε as function of $\hat{\omega} = \omega/\omega_\varepsilon$ it becomes ε independent for $T \rightarrow T_c$:

$$\lim_{\varepsilon \rightarrow \pm 0} \hat{\chi}_q''(\hat{\omega}) = h_q \hat{\omega} f_\pm''(\hat{\omega}) \ , \ \text{where} \ \hat{\chi}_q''(\hat{\omega}) = \chi_q''(\hat{\omega}\omega_\varepsilon) / [c_\varepsilon S_q / k_B T] \ . \tag{8b}$$

The range of rescaled frequencies $\hat{\omega}$, where the rescaled function $\hat{\chi}_q''$ becomes ε independent expands if the critical point is approached. This is illustrated by a series of curves in Fig.3, obtained for decreasing ε as solution of some model equations [12]. The preceding formulae are specified by a so-called exponent parameter λ. It can be evaluated from the decay vertex V(q;kp) in (2b) [9]. But since the latter depends on all microscopic details it cannot be predicted for realistic situations at present. Parameter λ fixes both masterfunctions $f_\pm(\hat{\omega})$, in particular it determines two critical exponents a and b [13]:

$$\Gamma^2(1-a)/\Gamma(1-2a) = \Gamma^2(1+b)/\Gamma(1+2b) = \lambda \ ; \ 0 < a < 1/2 \ , \ 0 < b < 1 \ . \tag{9}$$

Exponent a enters the first scale and it determines the so called critical spectrum:

$$\omega_\varepsilon = \Omega \, |\varepsilon|^{1/2a} \ , \tag{10a}$$

$$\hat{\omega} \, f_\pm''(\hat{\omega} \gg 1) \propto A \, \hat{\omega}^a \ . \tag{10b}$$

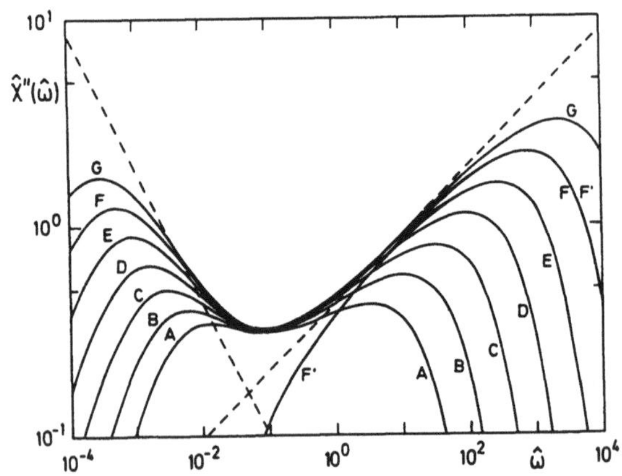

Fig.3 Rescaled susceptibility versus rescaled frequency. Taken from ref. [12]

These results imply power law decay of the intermediate scattering law

$$S_q(t\omega_\varepsilon \ll 1) / (\pi S_q) = f_q^c + Ah_q / t^a \ , \qquad (10c)$$

or power law increase of the scattering cross section

$$S(q,\omega \gg \omega_\varepsilon) / S_q \propto Ah_q / \omega^{1-a} \ . \qquad (10d)$$

This law is shown in Fig. 3 as the increasing dashed line. Within the time or frequency intervals specified, the right hand side of equations (10c,d) are independent of temperature; in particular the results do not discriminate between high and low temperature side of T_c. Exponent b enters the second scale and it determines another power law contribution to the spectrum

$$\omega'_\varepsilon = |\varepsilon|^{1/2b}\, \omega_\varepsilon \ , \qquad (11a)$$

$$\hat{\omega}\, f''_-(\hat{\omega} \ll 1) \propto B / \hat{\omega}^b \ . \qquad (11b)$$

These results imply another power law decay for correlations:

$$S_q(\omega_\varepsilon^{-1} \ll t \ll \omega_\varepsilon'^{-1}) / (\pi S_q) = f_q^c - h_q (t/\tau)^b \qquad (11c)$$

or another power law contribution to the cross section

$$S(q,\omega'_\varepsilon \ll \omega \ll \omega_\varepsilon) / S_q \propto \tau B h_q / (\omega\tau)^{1+b} \ . \qquad (11d)$$

This law is shown as the decreasing dashed line in Fig. 3. Notice, that the right hand side of (11c,d), as opposed to that of (10c,d), depends sensitively on temperature for $T \to T_c+$, since τ varies so rapidly in this region. Formulae (7,8a,10) were derived first for a schematic model [2,3]. The latter could not deal with wave vector dependencies, and it treated the non-generic case $\lambda = 1/2$ implying a trivial exponent b=1. Formula (11c) is the von Schweidler law, whose relevance for glass dynamics was discussed earlier by JONSCHER [14].

An experimental test of the scaling law (8b) is done most conveniently by studying double logarithmic plots of susceptibility spectra as function of frequency. Curves referring to different temperatures, T_1 and T_2 say, are related by parallel shifts. The shift parallel to the abscissa is $\Delta_x = \log[\omega_1/\omega_2]$

and the one parallel to the ordinate is $\Delta_y = \log[c_1/c_2]$. Redoing the shift the curves should coincide as shown in fig. 3.

Since there is a crossover from the von Schweidler law (11b) to the critical law (10b) the susceptibility spectra exhibit a minimum $\chi''(\omega_{min})$ at some frequency $\omega_{min} \propto \omega_\varepsilon$. The factorization property exhibited by (8b) implies, that the minimum position is the same for all susceptibilities. In particular it does not depend on wave-vector q and it is the same for coherent or incoherent scattering. This latter point is illustrated in Fig. 4; the full curve exhibits an incoherent spectrum [15] and the dashed one the coherent spectrum [12] for the same $\varepsilon \approx -0.0031$, evaluated for some schematic model. The curves differ for small as well as for large frequencies in many respects, but the minimum position is the same.

From (8a,b) one concludes, that the minimal spectral intensity decreases according to a $\sqrt{\varepsilon}$-law, if the critical point is approached:

$$\chi_q''(\omega_{min}) = Ch_q \sqrt{|\varepsilon|} + O(\varepsilon) \ , \ T \to T_c+0 \ . \tag{12}$$

Detection of this $\sqrt{|\varepsilon|}$ law would be the most convincing way to locate T_c from the high temperature side. The most convincing test on the existence of a critical temperature T_c would be the detection of the $\sqrt{T_c-T}$ decrease of f_q for $T < T_c$ and of the $\sqrt{T-T_c}$ decrease of $\chi_q''(\omega_{min})$ for $T > T_c$.

The same critical amplitude h_q enters (5b) and (12). Measuring the q dependence of the critical contribution to the Debye-Waller factor and of the minimum of $\omega S(q,\omega)$ would provide a test of the predicted connection

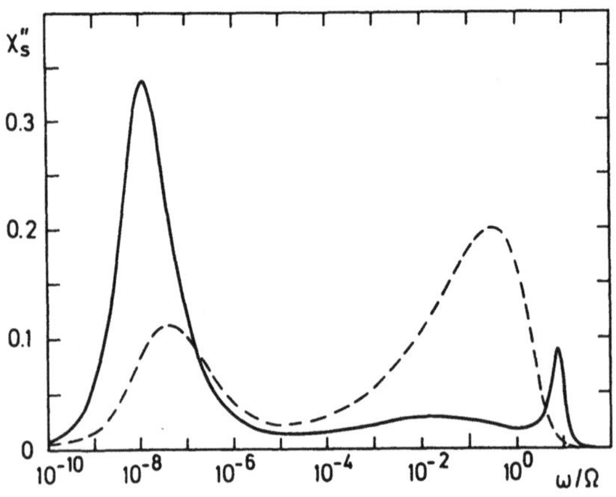

Fig. 4. Coherent and incoherent susceptibilities. Taken from refs. [12,15].

between a $T < T_c$ property with a $T > T_c$ property. The relevant q region can be judged from Fig. 2.

Checking of power law predictions and determining of exponents is most delicate and should be done only after the much more simple properties discussed above are verified. Let us consider some possibilities. Because of (8a,10a) the curve shifts of the susceptibility plots mentioned above are given by $\Delta_x = \frac{1}{2a} \log[\varepsilon_1/\varepsilon_2]$ and $\Delta_y = \frac{1}{2} \log[\varepsilon_1/\varepsilon_2]$. Hence the minimum of the various curves are predicted to fall on a straight line, $\Delta_y = a\Delta_x$. Verification of this property tests the prediction that the ratio of the scales c_ε and ω_ε follows a power law. The slope of the line is the exponent a as it enters the first scale in (10a). It is remarkable, that a critical exponent can be determined in this way without knowing the value of T_c. The wings of the susceptibility plots specified should be straight lines as indicated in Fig. 3 in dashed. The slopes are the exponents a and b respectively. Corrections to scaling can be big and their origin is also understood, but this point will not be discussed here. In Fig. 3 one clearly notices the approach of the curves towards the critical asymptote, (10b). On the other hand, ε is still so large that an unbiased fit would yield too small a value for the von Schweidler exponent b, (11b). If h_q is big, the critical spectrum is well developed and exponent a can be deduced more easily from the data. So it is advisable to work with a wave-vector off the peak position q_o if a shall be measured, see Fig. 2. The problem is reversed with the von Schweidler law. To determine b one chooses better a small value of h_q. This is illustrated in Fig. 5, reproducing BENGTZELIUS' results for a Lennard-Jones model [16]. The full curves refer to q_o and the large frequency wings of those curves obey the von Schweidler law with an exponent b related to the known λ via (9) [16]. The dashed curve is taken for a $q > q_o$ near the structure factor minimum. Deviations from scaling are so large that an unbiased fit to (11b) would yield too small a value for b.

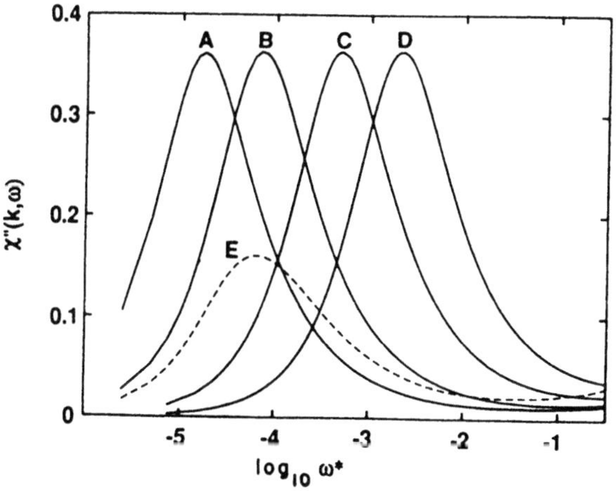

Fig. 5 Susceptibility spectra for a Lennard-Jones system. Taken from ref. [16]

Formulae (8,10) have similarity to results discussed in percolation theory. On the other hand GOLDSTEIN [17] has given arguments why percolation dynamics should be the genuine origin for the often observed β relaxation resonances. For these reasons it was proposed to call the whole dynamic regime, described by the first scaling law (8a), β relaxation. For large frequencies there must be a cut-off of the sublinear critical susceptibility increase (10b), and this produces a β relaxation peak at some frequency $\omega_\beta > \omega_\varepsilon$. For the model used for Fig. 4 the β peak for the coherent motion, shown by the dashed curve, is located in the region of microscopic oscillations. Similarly, the Lennard-Jones systems does not exhibit a β resonance in χ'' separated from the microscopic excitation band [16]. On the other hand, the tagged particle motion, whose spectrum is shown in Fig. 4 by the full curve, exhibits a β peak located about two decades below the microscopic resonance. The β peak does not show a critical temperature variation, in particular it remains non zero for $T = T_c$. This is shown in Fig. 6, where ω_β is plotted as function of ε together with the α peak position $\omega_\alpha \propto \omega'_\varepsilon$ and the microscopic excitation peak ω_{micr} [15]. The used model ignores regular temperature variations; the latter may imply normal activated behaviour of ω_β. A β peak may be present or not depending on the physics behind the deviations from scaling laws. If it is well developed it is described by a Cole-Cole shape. Further details will be presented elsewhere. The essential features of the β relaxation are however the critical spectrum (10b), the von Schweidler law (11b) and the minimum with its scaling properties.

The critical decrease of the form factors, observed for polybutadien [5] and TαNB [6], was accompanied by the appearance of a low frequency spectral

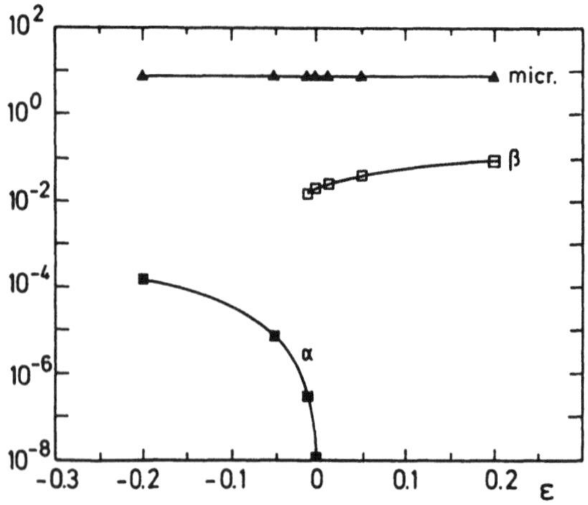

Fig. 6 Peak positions in the susceptibility versus ε. From ref. [15]

enhancement. The latter did not show a critical ε-dependence [5,6], a result which we consider as qualitative confirmation of the predicted critical spectrum (10d). It is a misunderstanding of our results that FRICK et. al. [5] declare their data as a contradiction to the mode coupling theory.

On scale ω'_ε the undercooled liquid dynamics is characterized by a second scaling law

$$\Phi_q(t) = F_q(t/\tau) \ . \tag{13a}$$

Here $F_q(\tilde{t})$ is a temperature independent master function obeying the scaling equation:

$$F_q(\omega) / [1+ \omega F_q(\omega)] = LT[\sum_{kp} V^c(q;kp)\, F_k t) F_p(t)](\omega) \tag{13.b}$$

The coupling coefficients are the mode coupling parameters from (2b) taken at the critical point. For short rescaled times one finds the asymptotic expansion

$$F_q(\tilde{t} \ll 1) = f_q^c - h_q \, \tilde{t}^b \ , \tag{13c}$$

with the same parameters f_q^c, h_q, b as used above in (11c) [18]. The non generic schematic models, studied in the first papers [2,3], imply Debye relaxation $F_q^D(\tilde{t}) = f_q^c \exp(-\tilde{t})$. The general case yields stretched relaxation [19]. In all cases studied so far in detail [20,16,19,21,15] the master function could be fitted well by a Kohlrausch law

$$F_q(\tilde{t}) = f_q^c \exp[-(\tilde{t}\Gamma_q)^{\beta_q}] \ . \tag{14}$$

The width Γ_q is estimated by $(h_q/f_q^c)^{1/b}$ and thus it should exhibit a minimum at the structure factor peak q_o. It is predicted to oscillate opposite in phase with S_q; compare Fig. 2. Similarly, β_q is predicted to have a maximum at q_o and to oscillate like S_q [19]. For the Lennard-Jones system one finds $\beta = 0.88$ for $q = q_o$, the full curves in Fig. 5, and $\beta = 0.68$ for q near the structure factor minimum, the dashed curve in Fig. 5 [16]. Scaling laws similar to (13a) hold for all quantities, which couple to density fluctuations as for example the incoherent scattering law, the elastic moduli, the dielectric loss [18]. The master functions in particular the Kohlrausch exponents, are different for different quantities. But the scale τ is predicted to be universal for all quantities measured at the same transition [18]. More explicitly, this means the following. If two quantities, say shear viscosity and coherent scattering rate for $q = q_o$ are studied, the respective time scales, τ_1 and τ_2 say, exhibit a non-Arrhenius increase if the critical point is approached from above. It is described by a power law divergency

$$\tau_{1,2} = B_{1,2}(T) + A_{1,2}(T) / [T-T_c]^{\gamma} , \gamma = (1/2a) + (1/2b) . \qquad (15)$$

The parameters A and B depend on microscopic details and cannot be predicted at present for realistic glass formers. They vary regularly near T_c, however $A_1(T)/A_2(T) \propto \exp(E/T)$ might occur. The second scaling law describes the well known α relaxation peak, illustrated clearly in figs. 4,5. Notice, that the von Schweidler law appears as low frequency part of the β relaxation due to equations (8a,11b) as well as high frequency part of the α relaxation [18]. In this manner the theory formulates the intimate relation between α and β relaxation processes. Of course, it is a matter of semantics, whether one wants to attribute the von Schweidler process to one or the other. Scaling and stretching have been discussed so often in the literature on glass dynamics, that no further remarks seem necessary. The neutron scattering experiment by MEZEI et. al. [4] demonstrated scaling law, universality of scale, stretching and Kohlrausch fit on microscopic length scales; the measured peak area $f^c_{q_o}$ was close to the value 0.85 calculated for the hard sphere system, Fig. 2.

A comment emphazising the relevance of α relaxation measurements by neutron scattering might be in order. Most of the classical measurements on structural relaxation are performed near the calorimetric glass transition temperature T_g, and therefore they test the dynamics on macroscopic time scales. The neutron scattering work [4] was done about 80° K above T_g and it tested the motion on a mesoscopic scale of about 10^{-9} sec. The measured α peak is the structural relaxation peak, discussed first by Maxwell. He described it by what is now usually called a Debye law and modern extensions of his work stick to that approximation. Upon undercooling the α relaxation becomes more and more pronounced till it dominates over the smooth background spectrum. For this situation the mode coupling theory predicted and the scattering experiment confirmed, that a description of the viscoelastic peak by a Debye law, or by a superposition of a few Debye resonances, is qualitatively wrong. Non linear couplings of density fluctuations are the microscopic origin of the α peak and of the strong temperature dependence of the time scale τ. Viscoelastic effects are glass transition precursors. Power law divergencies and stretched relaxation rather than activated processes and stochastic dynamics are the relevant features of strongly interacting simple liquids. Evidently, it is necessary to extend the scattering work in order to corroborate these statements.

It was shown for a number of examples, that formula (15) can fit viscosity data of simple systems quite well for a variation range of $\eta \propto \tau$ covering two to three decades [2,22]. The freedom for fitting is not so large as one might suspect, despite the four unspecified parameters involved in (15). One has to notice, that for simple liquids, as opposed to network systems or associated liquids, A and B should be treated as constants. Since mode coupling theories allow a correct first principle evaluation of viscosity and other transport coefficients under normal conditions [1], the temperature interval for the fit should include the normal liquid state. Let us consider TαNB as example. TABOREK et. al. [22] reported a fit with B = 0, T_c =424° K, $\gamma = 2.17$. The right full curve in Fig. 7 shows our fit with B > 0; the data [23] are shown as full dots, and we get $T_c = 410°$ K, $\gamma = 2.8$. By acknowledging $B \neq 0$ we could extend the fit interval a bit and decreased T_c in comparison to reference [22]. Dropping B our curves changes to the dotted one. The discrepancies between the two fits is an indication of the uncertainties involved, if one only looks at one set of experimental information. The debate gets a new turn, if one adds information on the dynamics for $T < T_c$ like the f_q measurements [6], reproduced as squares for $q = 2.6$ Å^{-1}. The dashed line is the regular variation (5a).The full line is a fit of the critical deviation by an ellipse, touching the regular curve and producing the $\sqrt{\varepsilon}$ law (5b) with $T_c = 410°$ K. As opposed to the authors of reference [6] we conclude that the data are compatible with the theoretical frame. We could not produce simultaneous fits for η and f_q with T_c < 400° K or T_c > 420° K. The figure is not shown as a support of the theory but merely to back up the following statements. If a quantity like f_q is measured for $T \leq T_c$, showing the predicted critical variation for temperatures approaching T_c from below, and if in addition another quantity like η is

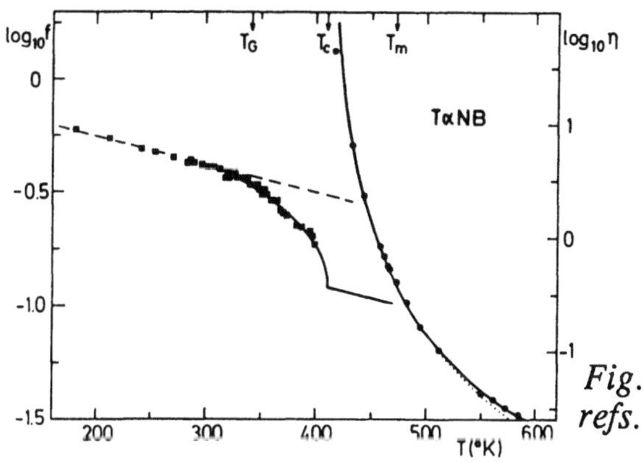

Fig. 7 As discussed in text. Taken from refs. [6] and [23].

measured for $T \geq T_c$ showing the predicted critical variation for temperatures approaching T_c above, the critical temperature can be located with an uncertainty of a few degrees. Trivial enough, if more quantities could be measured on either side of T_c, the existence of the glass transition singularity could be identified even more convincingly. In order to show that the theory does not apply, one has to find one of the quantities on either side, predicted to be critical, such that a fit by the theoretical formulae with the same T_c is impossible. From the shown comparison of the η and f_q data one concludes, that the neutron scattering experiments [6] have to be extended to larger temperatures if one wants to use them for an identification of T_c.

At the end of our discussion we would like to come back to a more theoretical issue. In a first order improvement of the most primitive version of the theory one has to replace kernel $M_q(\omega)$ in (2a) by a more complicated one called $\Gamma(q,\omega)$

$$\Gamma(q,\omega) = M_q(\omega) / [1 - \delta(q,\omega) M_q(\omega)] . \tag{16}$$

Here $M_q(\omega)$ is given by (2b) as before. For the new kernel $\delta(q,\omega)$ a similar two mode decay formula is obtained where, however, one of the decay correlators is formed with current rather than with densities. The former cannot exhibit a non ergodicity resonance; $\delta(q,\omega=0) = i\delta$ is finite and non-zero under all conditions [7]. Hence a non ergodic solution is impossible. The sharp dynamic phase transition can occur only in such approximation which ignores $\delta \neq 0$. The new kernel δ is the one phonon contribution to phonon assisted hopping processes and in this sense one can say that hopping restores ergodicity [7]. To characterize the realistic situation one can consider the solution of the nonlinear mode coupling equations as function of a parameter vector V, combining all the mode coupling vertices. Let us consider two dimensions and use one variable as the previous separation parameter ε, characterizing the strenght of the coupling to pairs of density fluctuations, and another variable δ characterizing the hopping processes, fig. 8. There is a subset S with boundary S_g in parameter space. In Fig. 8 S is the part of the abscissa ($\varepsilon>0$, $\delta=0$) and S_g is the point ($\varepsilon=0$, $\delta=0$). On S the solutions are non-ergodic. The corresponding states are called ideal glass states, since formally the viscosity is infinite there. The hypersurfaces of constant viscosity, indicated as dashed lines in Fig. 8, approach S if η becomes large. S_g is the set of glass transition singularities. Within the neighbourhood of S_g, indicated as circle in Fig. 8, the correlator behaves critically; it can be evaluated asymptotically since the distance from the singularity can be used as small parameter. For details we refer to the original papers [7,12]. Upon change of temperature the system moves on a path in parameter space. The

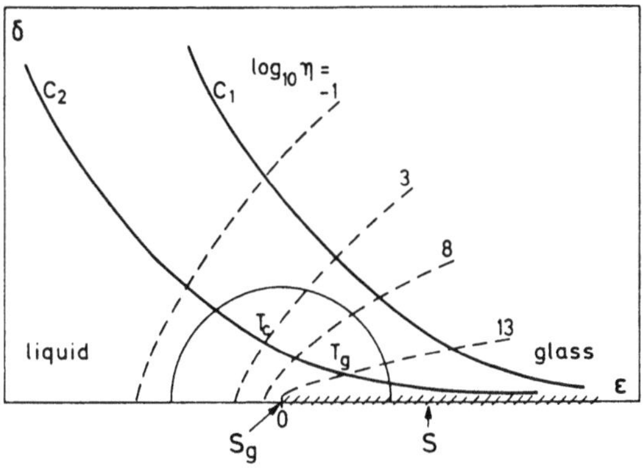

Fig. 8 As discussed in text. From ref. [7].

path avoids S_g and so there is no true phase transition. If the path like C_1 in Fig. 8 avoids even the region of influence of S_g, no particular anomaly is observed if the system is driven from the liquid to the glassy state. If, however, δ is sufficiently small as anticipated for C_2 in Fig. 8, there will be a region of transition anomalies centered around some critical temperature T_c. The main features of the experiments can be described then by the primitive version of the theory. But the results of the primitive theory become unvalid for $T = T_c$; $\delta \neq 0$ leads to a smearing of the idealized transition anomalies.

As most important example the effect of $\delta \neq 0$ is demonstrated for the α peak in Fig. 9 [12]. The curves A to D show the development of the α peak with ε increasing from negative values towards zero. For $\varepsilon = 0$ curve H does not show a peak of zero width; $\delta \neq 0$ implies a non-zero width. The latter shrinks if $\varepsilon > 0$ and curves D' to B' show the increase of the peak area according to (5b). The curves for $\varepsilon > 0$ can also be fitted well by a Kohlrausch law. So the range of applicability of this law is increased if the hopping

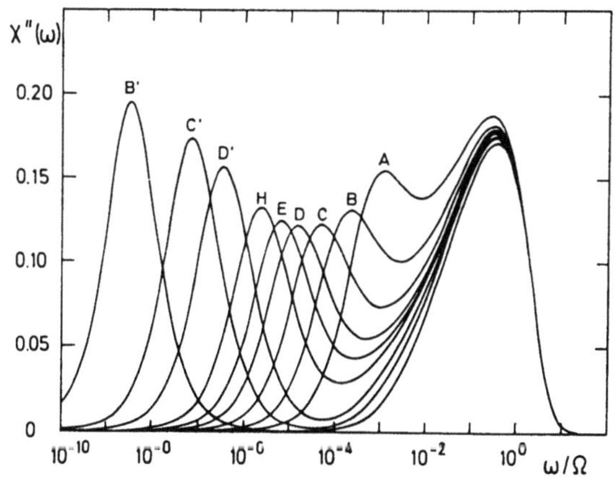

Fig. 9. Susceptibility for several ε-values. Taken from ref. [12].

phenomena are included. The smearing problem is similar to the one which one would face, if one would analyze a second order phase transition singularity for a system of small unknown size. Nothing is gained experimentally by monitoring the temperature T carefully close to T_c. To examine the glass transition singularity it is necessary to explore spectra extending over several decades below the typical microscopic excitation band, and for temperature intervals extending about 100 degrees above and below T_c. Since δ cannot be predicted at present for glass forming systems, only data analysis of the kind shown in references [4,5] or in Fig. 7 can reveal whether or not there are systems where the results of our theory apply.

We acknowledge gratefully helpful discussions with Prof. A. Sjölander. W.G. thanks Nordita and Chalmers University for making possible his visit in Göteborg as well as his colleques for their kind hospitality.

References

1. P. Resibois, J.L. Lebowitz: J. Stat. Phys. **12**, 483 (1975)
   P. Resibois: J. Stat. Phys. **13**, 393 (1975)
   J. Bosse, W. Götze, M. Lücke: Phys Rev A**17**, 447 (1978)
   J. Bosse, W. Götze, A. Zippelius: Phys. Rev A**18**, 1214 (1978)
   R.I. Cukier, J.R. Mehaffey: Phys. Rev. A**18**, 1202 (1978)
   L. Sjögren, A. Sjölander: J. Phys. C**12**, 4369 (1979)
   L. Sjögren: J. Phys. C**13**, 705 (1980)
   L. Sjögren: Phys. Rev. A**22**, 2866,2883 (1980)
   G. Wahnström, L. Sjögren: J. Phys. C**15**, 401 (1982)
   E. Leutheusser: J. Phys. C**15**, 2801 (1982)
2. U. Bengtzelius, W. Götze, A. Sjölander: J. Phys. C**17**, 5915 (1984)
3. E. Leutheusser: Phys. Rev. A**29**, 2765 (1984)
4. F. Mezei, W. Knaak, B. Farago: Phys. Rev. Lett. **58**, 571 (1987) - Physica Scr. T**19**, 363 (1987)
5. B. Frick, D. Richter, W. Petry, U. Buchenau: Z. Phys. B**70** 73 (1988)
6. F. Fujara, W. Petry: Europhys. Lett. **4**, 921 (1987)
7. W. Götze, L. Sjögren: Z. Phys B**65**, 415 (1987)
8. U. Bengtzelius: Phys. Rev. A**33**, 3433 (1986)
9. W. Götze: Z. Phys. B**60**, 195 (1985)
10. D.C. Champeney, D.F. Sedgwick: J. Phys. C**5**, 1903 (1972)
11. J. Bosse, J.S. Thakur: Phys. Rev. Lett. **59**, 998 (1987)
12. W. Götze, L. Sjögren: J. Phys. C**21**, 3407 (1988)
13. W. Götze: Z. Phys. B**56**, 139 (1984)
14. A.K. Jonscher: Nature **267**, 673 (1977)
15. G. Buchalla, U. Dersch, W. Götze, L. Sjögren: J. Phys. C**21**, 4239 (1988)
16. U. Bengtzelius: Phys. Rev. A**34**, 5059 (1986)

17. M. Goldstein: J. Chem. Phys. **51**, 3728 (1969)
18. W. Götze: In *Amorphous and Liquid Materials,* ed. by E. Lüscher, G. Fritsch and G. Jaccuci, (Martinus Nijhoff Publisher, Dordrecht, 1987) p.34
19. W. Götze, L. Sjögren: J. Phys. C**20**, 879 (1987)
20. H. DeRaedt, W. Götze: J. Phys. C**19**, 2607 (1986)
21. J. Bosse, U. Krieger: J. Phys. C**19**, L609 (1986)
22. P. Taborek, R.N. Kleinman, D.J. Bishop: Phys. Rev. B**34**, 1835 (1986)
23. J.D. Plazek, J.H. Magill: J. Chem. Phys. **45**, 3038 (1966)

Polymer Dynamics Near the Glass Transition

B. Frick and D. Richter

Institut Laue-Langevin, 156X Av. des Martyrs,
F-38042 Grenoble Cedex, France

Inelastic neutron scattering studies near the glass transition on a representative backbone polymer, polybutadiene $(-C_4H_6-)_n$, are discussed and are compared to results from other glass forming systems. Neutron Spin Echo measurements show the freezing of the structural relaxation above the calorimetric glass transition temperature T_g. The density-density correlation function $\phi(\tau)$ follows a universal stretched exponential function $\phi(\tau) \sim \exp(-t/\tau)^{\beta}$, with $\beta = 0.45 \pm 0.02$, if scaled by macroscopic shear viscosity data. A further faster spectral contribution is found on a time scale between structural relaxation and phonons, which does not freeze with the structural relaxation, but persists to temperatures below T_g. These results are discussed within the framework of mode-coupling theories and alternatively phonon like low energy excitations.

1. GLASS TRANSITION IN POLYMERS

Since a long time the phenomenon "glass transition" is well known for polymers /1/. Polymer glasses are easily prepared by modest cooling rates from the melt or "rubbery" state (order of 1K per second in comparison to ~ 10^6K/sec for amorphous metals). The glass forming ability often comes from an irregular microstructure, as e.g. atacticity for polystyrene or e.g. mixture of cis/trans/vinyl subunits in the case of polybutadiene. Furthermore, it is well known that glasses are not completely rigid at low temperatures, but show a large number of relaxation phenomena. Most of them are local relaxations due to side group motions of the polymer chain (e.g. $-CH_3$ methyl sidegroup rotations or phenylring flips). The number of these "secondary relaxations" can be drastically reduced if studying simple backbone polymers. The clearest case would be polyethylene $(-CH_2-CH_2-)$, which unfortunately is never completely amorphous, and thus one has to separate effects from the crystalline and amorphous regions. Another simple backbone polymer, which we will discuss below in detail is polybutadiene $(-CH_2-CH = CH-CH_2-)$. Using a backbone polymer we hope to concentrate on the class of relaxation processes which "freeze" at the glass transition and which are related to molecular transport processes (the α-relaxation) and probably to the "β-relaxation" /6/ in the sense of the mode coupling theory /2/.

A question which must be brought up is the universality of the glass transition. Can the freezing of polymers be representative for other glasses ? Intuitively one could think that this is not so. The covalent bonds along the polymer chain certainly restrict the motion of a polymer subunit. This restriction of the degrees of freedom could, however, simplify a theoretical description. Similar to simple liquids the interaction between polymer chains is of the van der Waals type. Both cases, simple liquid and polymer, are sketched in Fig. 1. At low temperatures, in the glass, an atom of a simple liquid can only vibrate within a cage, made up by his neighbours. Increasing the temperature leads to larger vibrational amplitudes and the chance to escape from the cage which is no longer rigid increases. At elevated temperatures, in the liquid, the atom will become mobile and diffusion sets in. Similarly for polymers we expect at low temperatures torsional vibrations, which go over to C-C-bond-rotations at higher temperatures and finally lead to a diffusion of

Springer Proceedings in Physics, Vol. 37 **Dynamics of Disordered Materials**
Editors: D. Richter · A.J. Dianoux · W. Petry · J. Teixeira

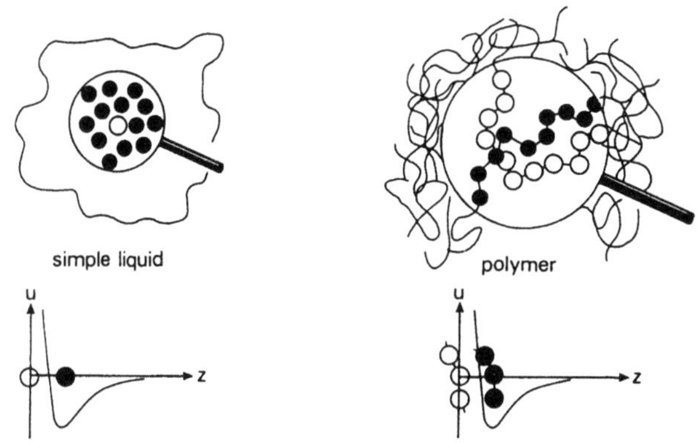

Fig. 1 : Sketch of the nearest neighbour surrounding and potential between the neighbours for a simple liquid and polymer, respectively.

the chain. We now shortly review features of the dynamics of polymers in the two extreme states : the melt or rubbery state and the glassy state.

1.1 Melt

If we consider a wider spatial scale the dynamics of a polymer chain in the melt is completely different from a simple liquid. The commonly accepted model is the "reptation model". It is assumed that the restrictions of neighbouring chains can be described by a dynamical tube /3/. The tube diameter is roughly 30-80 Å and thus, in order to study these effects by neutrons, a Q-range smaller than $Q \leqslant 10^{-1}$ Å would be necessary. Certainly all results which we will discuss below are not affected by reptation, as these measurements were performed at much larger Q, i.e. on a more local scale.

In the range, between the dimension of a tube diameter and a characteristic segmental distance, the polymer motion may be idealized by a so called "Rouse" motion /4/. The Rouse model considers the chain to be subject to random thermal forces and an entropic force related to the conformational entropy of the chain. Hydrodynamic interactions are thought to be screened out in the melt. /5/. The mean square displacement $\langle \Delta r_n(t) \rangle^2$ of a segment is expected to follow a $t^{1/2}$ law with $\langle \Delta r_n(t)^2 \rangle \simeq \sigma^2 \sqrt{\frac{kT}{\xi_o}\, t} \sim t^{1/2}$; where ξ_o is the local friction coefficient /6/. The Rouse behavior should be valid down to distances of the segmental length σ. Again we do not expect that Rouse behavior is valid in the Q-range where our measurements have been carried out. We would rather assume, that when reducing the scale further, the dynamics is influenced by details of the local monomer structure /7/. Furthermore let us mention, that the elementary process, which leads to conformational changes in the melt, is believed to be the C-C bond rotation which in solution occurs on a subnano-second timescale and which is probably slower in the melt.

1.2 Glass

The glass phase is characterized by the arrest of the large scale motions and thus only local motions are allowed. Under this category fall all side group motions, as e.g. methyl group rotations, which are most probably not directly related to the glass transition. More important with respect to the glass transition are local backbone motions. Several kinds of motions, where the ends of short chain segments are kept fixed (for example, crankshaft motion) or where only small translational motions are involved, have been discussed extensively in the literature /9/.

Of course in the glass we find vibrational motions of atoms or short chain parts within a local equilibrium configuration. The mean square displacement for the μ-th vibrational mode is given by : $\langle\Delta\bar{r}_{\mu}^{2}\rangle = \frac{2kT}{m\omega_{\mu}^{2}} n(\omega)$, where $n(\omega)$ is the Bose factor. At the low frequency end of the vibrational spectrum of polymers we find torsional vibrations around C-C bonds, for which the force constants are much lower than for example for valence angle deformations or bond stretching modes /10/. The relatively weak van der Waals interchain potential allows for different interchain modes as well. Large amplitude motions, with low frequencies, are quite unlikely in the glass.

1.3 Glass Transition

It is now of main interest to study the transition between these two extreme cases : the solid glass phase and the viscoelastic melt and how the dynamics changes there. Fresh motivation for this study comes from the mode-coupling theories /2/. These had been formulated for liquids and were extended to supercooled liquids and to the glass transition region. A great advantage of these theories are detailed predictions on the density-density fluctuations which can be scrutinized by scattering methods. This is discussed in detail by GÖTZE in this volume /11/. Less fundamental explanations, from a theoretical point of view, which have been used for the description of the glass transition (e.g. "free volume" model, e.g. /12/ and references therein) might explain as well some of the features, which are experimentally found by neutron scattering. A possible link between mode-coupling theories and the more phenomenological models is hoped to come in the future.

Let us shortly describe the behavior near the glass transition, which is experimentally found for polymers and which is similar to other glass forming substances. Common features of the transition in all these systems are a continuous change of volume, free energy and entropy, but discontinuities in the derivatives of the thermodynamical potentials, as e.g. specific heat, thermal expansion coefficient or compressibility. The transport coefficients like viscosity η or diffusion constant D and the relaxation times, show above the calorimetric glass transition temperature T_g a WLF (Williams-Landel-Ferry)- or equivalent VF (Vogel-Fulcher)-behavior : $\tau, D, \eta \sim \exp\{-A/(T-T_0)\}$, with a divergency at $T_0 \simeq T_g - \Delta$ ($\Delta \simeq$ 20-50K). The physical meaning of T_0 is not clear up to now. Furthermore one observes that the relaxation properties do not follow a simple exponential behavior (Debye relaxation), but follow a "KWW" (Kohlrausch-Williams-Watts) or "stretched exponential" behavior : $Y(\tau) \sim e^{-(t/\tau)^{\beta}}$ $0 < \beta < 1$, with an exponent β, which should be T-independent, if it is a universal parameter /13/.

Before we now report on the experimental results, we shall give some details on the neutron scattering methods used. Then we shall show results for polybutadiene and finally we shall compare the observations with the predictions of the mode coupling theory. We shall also discuss alternative explanations and compare our results with inelastic neutron scattering work on the glass transition of other glass formers.

2. EXPERIMENTAL DETAILS

With neutrons we probed molecular motions, which are faster than about 10^{-8} seconds and which take place in the spatial range 1 Å $\leqslant$ r $\leqslant$ 30 Å. The following measurements have been carried out : diffraction measurements on D1B with λ = 2.52 Å in the Q-range 0.3 - 3.3 $Å^{-1}$, neutron spin echo (NSE)experiments (with λ

= 6.0 Å and λ = 4.55 Å), where we measured the intermediate scattering law $S(Q,t)/S(Q,0)$ /14/ in the time regime 9×10^{-12} sec ⩽ t ⩽ 2.8×10^{-9} sec. Neutron backscattering (BS) measurements on IN10 and on IN13 with λ_i = 6.28 Å and 2.23 Å, respectively, giving an elastic resolution of δE ~ 1 μeV (~ 4×10^{-9} sec) and δE ~ 8 μeV (~ 3.2×10^{-10} sec). Time-of-flight (TOF) experiments on IN6 (λ_i = 5.1 Å; δE ~ 90 μeV (FWHM)), where we observe energy transfers up to at least 100 meV. The accessed Q-range on IN10 and IN6 lies roughly between 0.2 and 2 $Å^{-1}$ whereas between 0.2 $Å^{-1}$ and 5.3 $Å^{-1}$ on IN13. Here we report on measurements on polybutadiene : $-(CH_2-CH = CH-CH_2)_n-$ which is amorphous due to its irregular microstructure of cis-, trans- and vinyl units (about 47:46:7). The samples studied by incoherent scattering on the TOF- and BS-machines had a molecular weight of about 10^5 and a DSC glass transition of $T_g \approx 186K$. On the diffraction and spin echo machine we studied completely deuterated samples (C_4D_6) of molecular weight 9.3 x 10^5 and M_w/M_n = 1.03. The DSC glass transition temperature in this case was T_g = 181K. A further deuterated sample with M_w = 5 x 10^4 was studied on D1B.

3. EXPERIMENTAL RESULTS

3.1 Diffraction

Neutron diffraction experiments reveal the amorphous character of polybutadiene (Fig. 2). The diffraction pattern shows a main peak at Q_o ~ 1.4 $Å^{-1}$ and a second peak at Q ~ 2.9 $Å^{-1}$ /15/. The main peak is largely due to interchain correlations, i.e. monomer-monomer distances, whereas the second peak at Q ~ 2.9 $Å^{-1}$ is ascribed to intrachain distances. This has been studied in more detail for polyethylene /16/, but can also easily be seen from the temperature behavior of the main peak /17/ and the second peak. Whereas the second peak practically does not move with temperature, as one would expect for covalent bonds, the main peak shows a drastic shift to higher Q-values with decreasing temperature (Fig. 2b). The slope of the line, drawn through the peak position values at high temperatures, gives the linear thermal expansion in the melt; the line through the low temperature data shows the thermal expansion in the glass. The intersection point of the two curves is near T ~ 150K which is about 30K below T_g. In Fig. 2c we plot the temperature dependence of the width of the main peak, which is a measure of the sharpness of the nearest neighbour shell. With decreasing temperature the linewidth is diminishing and "freezes" with a width corresponding to a temperature of roughly T = 200K. The dip between T =150K and T = 200K is interpreted being due to a structural rearrangement towards the equilibrium value, setting in around 150K, when heating up the sample from low temperatures.

3.2 Approach to Tg from High Temperatures

We carried out spin echo measurements in order to study the change of the polymer dynamics in the melt, when approaching T_g /18/. In the investigated temperature range between T = 280K and T = 200K we observe relaxation in the time window of the instrument (Fig. 3). The relaxation slows down with decreasing temperature and at T = 200K the signal becomes elastic. A fit to the curves shows that the relaxation is not a simple exponential. A rescaling of the time scale, using the temperature dependence of the macroscopic shear viscosity, leads to a universal curve, as shown in Fig. 4. Thereby, we have used the relation between the viscosity and monomeric friction coefficient ξ, given by the Rouse model and the parameters determined from the shear viscosity by Berry and Fox /19/. The relevant microscopic time τ is related to the friction coefficient by $\tau = \frac{l^2}{6k_B}\,\frac{\xi_o(T)}{T}$; thereby k_B is the Boltzmann constant and l a microscopic jump length. The resulting universal curve can be described by a stretched exponential function $\frac{S(Q,t)}{S(Q,0)} \sim \exp-\left(\frac{t}{\tau}\right)^{\beta}$, with $\beta = 0.45 \pm 0.02$. For the jump length l we get : l = 2.07 ± 0.006 Å /18/. On the backscattering

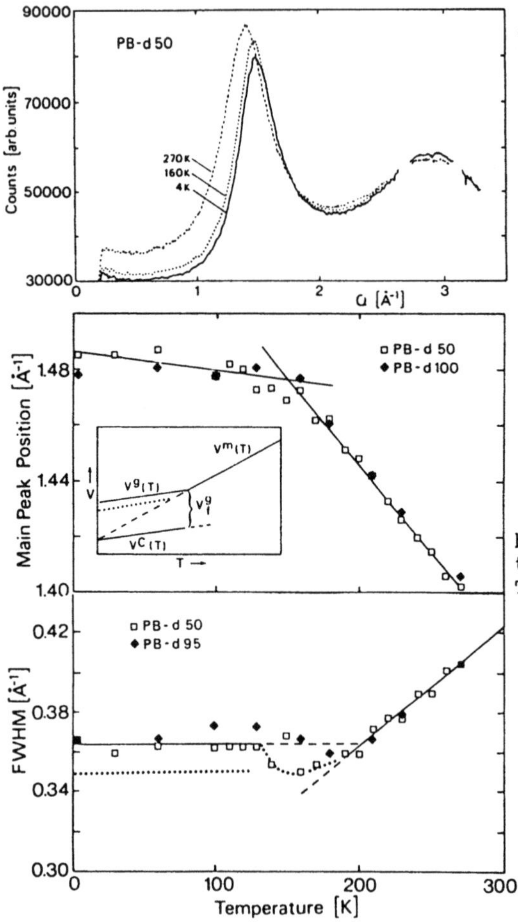

Fig.2 :

a) Diffraction patterns of polybutadiene at T = 4K, 160K and 300K measured on D1B

b) Main peak position as a function of temperature showing a crossover around T = 150K

c) Full width of the main peak as a function of temperature showing freezing near T = 200 K

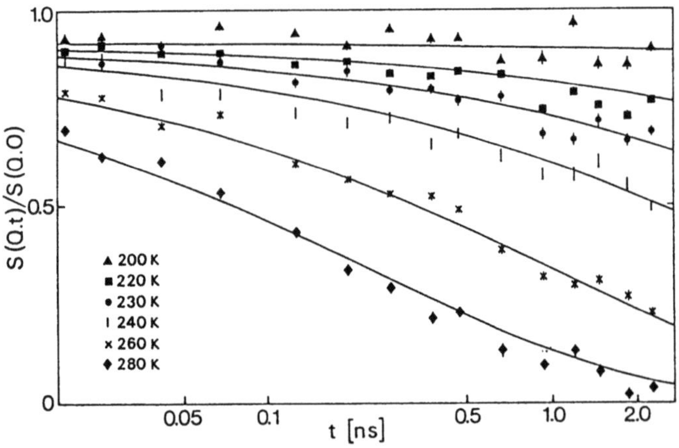

Fig. 3 : Spin echo measurements on polybutadiene. The intermediate scattering law is shown for several temperatures above T_g. The drawn lines correspond to stretched exponentials, where the parameters from the universal curve (Fig. 4) have been fixed /18/.

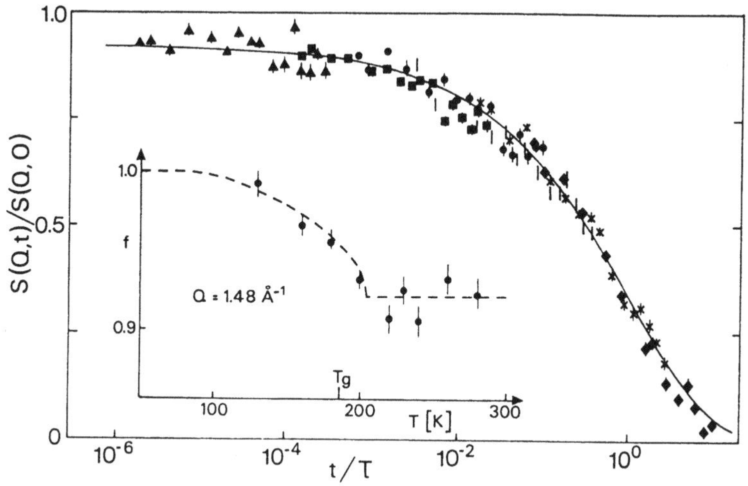

Fig. 4 : Intermediate scattering law of polybutadiene versus rescaled time. Rescaling has been done using viscosity data (see text) /18/. The insert shows the temperature dependence of the freezing non ergodic fraction in the scattering function.

instruments we also found quasielastic broadening, the lineshape of which cannot be fitted by a single Lorentzian, as it is expected for the Fourier transform of a stretched exponential. Another important finding is that (for $200 < T < 280K$) $S(Q,t)/S(Q,0)$ does not extrapolate to unity for $t \to 0$ (Fig. 4). The missing spectral component must be related to motional processes occurring on a time scale faster than the shortest measured time and slower than about $6 \cdot 10^{12}\ s^{-1}\ rad^{-1}$, given by the upper band pass of the spectrometer. The inset shows that this fast component is decreasing with decreasing temperature (i.e. f is increasing).

This fast component must be visible on spectrometers covering a broader dynamic range. Fig. 5 illustrates that this is the case. The spectrum, now measured in ω-space on a time-of-flight machine, shows a broad quasielastic signal, the wings of which go far out in energy (measured at T = 280K for an elastic Q of Q_o = 2.0 $Å^{-1}$).

3.3 Approach to T_g from Low Temperatures

3.3.1 Elastic Backscattering Measurements

Studying the glass, far below T_g we expect nearly every relaxation being frozen, but the vibrations will survive. Using high resolution backscattering spectrometers (IN10, IN13) and protonated samples we can study the self-correlation function of the protons (σ_{inc} = 486 b/monomer versus σ_{coh} = 22 b/ monomer for polybutadiene) and its change with temperature. We do this, using the "fixed elastic window" method, where within the instrumental resolution we analyse the elastic signal of the sample, scanning the sample temperature. Usually, these curves show a Q-dependent decrease, due to the Debye-Waller factor (DWF) : the observed elastic intensity for incoherent scattering should thus diminish with $I \sim I_o \cdot \exp(-1/3\ Q^2 \langle r_o^2 \rangle)$. For harmonic vibrations one would expect $\langle r_o^2 \rangle$ being proportional to temperature T and ln(I) decreasing with Q^2.

In Fig. 6 we compare the elastic scattering (resolution ~ 8 μeV) of the crystalline trans-modification of polybutadiene (~ 98% crystalline, Hüls, Marl) with the amorphous (cis/trans/vinyl) modification, measured on IN13 at a

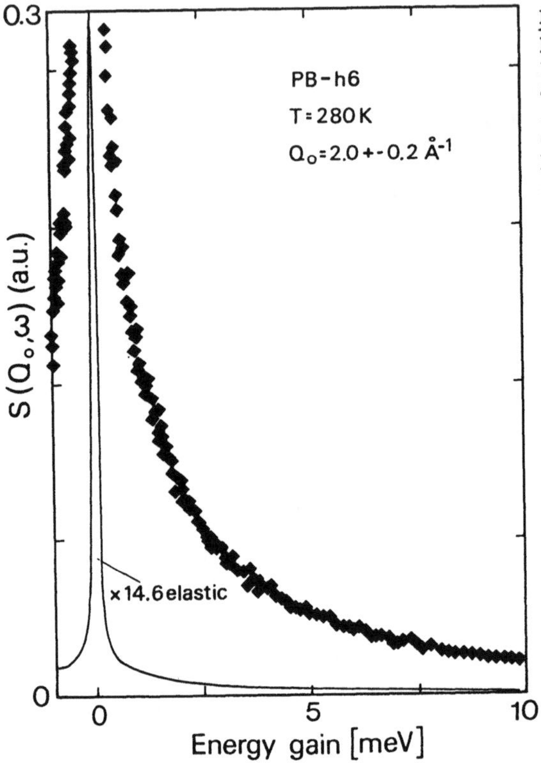

Fig. 5 : Inelastic intensity from polybutadiene, measured on the time-of-flight spectrometer IN6 at T = 280K (inset : elastic line). Q_0 is the Q-value for elastic scattering. The data have been taken at constant scattering angle.

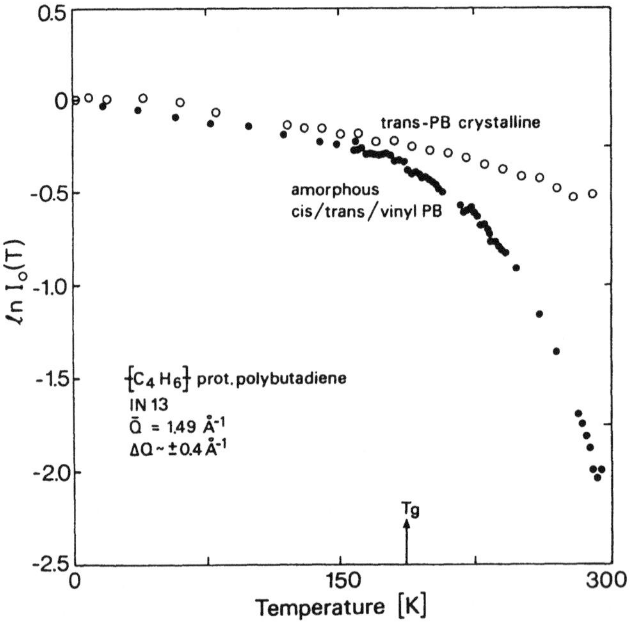

Fig. 6 : Comparison of the temperature dependence of the elastic incoherent scattering from highly crystalline trans-polybutadiene and from amorphous cis/trans/vinyl-polybutadiene, measured on IN13 (δE ~ 8μeV). The data are normalized to the lowest temperature for each sample.

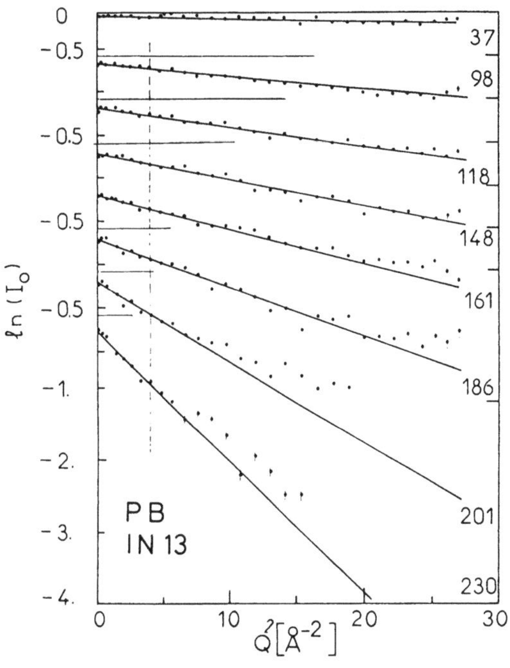

Fig. 7 : Logarithm of the normalized elastic incoherent scattering versus Q^2 from protonated PB ($M_w = 10^5$). Drawn lines are fit results for straight lines through both IN13 and IN10 results /20/. Temperatures are marked on the right side in K.

Q-value corresponding to the main peak of the amorphous species. For both samples the logarithm of the elastic intensity decreases in the same manner up to T ~ 150K. Then, at higher temperatures, we find an anomalous fast decrease of the elastic scattering for the amorphous polymer. Studying the Q-dependence in the low temperature range, we find perfect agreement with a Q^2 behavior (for $Q \leqslant 5.0$ Å^{-1}) up to roughly 150K (see Fig. 7). Slight deviations at large Q occur with increasing temperature; these become more drastic at $T \geqslant 200K$, where, however, the statistical error due to the low intensity at high Q values becomes important. From these curves we have calculated the mean square displacements r_o^2. This has also been done using similar scans performed at the backscattering spectrometer IN10. Furthermore we calculated the integrals over the intensity within the dynamic window of IN10 (± 12 μeV) and evaluated also there Q-dependence in terms of r_o^2. The thus obtained mean square displacement as a function of temperature is shown in Fig. 8 /20/. At low temperatures the mean square displacements follow well a linear temperature dependence as expected for harmonic vibrations. A stronger than linear temperature dependence is setting in around $T \simeq 150K$, where the mean square displacements start to diverge. We note that in the temperature range where $\langle r_o^2 \rangle$ is increasing, we do not find quasielastic scattering due to e.g. an activated process within the dynamic range of the instrument. The inset in Fig. 8 shows, that the spectral response with 1 μeV resolution is still elastic at T = 200K.

3.3.2 Time-of-Flight Measurements

TOF-experiments reveal the phonon spectra of polybutadiene at low temperatures : an inelastic hump centered around 2 meV is shown in Fig. 9 (note the ratio of the elastic (drawn line) to inelastic intensity). The increase of the inelastic scattering, being proportional to Q^2 at low temperatures, corresponds to the decrease found for the elastic intensity displayed in Fig. 7. The counterpart to the fast decrease of the elastic intensity (Fig. 6) at about T = 150K is found also in the inelastic range accessed by the TOF instrument, where at energies between the elastic line and the phonons rapidly

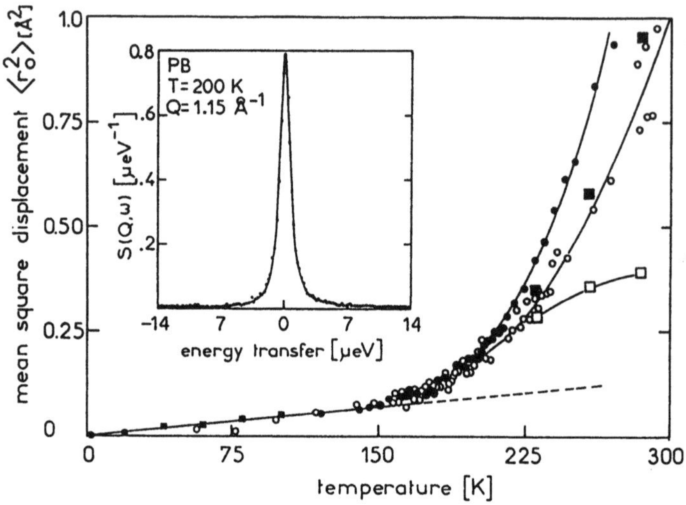

Fig. 8 : Mean square displacement as a function of temperature. It is obtained from fits to IN10 and IN13 elastic backscattering data (see Fig. 7 and /20/). The inset shows that the spectral response is still elastic at T = 200K.

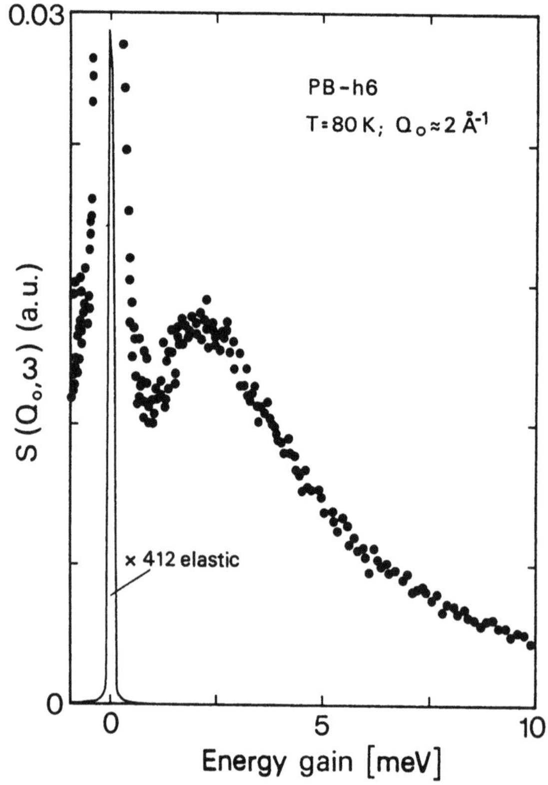

Fig. 9 : Inelastic intensity from protonated polybutadiene showing the phonon scattering at T = 80K in the glass. The drawn line corresponds to the elastic scattering. The data represent results from measurements at constant scattering angle. Q_0 corresponds to the momentum transfer at $\hbar\omega=0$.

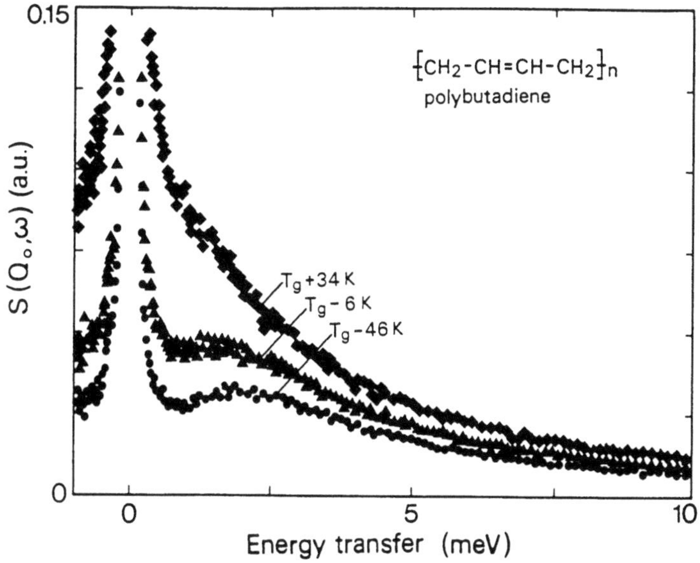

Fig. 10 : Temperature dependence of the low frequency scattering from the TOF-experiment IN6 showing the enhancement of the scattering intensity between the resolution function and the low temperature phonons.

additional intensity is arising (Fig. 10). This extra inelastic intensity also explains the missing fraction in S(Q,t)/S(Q,0), as observed by NSE (Fig. 4). A separation of this "low energy inelastic scattering" from the phonon scattering /21/ shows, that the width of this additional intensity does not increase with Q and roughly stays constant with temperature up to T = 280K, whereas the intensity increases dramatically with temperature. A separation of phonon scattering from the low energetic scattering, however, is not simple and to a certain degree arbitrary, as one has to assume that the phonons do not change (besides population factors) when passing the glass transition.

4. DISCUSSION

4.1 Approaching T_g from High Temperatures

The main results concerning the dynamics of the polymer above the glass transition temperature come from the spin echo experiment on deuterated samples /18/. The important findings were : i) a scaling of the relaxation on a microscopic scale corresponding to monomer monomer distances with the macroscopic viscosity relaxation; ii) a "stretching" of the relaxation process with a temperature independent exponent $\beta = 0.45 \pm 0.02$; iii) the indirect observation of a faster relaxation component, revealing itself as a missing fraction in S(Q,t)/S(Q,0) for $t \to 0$. Backscattering measurements on an incoherently scattering sample confirm the non-Debye behavior above T_g and TOF experiments on these samples show in detail the Q and temperature behavior of the fast component.

The intermediate scattering law, measured by the NSE method, is directly related to the density-density correlation function $\phi_q(t)$, which is of central importance for the description of the glass transition by the mode-coupling theory (see GÖTZE in this volume) : $\phi_q(t) = \langle \rho_q(t)\rho_q \rangle / S_q$; see eqn. (1a) in /11/. These density fluctuations $\phi_q(t)$ enter the mode-coupling expression (see /11/ eqn. (2b)) and are a necessary input for calculations of the viscosity or similar macroscopic quantities. Thus, anomalies found on a microscopic scale for density fluctuations $\phi_q(t)$ lead as a consequence to anomalies in the macroscopic transport coefficients. Above a critical temperature T_c ($T_c > T_g$) the density fluctuations decay to zero : $\phi(t \to \infty) = 0$, whereas they arrest spontaneously below T_c. Its spectral contribution is described by the "non ergodicity parameter" or "glass form factor" f_q (see /11/(3a,b)).

The theory predicts that the structural relaxation in the undercooled region is not described by a simple Debye law, but shows a stretched exponential behavior /11/. Our experiment shows this stretched relaxation behavior with a temperature independent β (200K < T < 280K) in good agreement with recent dielectric measurements at lower frequencies on polyisoprene /22/, where also a T independent β close to our value was found.Furthermore our data prove the strong interrelation between macroscopic transport properties and microscopic density fluctuations which is required by the theory.

A second, fast relaxation component, similar to GOLDSTEIN's β-relaxation/6/, is also predicted by the mode coupling theory. Assuming for the moment, that the observed missing fraction S(Q,t)/S(Q,O) for t = 0 is due to this β-relaxation, from the temperature dependence of this fraction f(Q = 1.4 $Å^{-1}$) we could try to find the critical temperature T_c, which is predicted by the mode-coupling theory to be located above the calorimetric T_g. The non-ergodicity parameter is then expected to exhibit a square root behaviour $[f-f(T_c)] \sim (\sqrt{T_c-T})$ below T_c, while above T_c the spectral contribution of the structural relaxation $f(T_c)$ should stay constant. The inset in Fig. 4 shows that f is constant within the error bars above roughly T = 200K and reaches f = 1 only far below T_g. If we assume a square root behavior for $[f-f(T_c)]$, then we could locate T_c roughly at $T_c \simeq$ 200-210K (note, that we find a freezing of the width of the main peak around this temperature (Fig. 2).

To further compare our results with the mode-coupling theory, we have investigated the Q-dependence of f(Q), which according to theory should show the same modulation with Q as the structure factor and thus should exhibit a maximum at Q_o for a given temperature. The relaxation rate τ is estimated to follow $\tau = \left(\frac{1-fq}{fq}\right)^{1/b}$ /11/ and thus should pass through a minimum at Q_o. Both features are in agreement with our recent results /18/ (Fig. 11).

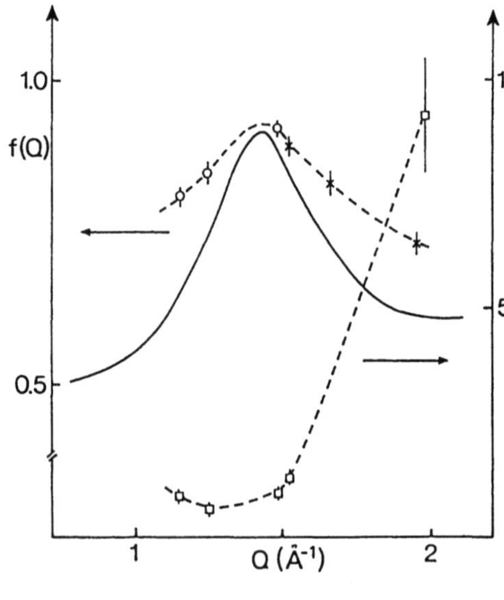

Fig. 11 : Q-dependence of the glass form factor f(Q) and relaxation rate τ /18/. The solid line represents the static structure factor at T = 230K.

From the point of view of the Rouse model the scaling of the microscopic data using the shear viscosity data is astonishing. Usually, it is assumed that the viscosity is dominated by the slower Rouse modes and the dynamics probed by neutrons by the fastest Rouse modes. Thus it is not obvious to find a freezing with the same temperature behaviour for all modes. A theoretical discussion on the way how the Rouse modes in polymers do freeze has recently been started /23/ and here our results might be a valuable information.

Comparing our findings with NSE measurements on an anorganic glass /24/ we find qualitative agreement, but a smaller β-exponent and a larger form factor f(Q) (Ca-K-NO_3 /24/ : $\beta = 0.65$; $f(Q \simeq 0.86)$). The different β-exponent could be due to a softer structure and weaker restoring forces in the case of polybutadiene (van der Waals forces) in comparison with Ca-K-NO_3, where the strong Coulomb forces could lead to a weaker non-linear coupling. This would also explain the stronger structural arrest ($f = 0.92$) in the case of polybutadiene near T_c.

4.2 Approaching T_g from Low Temperatures

From TOF- and from BS-experiments on incoherently scattering samples at low temperatures we have found a mainly harmonic temperature dependence of the phonons. The main result is the detection of deviations from this behavior below T_g, simultaneously with the appearance of additional intensity in the spectral range between the structural relaxation and phonons. Two different approaches for an explanation of these findings are discussed in the following : the interpretation of this fast component as i) the "β-relaxation", predicted by the mode coupling theory, and ii) the result from changes in the phonon density of states.

In the frame of the mode coupling ideas one can try to locate the critical temperature T_c, coming from the low temperature side. This has been done /20/, analysing our data in terms of the mean square displacement (see Fig. 8) which again is expected to follow a square root behavior. Our data appear to be in rather good agreement with this prediction, if $T_c = 220K$ is chosen (see Fig. 12). From NSE we had found $T_c \simeq 200 \pm 10K$ (note : T_g is in this case 5K lower). The exact determination of T_c from r_o, using incoherent scattering, is, however, rather difficult (separation of α- and β-relaxation) /20/.

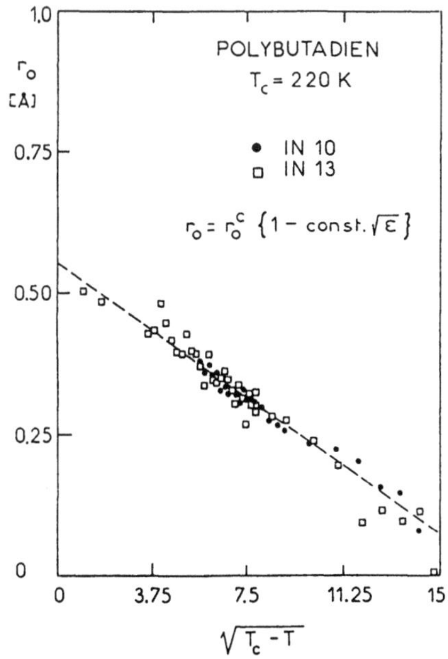

Fig. 12 : Displacement versus $\sqrt{T_c - T}$ /20/, which should be linear within this representation according to Ref. /11/.

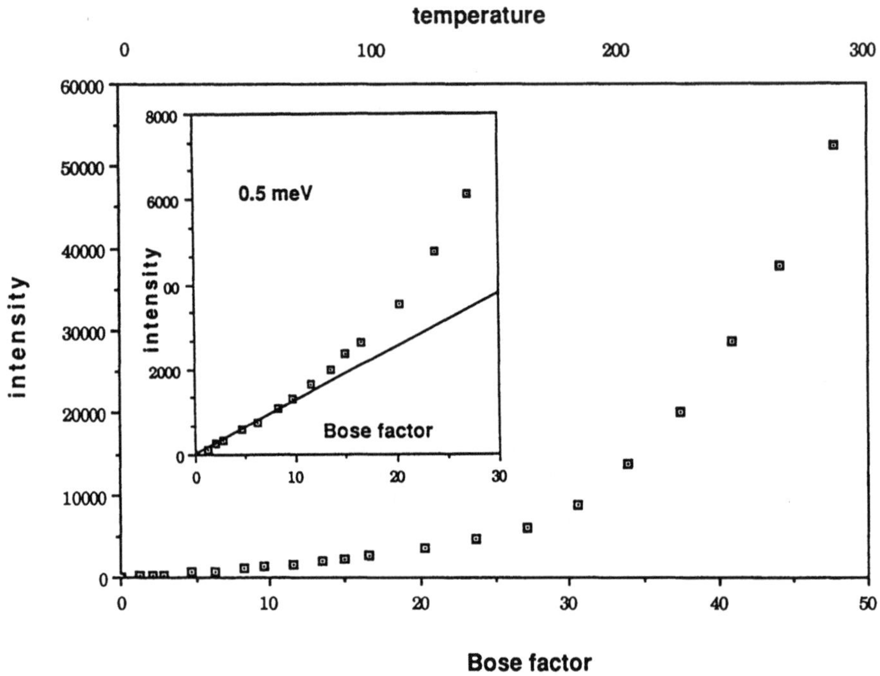

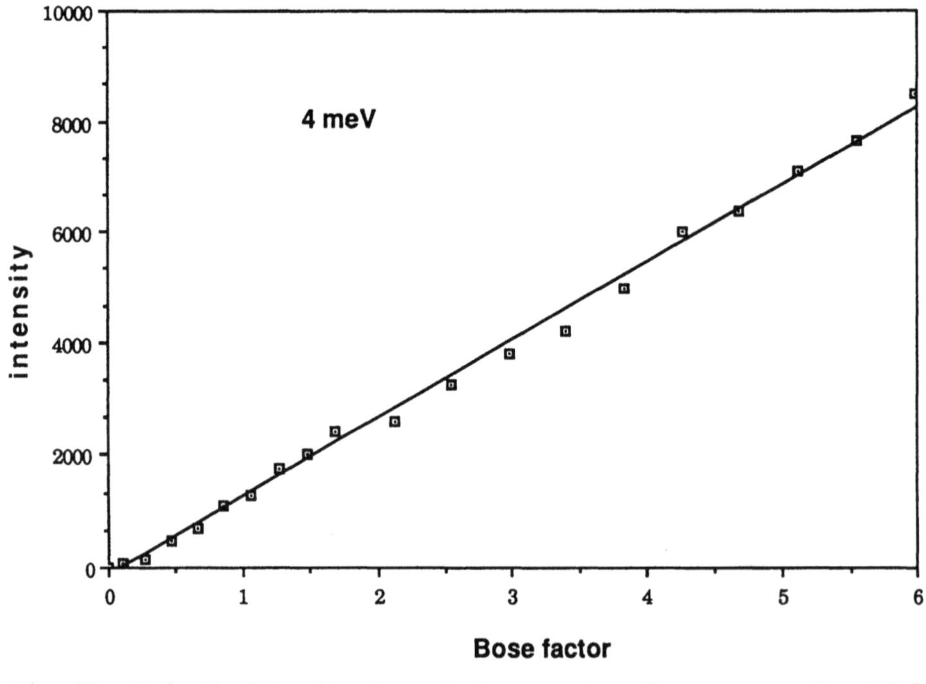

Fig. 13 : Inelastic intensity at constant energy transfers, upper part, at 0.5 meV and, lower part, at 4 meV versus Bose factor (0 < T < 300K in both figures) summed over several high angle detectors. For 0.5 meV energy transfer we already observe clear deviations from a straight line near T = 50K (see insert), whereas at 4 meV the inelastic intensity follows the straight line, as expected for phonons. These low energy excitations dominate above T_g.

Analysing for the fast component, we find near T_g a strong increase in intensity at low energies. The width of the additional intensity distribution stays roughly constant in the observed temperature range. If we regard this as the β-relaxation, we are not yet able to perform a reliable comparison with the critical exponents predicted by the mode coupling theory. A preliminary approach has been tried in /21/. One of the main problems is the separation of phonon scattering from a possible relaxation component. This must rely on the assumption, that the phonon spectrum does not change when varying the temperature through the calorimetric glass transition temperature T_g up to T_g + 100K. This is a strong assumption if we imagine that the structure softens and becomes more and more liquid like. Therefore, we want to discuss the experimental finding of this low frequency intensity, observed in a frequency range between phonons and structural relaxation, also from another side.

A detailed discussion of low frequency excitation in glasses is given by BUCHENAU in this volume /25/. The attraction of these ideas lies in a common description of two level excitations, relaxational and vibrational phenomena observed in glasses. For vitreous silica inelastic neutron scattering results strongly support the coexistence of these phenomena. A recent attempt to follow the observed low energy excitations in polybutadiene down to T =1.5K /26/ has shown that we find already around T = 50K = T_g - 136K deviations from the normal vibrational properties in the low frequency range (below 2 meV) (see Fig. 13). This is only detectable on the high intensity TOF-machine IN6 with very long counting times and substraction of the elastic peak. The deviation from Bose factor behavior sets in already around T = 50K and a strong divergency is found around T = 150K (note : from diffraction (Fig. 2) the change in linear expansion is located near T = 150K). In the concept of Ref. /25/ we could think about vibrations at low temperatures, which lead with incrasing temperature to a growing number of relaxation processes (probably in local double-well potentials). Finally, due to the strong softening of the structure the restoring forces are weakened and near the glass transition the appearance of resonant modes may be expected. This idea should be discussed in connection with the low energy modes observed now in many glasses near T_g /27/ and could be considered also for the case of polybutadiene. A relation between this concept and the β-relaxation of the mode coupling theory may be possible but is not yet clear.

To summarize, at low temperatures we have presented experimental results on the vibrations in a polymeric glass and its change near the glass transition. We have found a strong intensity enhancement at low frequencies already below T_g. At higher temperatures structural relaxation was observed which scales with macroscopic shear viscosity data. The results of the mode coupling theory for the structural relaxation agree qualitatively with our experimental results. The observed fast relaxation component could probably be explained by the predicted "β-relaxation" as well, but up to now the experimental evidence is not yet conclusive. Alternatively, phenomenological models for the low frequency relaxation may be applied for an interpretation.

REFERENCES

1. R.N. Harward: In The Physics of Glassy Polymers (Applied Publisher LTD, London 1973);
   J. Ferry: In Viscoelastic Properties of Polymers (Wiley Inc. 1980)
2. E. Leutheusser: Phys. Rev. A29, 2765 (1984);
   U. Bengtzelius, W. Götze, A. Sjölander: J. Phys. C 17, 5915 (1984);
   L. Sjögren, W. Götze: J. Phys. C 20, 879 (1987) and this volume
3. P.G. de Gennes; J. Chem. Phys. 55, 572 (1971)
4. P.E. Rouse: J. Chem. Phys. 21, 1272 (1953)
5. D. Richter, K. Binder, B. Ewen, B. Stühn: J. Phys. Chem. 88, 6618 (1984)
6. M. Goldstein: J. Chem. Phys. 51, 3728 (1969)
7. B. Stühn, B. Ewen, D. Richter: Z. Phys. B 58, 305 (1985)
8. E. Helfand: Science 226 (1984);
   J.L. Viovy, L. Monnerie, F. Merola: Macromolecules 18, 1130 (1985)

9. L. Monnerie, J.L. Viovy, R. Dejean de la Batie, F. Lanprêtre: In Polymer Motion in Dense Systems, ed. by D. Richter, T. Springer 1988, Springer Proc. Phys. 29
E. Helfand: J. Chem. Phys. 54, 4651 (1971)
10. N. Saîto, K. Okano, S. Jwayanagi, T. Hideshima: In Solid State Physics, ed. by F. Seitz, D. Turnbull 1963
11. W. Götze, L. Sjögren: this volume
12. M.H. Cohen, G.S. Grest: Phys. Rev. B 20, 1077 (1979)
13. J. Jäckle: Rep. Prog. Phys. 49, 171 (1986);
C.A. Angell: J. Non-Crystalline Solids 102, 205 (1988)
14. F. Mezei : In Neutron Spin Echo, ed. by F. Mezei (Springer Verlag, Berlin 1980)
15. B. Frick, D. Richter, C. Ritter: submitted to EPL
16. G.R. Mitchell, R. Lovell, A.H. Windle: Polymer 23, 1273 (1979);
I. Voigt-Marlin, F.C. Mijhoff: J. Appl. Phys. 46, 1165 (1975)
17. H.G. Kilian, K. Boueke, J. Polym. Sci. 58, 311 (1962)
18. D. Richter, B. Frick, B. Farago: Phys. Rev. Lett. 61, 2465 (1988)
19. G.C. Berry, T.G. Fox: Adv. Polymer Science 5, 261 (1968)
20. B. Frick, D. Richter, W. Petry, U. Buchenau: Z. Phys. B 70, 73 (1988)
21. B. Frick: In Polymer Motion in Dense Systems, ed. by D. Richter, T. Springer, Springer Proc. 29 (1988)
22. Kremer, Boese: priv. comm.
23. S.F. Edwards, T.A. Vilgis: In Time Dependent Effects in Disordered Materials, ed. by R. Pynn, T. Riste, NATO ASI series B167 (Plenum Press, New York 1987)
24. F. Mezei, W. Knaak, B. Farago: Phys. Rev. Lett. 58, 571 (1987)
25. U. Buchenau: this volume
26. B. Frick, U. Buchenau, D. Richter: unpublished results
27. F. Fujara, W. Petry: Europhys. Lett. 4, 921 (1987);
F. Mezei, W. Knaak, B. Farago: Physica Scripta T 19, 363 (1987);
U. Buchenau, M. Monkenbusch, M. Stamm, C.F. Majkrzak, N. Nücker: In Polymer Motion in Dense Systems (Springer Verlag 1988) and several contributions in this volume

Dielectric Relaxation Behaviour Around the Glass-Transition of Polar Polymeric Systems

J. Colmenero, A. Alegría, J.J. del Val, and J.M. Alberdi

Departamento de Física de Materiales,
Universidad del País Vasco. Facultad de Química, Apdo. 1072,
E-20080 San Sebastián, Spain

In this work we have investigated the changes in the dielectric behaviour of a glass-forming polymeric system (PVME) through the experimental glass-transition temperature range. The results obtained have been interpreted in the framework of a kinetic free-volume model previously developped by us.

1. INTRODUCTION

When a liquid is supercooled sufficiently far below its equilibrium freezing temperature without crystallizing, it leaves out their equilibrium liquid-like behaviour and becomes an unstable solid-like glass. This process, known as the glass-transition, has been the subject of a great interest during recent years from both, theoretical and experimental points of view. However, the exact nature of the glass-transition, as well as the related problem of the dynamics of a supercooled liquid, are questions that still remain open to controversy.

In this work we have been focussed in the dielectric relaxation behaviour around the glass-transition of polar polymeric glass-forming systems. In this framework, the word around refers to the temperature range at which the actual glass-transition process above described takes place. This temperature range will be characterized for simplicity by a temperature T_g. Other likely hidden glass-transition temperatures associated to possible sharp thermodynamic or dynamical transitions, which are predicted by theoretical approaches like Free-Volume-model or mode coupling ones, have not been considered here.

From an experimental point of view, the standard dielectric techniques like for example, frequency isothermal measurements (frequency range 10-10^5 Hz) have been proved to be very appropriate in order to study the dielectric relaxation behaviour in the supercooled liquid-like state, i.e., above T_g. However, at the temperature range just below T_g, the frequency range we have to use (~10^{-3} Hz) implies practical problems. In addition, at this range, the glassy polymeric system evolves towards the liquid like equilibrium state during experimental times. As a consequence, the dielectric relaxation obtained in such conditions by these classical methods should correspond to the liquid-like equilibrium state instead of the actual glassy state. There is, however, a thermally stimulated technique, known as Thermally Stimulated Depolarization Current (TSDC), which could be succesfully used with an adequate experimental procedure in order to monitor the dipolar motions which occur just below T_g. This technique is a particular time domain technique with an equivalent frequency range of 10^{-2} - 10^{-5} Hz.

In this work we have tried to use both, TSDC technique and standard dielectric spectroscopy in order to follow the change of the dielectric relaxation behaviour through the glass-transition of a glassy polymer: poly(vinyl methyl ether) (PVME). The glass transition temperature of this polymer, measured by differential scanning calorimetry (DSC), was $T_g = 240$ K.

2. EXPERIMENTAL

The dielectric relaxation behaviour above T_g has been studied by means of standard dielectric frequency measurements (frequency range 10 to 10^5 Hz). Both real and imaginary parts of the dielectric susceptibility were measured as a function of the temperature and frequency. Both the experimental system and procedure used by us have already been described in a previous work [1].

At the temperature range just below T_g the dielectric response was measured by means of the TSDC technique. Both, the basis of TSDC technique and the general operating principle, have been widely described in previous papers [2,3]. The measurement procedure used in this work can be summarized as follows. The glassy PVME is formed by cooling from the equilibrium liquid-like state ($T >> T_g$) at a constant cooling rate q_c, of 20 K min^{-1}. An electrical polarizing field, E_p (10^6 Vm^{-1}) is applied to the sample during this glass-forming process. The field is set-on at a temperature $T_{p,on} > T_g$ and set-off when the sample reaches a temperature $T_{p,off} < T_g$ where the dipolar motions associated to the glass-transition process are completely hindered.

This procedure allows dipoles with a relaxation time smaller than the experimental time (of the order of the time that sample remains at each temperature, ~$1K/q_c$), at any temperature between $T_{p,on}$ and $T_{p,off}$, to orientate. Those dipoles whose relaxation time at $T_{p,off}$ is higher than the experimental one can remain oriented in the field direction when it is removed and will contribute to the final polarization of the sample. The sample is subsequently heated at a constant rate, q_h, of 10 K min^{-1} so that, dipoles can return to random orientation. The corresponding depolarization current, I, is recorded versus sample temperature. If the interval ($T_{p,on}$, $T_{p,off}$) covers a wide temperature range, TSDC response is known as the whole TSDC spectrum. In polymers, whole TSDC spectra are in general too broad to be associated to only one Debye-like relaxation process. However, it is well known that the TSDC technique can separate different elementary processes that contribute to such a complex relaxation by means of the fractional polarization procedure. In this procedure the polarization field, Ep, is only applied to the sample in a narrow temperature window, $T_{p,on}-T_{p,off}=5K$ in our case, during cooling at constant rate. By varying the interval ($T_{p,on}$,$T_{p,off}$) in the temperature region where global TSDC spectrum appears, this one can be resolved into quasi-elementary peaks. So obtained peaks can be considered as thermally activated Debye-like TSDC peaks in a first approximation. Therefore, the temperature behaviour of the relaxation time, associated to each of one of the quasi-elementary peaks $\tau_i(T)$ can be obtained in the temperature range, where the i-peak appears from the basic relationship

$$\tau_i(T) = - \frac{P_i(T)}{J_i(T)} \tag{1}$$

where $P_i(T)$ is the polarization which remains in the sample at temperature T, and $J_i(T)$ the depolarization current density at the same temperature.

On the other hand, the empirical distribution function of the relaxation times $g(\tau_i)$ at different temperatures in the experimental range where TSDC peaks appear, can be obtained from $\tau_i(T)$ behaviour as follows. Taking a constant temperature in this range, T_c, we can calculate $\tau_i(T_c)$ values. Moreover, it is well known from the basic Debye theory that the number of dipolar units, which contribute to a Debye TSDC peak characterized by a single relaxation time τ_i, becomes proportional to the area of this peak. Then, $g(\tau_i)$ at the temperature considered will be proportional to the area of the quasi-elementary i-peak.

3. RESULTS

3.1 Frequency measurements above T_g

The complex dielectric susceptibility $\chi^* = \chi' + i\,\chi''$ as a function of the frequency was measured at different temperatures above T_g. The values of the slopes, corresponding respectively to the high and low frequency linear assymptotic behaviour of $\lg[\chi''(\omega)]$ vs. $\lg\omega$, were obtained at each measuring temperature. These values proved to be temperature independent within the experimental error. Then, all the $\chi''(\omega)$ curves corresponding to different temperatures could be put on a single master curve by scaling the frequency with a temperature dependent shift factor a(T). This master curve, built at a reference temperature of 250 K, is shown in Fig. 1 together with the values of the slopes corresponding to the assymptotic behaviour. Figure 1 also shows a Havriliak-Negami [4] fit of the experimental behaviour.The inset in Fig.1 shows the temperature behaviour of the shift factor a(T). This has been fitted by a standard Vogel-Fulcher expression: $a(T) \propto \exp[A/T-T_o]$. The values of the Vogel-Fulcher parameters, A and T_o, are also shown in Fig.1. Due to the scaling property abovementioned, the temperature behaviour of the shift factor a(T) results to be the same that the corresponding one to the average relaxation time which can be defined as $\tau(T) = 1/\omega_{max}(T)$. Here $\omega_{max}(T)$ is the frequency at which the maximum of the $\chi''(\omega)$ appears at each temperature.

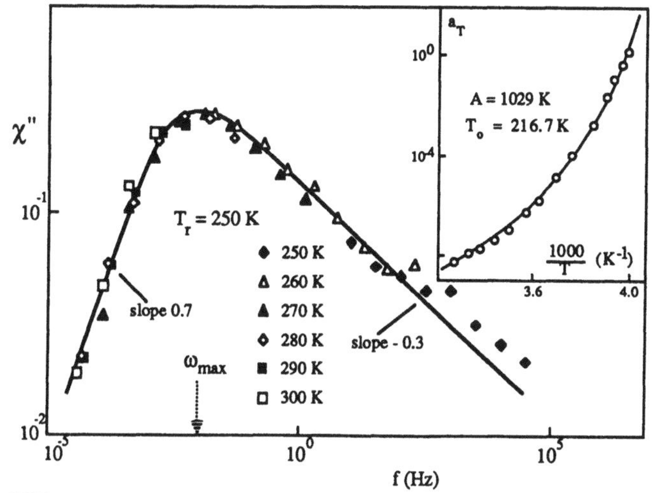

FIG. 1. Logarithmic $\chi''(\omega)$ master curve built starting from curves corresponding to different measuring temperatures. These temperatures are indicated in the figure. The continuous line correspond to a Havriliak-Negami fit. The inset shows the shift factor temperature behaviour a(T). In this case the continuous line is a Vogel-Fulcher fit. A and To are the Vogel-Fulcher parameters (see the text).

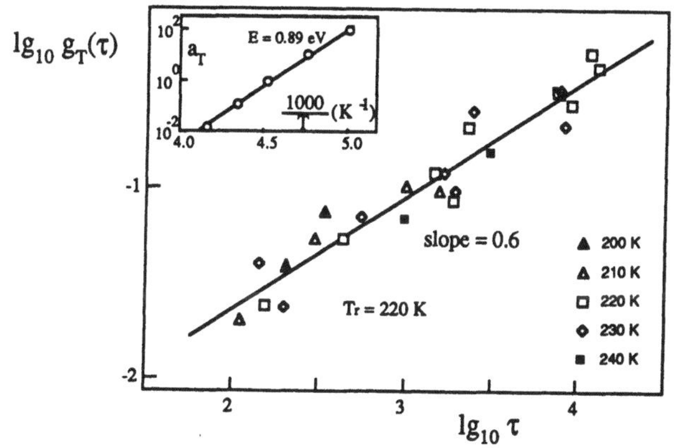

FIG. 2. Logarithmic plot of the master curve corresponding to the distribution function of the relaxation times obtained from TSDC measurements. Tr is the reference temperature at which the master curve has been built. The continuous line is a linear regression fit. The inset shows the shift factor temperature behaviour a(T). In this case the continuous line corresponds to an Arrhenius fit with an apparent activation energy E.

3.2 TSDC measurements below T_g

In this case, the empirical distribution function of the relaxation times at different temperatures can be directly obtained from the experimental results, as has been described in section 2. However, only the low times tail of the distribution function proved to be experimentaly accessible. The logarithm of this tail behaves linearly with respect to the logarithm of the relaxation time values. The slope of this straight line was temperature independent, although in this case the experimental indetermination was greater than in the case of $\chi''(\omega)$ curves. A *master curve* was built by scaling the relaxation time axis with a temperature dependent shift factor a(T).The master distribution function curve obtained, as well as the a(T) behaviour are shown in Fig. 2. As can be seen in this figure, the temperature dependence of a(T) is Arrhenius like, in this case. The value of the apparent activation energy, E, obtained from the fit is also shown in Fig. 2. From these data, an average relaxation time can be defined as: $\tau(T) = \tau_a \exp[E/k_B T]$ where τ_a can be calculated as [2]

$$\tau_a = \frac{k_B T_m}{q\, E \exp[E/k_B T_m]} \tag{2}$$

where T_m is the temperature at which the maximum of the global TSDC peak appears.

4. DISCUSSION

The temperature behaviour around the glass-transition of the average dielectric relaxation time, $\tau(T)$, as can be calculated from both, TSDC and frequency measurements, is shown in Fig. 3. As can be seen $\tau(T)$ displays a crossover from a Vogel-Fulcher behaviour (liquid-like zone above T_g) to an Arrhenius-like one (glassy zone just below T_g). This kind of behaviour has also been found in several structurally different polymeric systems [5]. Moreover, molecular transport characteristic magnitudes other than relaxation times (viscosity and diffusion coefficient) also show this temperature dependence around the glass-transition of non-polymeric glass-forming systems [6]. This *universal* behaviour can be understood in the frame of a kinetic free-volume approach which describes the equilibrium loss of a supercooled liquid through the experimental glass-transition range. A general description of this phenomenological approach can be found in our previous papers [5,6 and 7] but the main features can be summarized as follows. The classical free-volume model for the molecular transport in a dense supercooled liquid [8,9] predicts that the molecular transport characteristic magnitudes like relaxation times, are driven by the free-volume, and they behave according to the empirical Vogel-Fulcher equation. In this framework the Vogel-Fulcher parameters, A and T_o, are interpreted as the reciprocal of the expansion coefficient of the free-volume ($A=1/\alpha_f$) and as the temperature at which the free-volume vanishes (T_o) respectively. However, this Vogel-Fulcher behaviour is modified during the glass-forming process when the supercooled liquid falls out of the metastable equilibrium state. Our main assumption is to consider that the relaxation time in this region (just below the glass-transition) is still driven by free-volume. However the fractional free-volume in this region is defined as $f=\alpha_f(T-T_o) + \delta$, where δ is the average fractional excess free-volume which was first introduced by Kovacs [10]. The kinetic evolution of δ is governed by Kovacs' equation

$$\frac{d\delta}{dt} = -\Delta\alpha\, q - \frac{\delta}{\tau(\delta,T)} \tag{3}$$

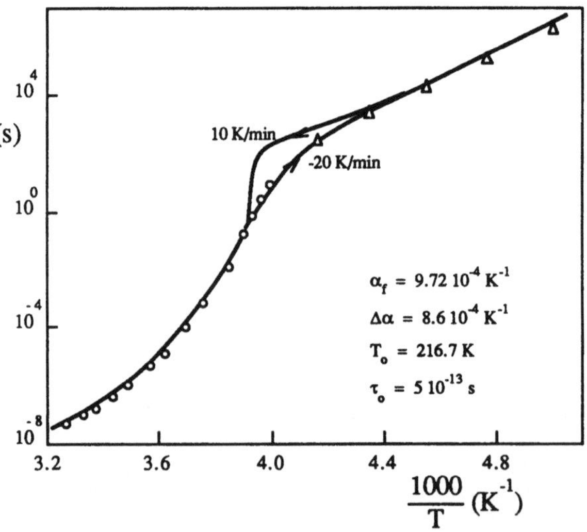

FIG. 3. Temperature behaviour of the average dielectric relaxation time around Tg. The continuous lines are theoretical simulations performed by means of eq. (3) and (4) (see the text). The values of the paremeters used in the simulation are also indicated.

where q is the rate of temperature variation, $\Delta\alpha$ is the difference between expansion coefficients in supercooled liquid and glassy states and τ is the relaxation time given by expression

$$\tau(\delta,T) = \tau_o \exp\frac{1}{\alpha_f(T - T_o) + \delta} \tag{4}$$

These two equations describe both the equilibrium (δ=0) and the non-equilibrium ($\delta\neq$0) behaviour of the time scale for the relaxation process. The values of the paramcters α_f, T_o and τ_o can be obtained from the Vogel-Fulcher fit of τ(T) behaviour above T_g. $\Delta\alpha$ value can be obtained from the Arrhenius fit of τ(T) just below the glass transition (see for example ref. 7). Equations (3) and (4) have been numerically solved in order to simulate the experimental behaviour shown in Fig. 3. The values of the model paremeters used are also shown in this figure. The two continuous lines correspond to two different values of the rate of temperature variation q. One case corresponds to the glass-forming process (q_c=20 K/min.) and the other one to a subsequent heating (q_h=10K/min.) from the glassy state. As can be seen in Fig. 3 there is a good agreement between the experimental behaviour of τ(T) and the theoretical simulation.

On the other hand, the non-Debye character of the dielectric response around T_g obtained can be expressed in terms of the well known Kohlrausch law behaviour ($\Phi(t) = \exp[-(t/\tau_p)^\beta]$). The slope of the high frequency assymptotic behaviour of χ'' as well as the slope of the low-times tail of the relaxation distribution function obtained from TSDC measurements correspond to the β value of the Kohlrausch law. Taking into account this fact, we have obtained a variation of the β value around T_g from β=0.3 (above T_g) to β=0.6 (below T_g). If the Kohlrausch behaviour is considered as consequence of a distribution of Debye-like processes, the change of the β value obtained through T_g can be interpreted as a narrowing of the distribution (glassy state) function of the Debye relaxation times. This result can also be understood, at least qualitatively, in the frame of an extension of the kinetic free volume model abovementioned [5,11].

ACKNOWLEDGMENTS

We thank Gipuzkoako Foru Aldundia for partial financial support.

REFERENCES

1. A. Alegría, J. Colmenero, J.J. del Val and J.M. Barandiarán, Polymer 26, 913 (1985)
2. J.J. del Val, A. Alegría, J. Colmenero and J.M. Barandiarán, Polymer 27, 1771 (1986)
3. J. Colmenero, A. Alegría, J.M. Alberdi, J.J. del Val and G. Ucar, Phys. Rev. B35, 3995 (1987)
4. S. Havriliak and S. Negami, Polymer 8, 161 (1967)
5. J. Colmenero, A. Alegría, J.J. del Val and J.M. Alberdi, Makromol. Chem., Macromol. Symp. 20/21, 397 (1988).
6. J. Colmenero, A. Arnau and A. Alegría In Current Topics on Non-Crystalline Solids p. 183, M.D. Baró and N. Clavaguera (eds.), World Scientific, Singapore 1986
7. A. Alegría, J.M. Barandiarán and J. Colmenero, Phys. Stat. Sol. (b) 120, 349 (1983)
8. M.H. Cohen and D. Turnbull, J.Chem. Phys. 31, 1164 (1959)
9. M.H. Cohen and G.S. Grest, Phys.Rev. B20, 1077 (1979)
10. J.M. Hutchinson and A.J. Kovacs, J.Polym. Sci. 14, 1575 (1976)
11. A. Alegría, J.M. Barandiarán and J. Colmenero, Phys. Stat. Sol. (b) 125, 409 (1984)

Primary Glass Tranisition in a Molecular Glass as Observed by Incoherent Neutron Scattering

W. Petry[1], *M. Kiebel*[2], *and H. Sillescu*[2]

[1]Institut Laue-Langevin, 156X Av. des Martyrs, F-38042 Grenoble Cedex, France

[2]Institut für Physikalische Chemie, Universität Mainz, Postfach 3108, D-6500 Mainz, Fed. Rep. of Germany

Significant progress in the understanding of the dynamics of the glass transition has been obtained in the last four years by new microscopic theories. In the mode coupling approach the freezing of density fluctuations is seen as the process defining the glass transition. Among others, this theory makes two characteristic predictions for the glass transition [1-6] : 1) The transition from non ergodicity to ergodicity of the density fluctuations occurs at a critical temperature T_0 well above the thermodynamic T_g, (T_g may be defined in a jump of the specific heat c_p). These density fluctuations split up into two frequency regions both of them following different scaling laws and separated by a critical frequency ω_c.These fluctuations are expected to be of low frequency, i.e. below a certain phonon cut off of roughly 10^{12} Hz.

The scattering law as measured by coherent or incoherent neutron scattering is a direct measure of these density fluctuations. Using the available quasielastic scattering instruments a frequency range of roughly 10^8 to 10^{12} Hz is achievable and in the last two years several neutron scattering experiments on different glass formers have been reported [7-13]. Independent of the particular glass all these experiments confirm at least qualitatively the essential predictions of the recent microscopic theories.

Here we report on incoherent quasielastic neutron scattering (IQNS) on the low molecular van der Waals glass 1,1-di-(p-methoxyphenyl)-cyclohexane (DMPC). This glass has been chosen for practical purposes because it can be kept easily for several days in the supercooled or viscous state without crystallization. In spite of their complex structure van der Waals glasses are considered as a reasonable approach to the simple hard sphere model of the microscopic theory because the van der Waals interaction between the molecules is essentially a short range interaction $V(r) \propto C/r^6$. In order not to be misled by internal high frequency motions the rather free CH_3 ligands of DMPC have been replaced by deuterated ones thus being almost invisible for IQNS. 15 % of the CD_3 groups are statistically replaced by H [14]. Such a modification of part of the molecules very effectively prevents recrystallization but results in only slight changes of T_g and viscosity with respect to the pure system.

The system is a special make for our purposes [15] and has been characterized by DSC calorimetry [14] (T_g = 249 K). Using dielectric relaxation and photon correlation spectroscopy [16], the α process has been studied in the Hz to MHz region. There, the α process is found to yield stretched exponential relaxation functions of type $\sim \exp(-(t/\tau_0)^{\beta})$ with a temperature independent $\beta \approx 0.50$ (0.05).

The neutron experiments were carried out at the thermal neutron back-scattering spectrometer IN13 at the Institut Laue-Langevin in Grenoble. The energy resolution was set to 8 μeV FWHM allowing to look for dynamics faster than 10^{-9}s. The accessible range of momentum transfer Q was 0.3Å^{-1} to 5 Å^{-1}.

As a starting point we performed elastic window scans as shown in Fig. 1. At temperatures sufficiently below T_g the elastic intensity can be described by a

Springer Proceedings in Physics, Vol. 37 Dynamics of Disordered Materials
Editors: D. Richter · A.J. Dianoux · W. Petry · J. Teixeira

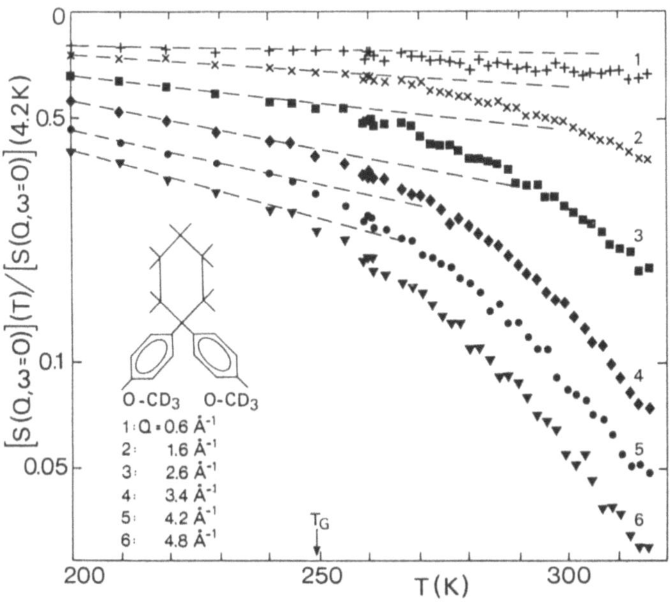

Fig. 1 : Incoherent structure factor of DMPC. $S(Q,\omega = 0)$ normalized to $S(Q,\omega = 0)$ at T = 4 K vs. temperature T (elastic window), plotted for selected Q-values. Each curve represents the sum of five independent measurements within an interval of $\pm$ 0.4 $Å^{-1}$ about the indicated Q-values. Far below T_o the data obey the Debye-Waller factor $\ln(I^{DWF}) \propto - T$ indicated by the dashed lines. Around T_g a strong Q-dependent onset of an extra decrease is observed.

normal Debye-Waller factor predicting a linear temperature decrease of $\ln[S(Q,\omega = 0)]$. Around T_g, however, significant anomalies occur. The observed strong extra decrease of the elastic intensity is fully consistent with our finding in tri-α-naphtylbenzene (TNB) [8] and ortho-terphenyl (OTP) [12] and can be analyzed in terms of a local motion with strongly increasing amplitudes as T is raised above T_g. This local motion seems to be a precursor phenomenon of the glass transition or a kind of secondary relaxation process. The detailed presentation of the full analysis of this anomalous Debye-Waller factor is omitted in this paper. It supports the ideas expressed in Ref. 8 and 12 but does not add further insight into the physical nature of the observed local modes.

In this paper we would rather like to draw the attention on phenomena at even higher temperatures where gradually we start to see molecular diffusion in addition to local motion. To this end, quasielastic spectra $S(Q, \omega)$ have been taken at T = 280 K, 320 K, 350 K and 365 K. A dynamic window of $\Delta E = - 110...$ + 210 μeV has been set. The resolution profile of the spectrometer has also been measured by a run at 2 K. Fig. 2 shows typical experimental spectra for a) a fixed momentum transfer and varying temperatures and b) a fixed temperature and varying values of momentum transfer. From Fig. 2a we see that at 280 K (= T_g + 31 K) the spectrum is essentially identical with the instrumental resolution function. For T $\geq$ 320 K, however, a gradually increasing quasielastic broadening occurs. From Fig. 2b we qualitatively see that the quasielastic broadening shows a Q-dependence typical for long range diffusive type motion.

In order to visualize lineshape characteristics we Fourier transform the spectra and divide by the Fourier transform of the resolution function. In this way we obtain the intermediate scattering function S(Q,t). Fig. 3 shows data

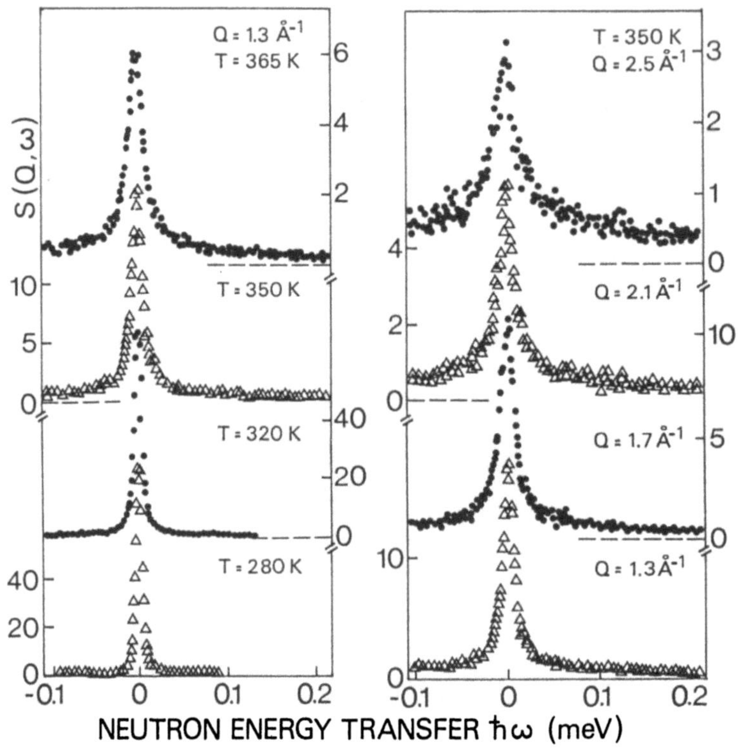

Fig. 2 : Incoherent scattering law $S(Q,\omega)$ of DMPC for a) different T at fixed scattering vector Q and b) at different Q for fixed T. The spectrum measured at T = 280 K and Q = 1.3 $Å^{-1}$ corresponds to the instrumental resolution in energy.

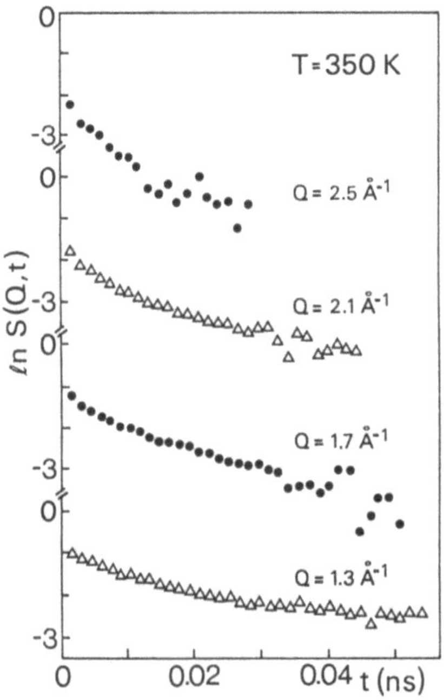

Fig. 3 : Intermediate scattering function $S(Q,t)$ of DMPC. Spectra in Fig. 2b have been Fourier transformed in t-space and deconvoluted by the instrumental resolution.

(same as those shown in Fig. 2b) treated in this way, independent from any theoretical preassumptions. It is apparent that ln[S(Q,t)] deviates significantly from a linear behaviour with t : the decay is not described by a single exponential but - as will be shown below- is well described by a stretched exponential law.

For a quantitative analysis we prefer to go back to the ω-domain in order to avoid cut-off problems originating from Fourier transforming the experimental data in the narrow dynamic range. We rather transform a Kohlrausch law obtaining the fitting curve

$$S(Q,\omega) \sim \int e^{-i\omega t}\, e^{-(t/\tau_0)^{\beta}}\, dt \qquad (1)$$

This numerical Fourier tranformation is then numerically convoluted with the measured resolution function and placed inside a least square fit procedure [17]. The fits (solid line in Fig. 4) work perfectly well yielding results for τ_0 and β as summarized in the following Table :

Table 1 : Results of fits of the measured $S(Q,\omega)$ by means of the Fourier transform of a stretched exponential. τ_0 in parenthesis correspond to fits with a fixed $\beta = 0.6$.

τ_0(ns)		Q($Å^{-1}$)				
		1.3	1.7	2.1	2.5	2.8
	320	1.7 (2.7)	0.52 (0.49)	0.30 (0.34)	0.22 (0.23)	0.12 (0.24)
T(K)	350	0.060 (0.10)	0.040 (0.056)	0.023 (0.041)	0.020 (0.020)	0.0082 (0.026)
	365	0.045 (0.041)	0.027 (0.028)	0.016 (0.018)	0.0075 (0.015)	-

β		Q($Å^{-1}$)				
		1.3	1.7	2.1	2.5	2.8
	320	0.65	0.60	0.50	0.45	0.39
T(K)	350	0.43	0.50	0.46	0.58	0.39
	365	0.63	0.58	0.53	0.44	-

The uncertainty of β and τ_0 is estimated to roughly 15 %. As is seen from the Table, τ_0 is clearly T- and Q-dependent (diffusion, → α-process). β is T-independent in our (narrow) T-domain and indicates some decrease with increasing Q. Because of the strongly decreasing Debye-Waller factor the analysis of the α-process is not extended beyond 2.8 $Å^{-1}$. Analysis of the higher Q-range is in principle possible but would demand significantly better statistics.

The Q- and T-dependent relaxation time τ_0 can be compared with other dynamics measurements (dielectric relaxation, photon correlation [16]) recently performed with DMPC. Taking the τ_0 at Q = 1.3 $Å^{-1}$ they fit to a continuous

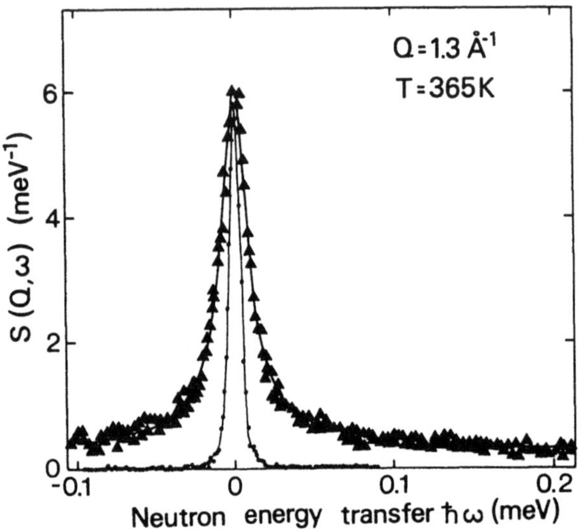

Fig. 4 : Typical spectra of DMPC perfectly fitted in the ω-domain by the Fourier tranform of the Kohlrausch-function. For comparison the measured resolution is shown.

curve of relaxation times extending from 10 to 10^{-11} sec. As shown in the Table values of τ_o at larger Q are smaller and therefore slightly deviate from the relaxation times measured by dielectric relaxation or photon correlation.

The T-independence of β, i.e. the predicted scaling of the α-process can be demonstrated in terms of a master plot. Fig. 5 shows the scaled intermediate scattering law $f(t/\tau_o) = S(Q,t/\tau_o) = A.\exp(-(t/\tau_o)^\beta)$. The data points stem from the spectra measured at Q = 1.7 Å^{-1}, Fourier transformed as described above and scaled with the fitted τ_o (see table). This scaling is certainly justified by the results itself, i.e. all measured points fall on one master curve but also by the fact, that τ_o compares to what one measures with other dynamic methods. The starting value of $f(t/\tau_o)$ of less than 1 describes the fast β-relaxation as obtained from the elastic scan (Fig. 1).

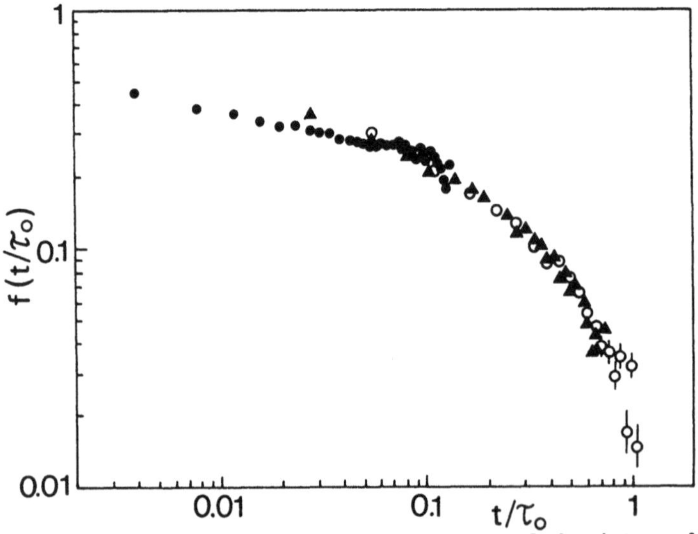

Fig. 5 : Master plot of the scaled intermediate scattering law $f(t/\tau_o) = S(Q,t/\tau_o)$.

Summarizing, we see the α-process in a molecular glass by observing a T- and Q-dependent spectral shape of the incoherent scattering function. The lineshape is well described by a Kohlrausch exponent β = 0.65...0.4 for Q = 1.3 ... 2.8 $Å^{-1}$ which compares to those measured by other methods in completely different Q-ranges [16] but on the same material. The T independent spectral shape can be nicely visualized by a plot of the intermediate scattering function $S(Q,t/\tau_0)$ measured at different temperatures and scaled by the relaxation time τ_0.

References

1. E. Leutheusser, Phys. Rev. A29, 2765 (1984).
2. U. Bengtzelius, W. Götze and A. Sjölander, J. Phys. C 17, 5915 (1984).
3. W. Götze, Physica Scripta 34, 66 (1986).
4. U. Bengtzelius, Phys. Rev. A34, 5059 (1986).
5. U. Krieger and J. Bosse, Phys. Rev. Lett. 59, 1601 (1987).
6. W. Götze and L. Sjögren, this conf.
7. F. Mezei, W. Knaak and B. Farago, Phys. Rev. Lett. 58, 571 (1987).
8. F. Fujara and W. Petry, Europhysics Lett. 4, 921 (1987).
9. B. Frick, D. Richter, W. Petry and U. Buchenau, Z. Physik B 70, 73 (1987).
10. W. Knaak, F. Mezei and B. Farago, Europhysics Lett. 7, 529 (1988).
11. D. Richter, B. Frick and B. Farago, Phys. Rev. Lett. 61, 2465 (1988).
12. E. Bartsch, M. Kiebel, F. Fujara, H. Sillescu and W. Petry, this conf.
13. W. Doster, S. Cusack and W. Petry, this conference.
14. B. Gerharz, diploma thesis, Universität Mainz, 1989.
15. M. Antonietti, T. Wagner, unpublished synthesis work.
16. J.U. Hagenah, Ph.D. thesis, Universität Mainz, 1988; F. Kremer, unpublished.
17. J. Wuttke, ILL technical report 87WU12T, 1987.

A Contribution of Neutron Scattering to a New Understanding of the Glass Transition

W. Knaak

Hahn-Meitner-Institut, D-1000 Berlin 39, Germany

While passing the liquid-glass-transition only some gradual changes in the *static* structure factor S(Q) are observed, characteristic changes occur in the (coherent) *dynamic* structure factor S(Q,ω) of the ionic glass $Ca_{0.4}K_{0.6}(NO_3)_{1.4}$. There is an anomalous increase of inelastic intensity setting in at T_g simultaneously at all energies below ~2meV, which, by analysing four orders of energy transfer, shows signs of critical behaviour as predicted by the mode coupling theory. As opposed to earlier phase transition models of the liquid-glass transition the critical temperature of this purely dynamical instability lies distinctly (~30K) *above* T_g.

Introduction

It is well known, that the change of the apparent activation energy E_A for describing the viscosity of a glass forming liquid by an Arrhenius law correlates with the magnitude of the specific heat step ΔC_p at T_g. The observation that a big change of E_A is connected with a big step ΔC_p, i.e. a big change of configurational entropy in the liquid leads to the suggestion, that subtle structural changes are responsible for a change in transport behaviour /1/. Although these structural changes can not directly be seen in S(Q), it seems that their consequences on the dynamics of the system show up in the dynamic scattering function S(Q,E). What has now been observed in some quite different glasses at T_g /2-4/ is an anomalous decrease of the Debye-Waller factor and a corresponding anomalous increase, i.e. above thermal occupation, of the inelastic intensities around and below 1meV. The latter feature is shown in fig.1a for a glass of another class, the ionic glass $Ca_{0.4}K_{0.6}(NO_3)_{1.4}$, measured on the time-of-flight spectrometer IN5, ILL /5/. The difficulty is to make a meaningful interpretation of this observation and it is not anticipated that all glass forming systems behave equally. What qualifies this ionic system in this context, is that is has among other properties a big change in E_A, a big specific heat step ΔC_p and very little frozen in entropy and therefore all features connected with the liquid-glass-transition (LGT) should be observable in a pronounced manner.

Results and Discussion

The sharp onset of the increase of the inelastic intensities at T_g (fig.1b) simultaneously at all frequencies up to 0.5 THz measured at a characteristic correlation length of $2\pi/Q_0$ = 3.3Å may be regarded as a "microscopic high frequency signature" of T_g. There are other "high frequency signatures" of T_g, e.g. the changing temperature coefficient for the index of refraction n(T), for stress induced double refraction or for the ultra sound velocity. But these quantities probe the glass on a macroscopic length scale and the changes can be traced back to the temperature dependence of the density ρ(T) or of the mode Grüneisen parameters $\Gamma_i = -d(\ln\omega_i)/d(\ln V)$ introduced to describe anharmonic effects. The macroscopc quantities however give no information which modes have a changing Γ_i. In any case an increase of inelastic intensity is clearly due to increased motion favoured by an increase of the looseness of the structure above T_g or of "free volume". The free volume of a molecular cell as introduced by Cohen and Grest /6/ is that above a critical volume and by entropy arguments it is shown, that cells having free volume agglomerate into liquid-like clusters, which above the percolation threshold allow viscous flow. To avoid a first order phase transition at percolation

Springer Proceedings in Physics, Vol. 37 **Dynamics of Disordered Materials**
Editors: D. Richter · A.J. Dianoux · W. Petry · J. Teixeira

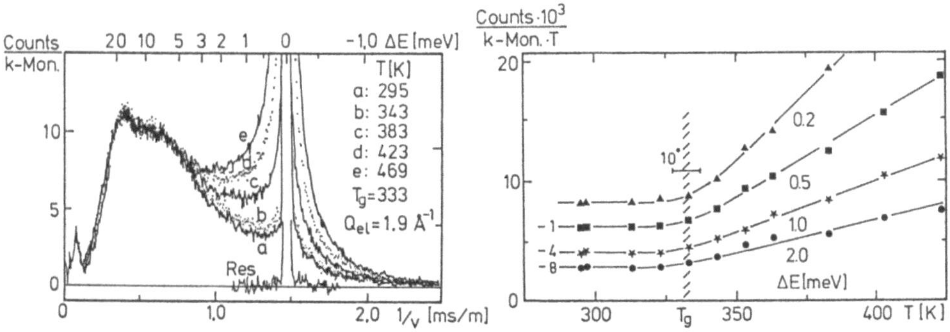

Fig.1: a) Time-of-flight neutron scattering spectra of $Ca_{0.4}K_{0.6}(NO_3)_{1.4}$ at different temperatures around the glass transition corrected for background, detector efficiency and self shielding. All spectra are normalized to the same temperature by the Bose-factor, in order to indicate the behaviour of the spectral density itself. The buttom curve shows the resolution measured with a vanadium standard and scaled to the same peak intensity.

b) Change of the inelastic intensities as a function of the temperature for different energy transfers. The T denominator roughly reflects the Bose factor. For clarity some curves have been shifted downards by the amount indicated. (from ref.5)

threshold it is assumed, that due to increasing relaxation times the system freezes at T_g into a fixed state above the threshold. This means, the total amount of free volume is fixed below T_g and increases above, a finding which would be consistent with the temperature variation of the inelastic intensities at ~1meV. The above mentioned T_g-signatures are of course cooling rate dependent and we would guess, this is also the case for the n-scattering signature, which was gained with low cooling rates (~1K/min). Although the concept of free volume was introduced to explain viscous flow, i.e. slow structural relaxation, this reasoning may intuitively also account for localized motion on a time-scale between the structural relaxation and the microscopic oscillation like the dielectric ß-relaxation, which however is not present in $Ca_{0.4}K_{0.6}(NO_3)_{1.4}$ /7/.

One way to characterize this motion further is to study the inelastic structure factor $S_E(Q)$. Fig. 2a shows a contourplot of the dynamic cross section $\partial^2\sigma/\partial\Omega\partial E$ for the glass at room temperature. Although the peak of the structure factor S(Q) is rather broad clear signs of diffuse Umklapp-scattering are observed. This feature is completely absent in the difference between the Bose-scaled room temperature and the 383K spectrum, containing only the anomalous intensity. To extend the Q-range, a further, preliminary TOF-experiment was performed at the SV5 spectrometer at the reactor Dido, KFA Jülich. The resulting inelastic structure factor, $S_E(Q)$, for room temperature and E=(1.5±0.5)meV, shown in fig.2b, is proportional to $S(Q)\cdot\exp(-2W)\cdot Q^2 = I_{el}(Q)\cdot Q^2$, $I_{el}(Q)$ being the elastic or quasielastic intensity. This leads to the conclusion, that acoustic waves are dominant in the glass in this Q-E-region, not very much above the Debye-limit (see fig.3), calculated from /9/:

$$S(Q,E\to 0) = \frac{\hbar^2\cdot Q^2}{2\cdot\bar{m}}\cdot k_B T\cdot\frac{3\cdot\pi}{(\hbar\omega_D)^3}\cdot I_{el}(Q) \quad \text{with } \frac{\hbar^2\cdot Q^2}{2\cdot\bar{m}} = 0.119\text{ meV} \quad \text{and}$$

$\hbar\omega_D$ = 12.95 meV (≙ v_D=1.72 km/s). The inelastic structure factor for 513K is much more difficult to interprete. For high energy transfer, say E=10meV, both spectra differ only by a Q-independent Bose-Einstein temperature factor of approximately 1.7. Applying the same factor to the room temperature inelastic structure factor at E=1.5meV gives an estimate of the phonon background, which has only little structure. Therefore the inelastic stucture factor of the anomalous intensity also seems to be rather flat for Q > 2Å^{-1}, i.e. it doesn't drop like $I_{el}(Q,T=513K)$, also shown in

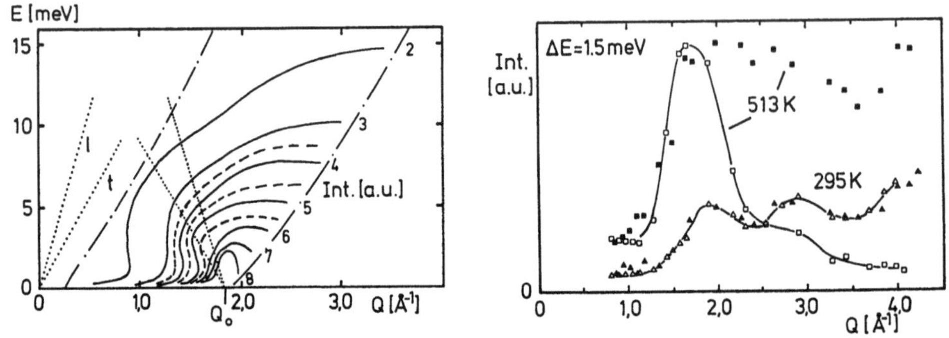

Fig.2: a) Contourplot of the dynamic cross-section ($\partial^2\sigma/\partial\Omega\partial E$) at room temperature measured on IN5, ILL, with λ=5.8Å /8/. The numbers at the solid lines give the intensity in a.u. and the dotted lines show the longitudinal and transverse ultrasound velocities.

b) Inelastic structure factor (ΔE=(1.5±0.5)meV) at room temperature (triangles) and 513K (squares), measured on SV5, KFA Jülich with λ=3.0Å. The full symbols show the data on the same intensity scale, while the open triangles show $Q^2 \cdot I_{el}(Q,T=295K)$ and the open squares $I_{el}(Q,T=513K)$ scaled properly.

fig.2b, as would be expected if there would be only one quasielastic component. On the other hand, S(Q,E>0) should have a Q^2-factor as long as we are dealing with some oscillatory motion. Although its impossible to draw a firm conclusion from these preliminary data, they are in favour of a picture of some localized oscillations. However no model exists explaining the change of the density of states near the glass transition.

Another framework to analyse the data is offered by the now well known mode coupling (MC-) theories of supercooled liquids /10,11/. For this purpose it's essential to examine a very wide energy range of S(Q,E), which has been done using two different techniques (NSE and TOF) /5/. The data of both experiments were put together using standard Fast Fourier Transform techniques and the results of this calculations are shown in fig.3 in the ω-domain (solid lines) together with the original TOF-data (points), transformed into the symmetrized (classical) scattering function:

$$\hat{S}(\theta,E) = \left(\frac{\partial^2\sigma}{\partial\Omega\partial t}\right) \cdot \left(t/t_0\right) \cdot \left(\frac{t^3}{2t_0^2 \cdot E_0}\right) \cdot \exp\left(-\frac{E(t)}{2kT}\right)$$

There, t denotes the time-of-flight ($t=t_0$ for elastically scattered neutrons) and E_0 the incident energy of 2.42meV. The scattering functions thus measured over more than four decades in energy show two adjacent power law sections and a crossover frequency $\omega_\varepsilon(T)$. These experimental spectra look very much the same as those numerically calculated within the MC theory for a Lennard-Jones liquid by Bengtzelius /12/. On lowering the temperature the whole spectrum narrows, but the slowing down of the half width of the central peak (ω_ε^l) is much stronger than that of the crossover frequency ω_ε. These are exactly the generic features predicted by the mode coupling theories: $\Phi_q(t)$:=S(Q,t)/S(Q) is expected to decay in two steps, where the faster one shows up in the Fourier Transform S(Q,ω) as a high energy tail of the slower ("primary") relaxation. The splitting of timescales of these two relaxation steps produces a minimum in the susceptibility $\chi_q''(\omega) = \omega \cdot S(Q,\omega)/kT$, which indeed is observed in the experimental spectra /5/. Inherent to the theory is the property, that the frequency of this minimum should not depend on Q /13/. Fig.4 shows $\chi_q''(\omega)$ for different Q-values below Q_0 and some small but insignificant changes are observed. The biggest change is in the phonon intensities above 1meV, which however has only minor effects on ω_ε. A direct measurement of the spectrum (not via S(Q,t), e.g. on instrument IN13) should be helpful to clarify this point.

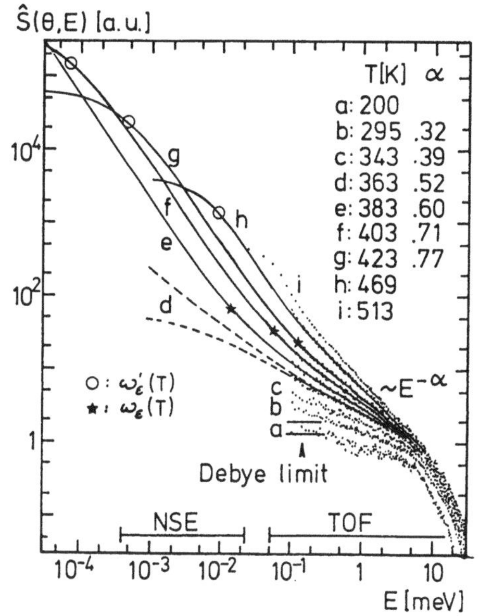

Fig.3: Symmetrized (classical) scattering functions at constant scattering angle ($Q_{el} \approx 1.9$ Å^{-1}) obtained from the combination of NSE and TOF experiments by Fourier Transformation (solid lines). The dots represent the TOF data as measured. The two broken lines for 363K illustrate how the error bars of the NSE experiment transform into the ω-domain due to the fact that the NSE-signal is mainly elastic there. At higher temperatures there is no such ambiguity. (from ref.5)

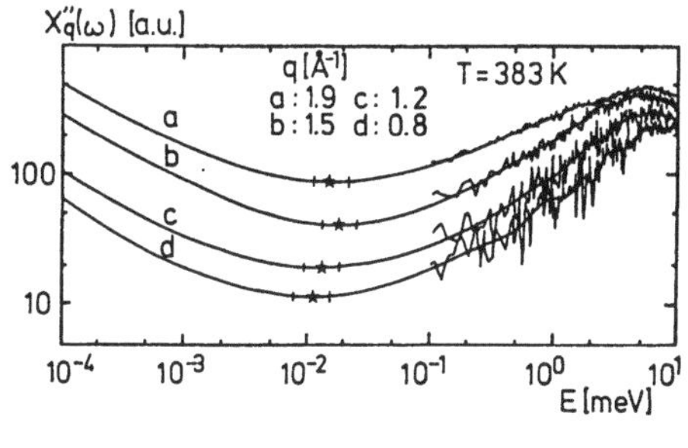

Fig.4: Dynamic susceptibility $\chi''_Q(\omega)=\omega\cdot S(Q,\omega)/kT$ for T=383K and different Q-values below Q_0 /8/. The stars mark the minimum defining the frequency-scale ω_ε.

Central to the understanding of the LGT in the MC theory is the prediction of a critical dynamical instability at some temperature T_0 at which ω_ε (and ω'_ε) should drop to zero opening up a window to the Enskog-collision-frequency ω_E ($\approx$ some meV) in which the critical scattering can be observed. Our previous NSE-results /14/ show, that ω'_ε scales with the viscosity $\eta(T)$ which in turn can be fitted by a powerlaw $\eta(T)=(T/T_0-1)^{-4.2}$ in the range of low viscosities ($\eta < 10^4$dPas) with $T_0=(366\pm4)$K. The temperature dependence of ω_ε is also compatible with a powerlaw with the same choice of T_0. Again a direct measurement of the spectrum in this energy range should give better significance. The crucial point is now, that T_0 lies distinctly above the real glass transition temperature of T_g=333K of this system. There have been several attempts in the past to describe an "ideal" glass transition as a structural phase transition /6,15/ which should occur when a supercooled liquid is driven through the LGT infinitely slowly and which therefore should lie below the conventional T_g. This was thought to be in line with the Kauzmann paradoxon /16/ which provides very generally a lower temperature limit for the existence of a supercooled liquid. Provided the observation of $T_0 > T_g$ has some general validity /17/, the Kauzmann paradoxon is solved by the MC theory in a quite different way: the LGT is driven by some purely dynamical processes *above* T_g and nothing anomalous happens below. Again, these dynamical processes are in the theory triggered by a small, smooth change in the structure S(Q).

Another prediction of the theory is that the spectra should scale like $\omega_\varepsilon(T)$ near to T_0 in the range $\omega_\varepsilon' \ll \omega \ll \omega_\varepsilon$, i.e. around ω_ε. Such a scaling is also observed /5/ although the spectra for 423K and up may already be too far from T_0. The critical spectrum itself, which in the ideal transition should extend over the whole ω-range leading to a power-law divergency, is observed only over 1.5 decades due to lacking resolution for the temperatures below 383K and perhaps due to the nonideality of the transition of real systems /18/.

Although the general features of the observed spectra lend strong experimental support to the MC theory, the question of the nature of observed anomalous low frequency density fluctuations remains open. The preliminary results on the Q-dependence (fig.1) suggest some local modes of slow oscillation, but, although the MC theory always speaks of "relaxation" this needs not to be, a priory, a contradiction because in the theory no explicit statement about the nature of the critical motion is given. Instead, an analysis of the Q-dependences should be tried using the MC theory, but for this purpose only a very small amount of theoretical data is available. On the other hand it is clear, that local modes particularly depend strongly on the short range structure of the liquid which is not accounted for in the one component MC-theory.

Conclusion

The most important finding of this study is the observation of anomalous spectral intensity in the range below ~2meV exhibiting signs of critical behaviour predicted by the MC-theory. This is the first theory, which together with experimental data, leads to a hidden dynamical phase transition at a temperature *above* T_g. The observed signs of criticality are the powerlaw for the anomalous spectral density itself and for the temperature dependence of its frequency scale ω_ε as well as for ω_ε'. Whether this spectral density arises from some local oscillations, suggested by preliminary results on the dynamic structure factor, remains an open question. However, we are confident, that the increasing looseness of the structure above T_g like the evolution of "free volume" is responsible for the increased inelastic intensities.

Acknowledgements

I would like to thank U. Buchenau for giving me beam-time on his instrument SV5 as well as F. Mezei and B. Farago for good collaboration and J. Bosse and W. Götze for stimulating discussions.

References

1. for a recent rewiew of "Models of the Glasss Transition" see: J.Jäckle Rep.Prog.Phys. 49,171,(1986)
2. Phillips, Nücker, Buchenau, Petry, to be published G.Galli, P.Migliardo, R.Bellisent, W.Reichardt, Sol.Stat.Comm.57,196,(1986)
3. B.Frick, D.Richter, W.Petry, U.Buchenau, Z.Phys.B70 ,73,(1988)
4. F.Fujara, W.Petry, Eur.Phys.Lett.4,921(1987)
5. W.Knaak, F.Mezei, B.Farago, Eur.Phys.Lett., in press
6. M.H.Cohen, G.S.Grest, Phys.Rev.B24,4091,(1981)
7. J.Wong, C.A.Angell, "Glass Structure by Spectroscopy", Marcel Dekker, New York, (1976), p.756
8. same measurement as ref.5
9. deduced from equ!s 42,43 of: U.Buchenau, Z.Phys.B58,181,(1985)
10. W.Götze, this volume and references therein
11. J.Bosse, this volume and references therein
12. U.Bengtzelius, Phys.Rev.A34,5059,(1986)
13. W.Götze, Verh.d.DPG VI,23,64,(1988)
14. F.Mezei, W.Knaak, B.Farago, Phys.Rev.Lett.58,571,(1987)
15. E.A.DiMarzio, Ann.NY Ac.Sci.371,1,(1981)
16. W.Kauzmann, Chem.Rev.43,219,(1948)
17. P.Taborek, R.N.Kleimann, D.J.Bishop Phys.Rev.B34,1835(1986)
18. W.Götze, L.Sjögren, Z.Phys.B65,415,(1987)

β-NMR on $LiCl{\cdot}7D_2O$ in the Glassy and the Liquid State

W. Faber, P. Heitjans, and A. Schirmer

Institut für Physikalische Chemie, Universität Hannover, Callinstr. 3-3A, D-3000 Hannover 1, Fed. Rep. of Germany, and
Fachbereich Physik, Universität Marburg, Renthof 5, D-3550 Marburg, Fed. Rep. of Germany

The glass forming system $LiCl \cdot 7D_2O$ (T_g = 144 K) was studied by the method of β-radiation detected NMR. The spin-lattice relaxation rate $1/T_1$ of ^8Li was measured as function of temperature (4 K $\lesssim$ T $\lesssim$ 235 K) and magnetic field (50 mT $\lesssim$ B $\lesssim$ 600 mT). $T_1^{-1}(T)$ shows a motion induced peak with its maximum near 160 K. Down to about 80 K $T_1^{-1}(T,B)$ is described by the sum of two spectral densities with different correlation times and activation energies belonging to two motional processes. Below about 80 K the temperature and field dependence is weak and proportional to $T_1^{-1} \propto T^{0.3}/B^{0.5}$, which may be ascribed to the dynamics of two-level-systems.

1. Introduction

Concentrated aqueous electrolyte solutions of LiCl are the subject of considerable current interest, because when vitrified they allow the study of the low-temperature range with its glassy properties as well as the glass transition and the liquid state. The solutions $LiCl \cdot n\, D_2O$ are easily vitrified for the concentration ranges n $\simeq$ 4 and n $\simeq$ 7 /1,2,3/. Whereas the liquid state has been investigated by NMR for several years /4,5,6/, the glassy state was covered only recently /7,8/ apart from an early NMR work at 100 K /9/. Here we present first results by β-NMR on ^8Li in $LiCl \cdot 7D_2O$. Among the various features of the β-NMR method being relevant for spin relaxation studies /10/ we emphasize the fact that, contrary to conventional NMR, the isolated β-active probe nuclei used here are generally not sensitive to paramagnetic impurities. Furthermore the magnetic field can easily be varied continuously down to low values without loss of sensitivity.

2. Experimental

The measurements were performed on the new β-NMR spectrometer at the research reactor DIDO in Jülich. Cold neutrons ($\lambda \gtrsim 4$ Å) are extracted with a bender out of a straight neutron guide, polarized by reflections from a magnetized Co-Ti supermirror and led to the sample with another neutron guide, whose final section is a focussing guide in order to enhance the flux at the sample. To conserve polarization a solenoid surrounds the guide. By capture of the polarized neutrons in the $^7LiCl \cdot 7D_2O$ sample polarized β-active ^8Li ($T_{1/2}$ = 0.8 s) nuclei are produced. The probe polarization is monitored via the asymmetry of the β-radiation, which is detected by two pairs of plastic scintillation counters. To eliminate instrumental asymmetries the neutron polarization is changed periodically with a gradient radio-frequency spin flipper. With a cold polarized neutron flux of about $2.5 \cdot 10^7$ cm^{-2} s^{-1} we obtain a ^8Li β-counting rate of about 10^3 s^{-1}.

3. Results and Discussion

The relaxation time T_1 was obtained by measuring the β-counting rate asymmetry as function of time. In the liquid as well as in the glass single exponential decay of the asymmetry, i.e. of the nuclear polarization, was found within the accuracy of the data. Figure 1 shows the T dependence of T_1^{-1} for B = 500 mT in the whole temperature range covered in this experiment. The overall T dependence is characterized by a pronounced motion-induced maximum in the liquid state and a flat continuous slope in the glassy state. The caloric glass transition point (T_g = 144 K /11/) falls into the range of the low-T wing of the maximum.

For low temperatures in the glassy state, say below about 80 K, the T dependence can be described by a power law $T_1^{-1} \propto T^{\alpha}$ with $\alpha \simeq 0.3$ for B = 100...600 mT. On the other hand, the B dependence of T_1^{-1} for $T \lesssim 80$ K approximately follows $T_1^{-1} \propto B^{-\beta}$ with $\beta \simeq 0.5$. Similar weak T and B dependences were found in earlier β-NMR and NMR studies of nuclear spin relaxation in glasses at low temperatures /7,12,13/ and can be ascribed to the dynamics of disorder modes typical of glasses, i.e. two-level-systems.

It should, however, be mentioned that at $T \lesssim 20$ K a broad shallow maximum in $T_1^{-1}(B)$ seems to occur round about 300 mT as illustrated in Fig. 2 for T = 10 K. The other examples for $T_1^{-1}(B)$ at 150 K and 100 K, given in Fig. 2, again show different behaviour each and belong to temperature regions which may be defined by $T \gtrsim 140$ K (region I) and $140 \text{ K} \gtrsim T \gtrsim 80$ K (region II), respectively.

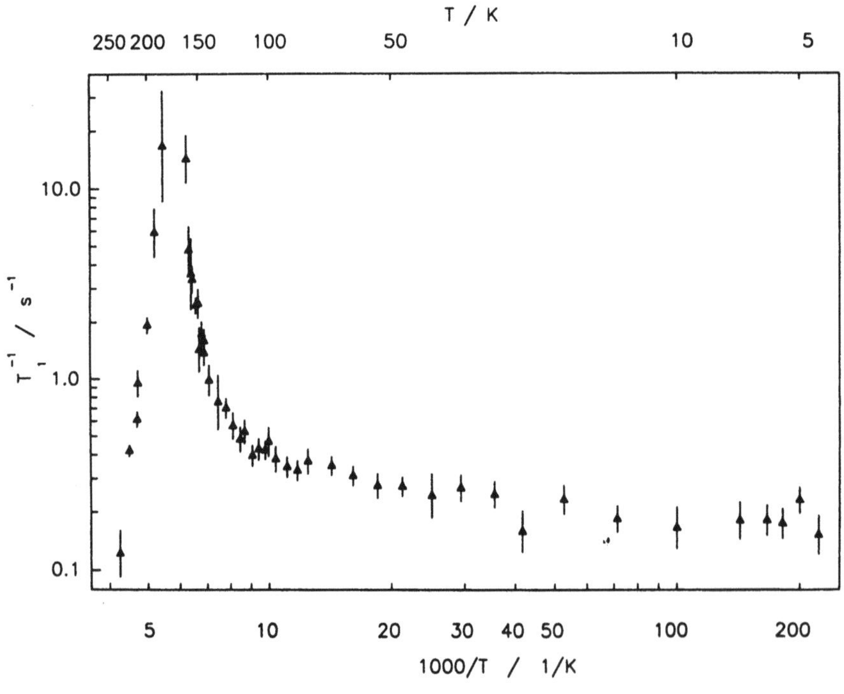

Figure 1: ^{8}Li relaxation rate T_1^{-1} as function of temperature for B = 500 mT

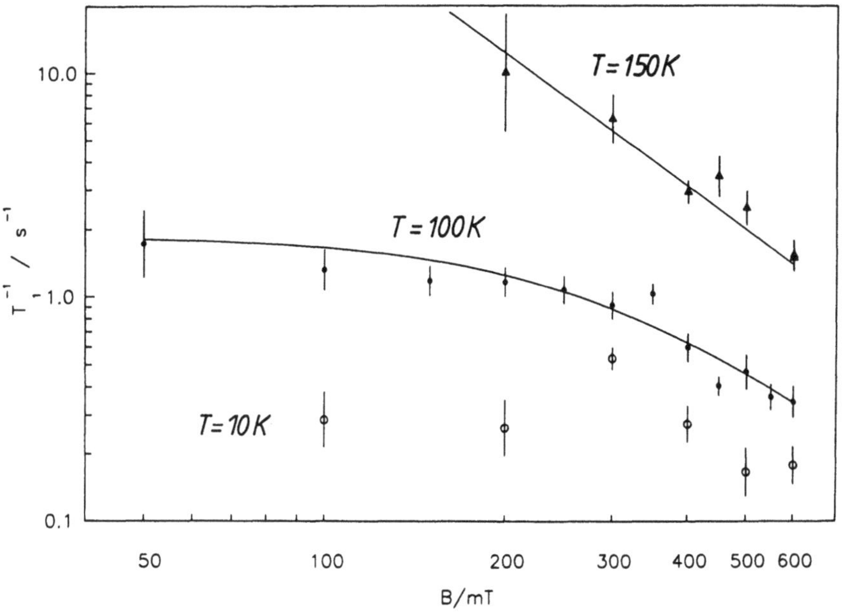

Figure 2: ^{8}Li relaxation rate T_1^{-1} as function of magnetic field in various T regimes. The lines through the data for T = 150 K and 100 K represent fits explained in the text.

In the high-temperature region (I) $T_1^{-1}(B)$ for T = 150 K is proportional to B^{-2} (see Fig. 2), so in a preliminary analysis we used the simple ansatz

$$T_1^{-1} = a\,\tau_c/(1+\omega^2\tau_c^2) \qquad (1)$$

$$\text{with} \qquad \tau_c = \tau_0 \exp(E_A/kT)$$

where τ_c is the correlation time of the dominating motional process. With the parameters $E_A = 0.28$ eV, $\tau_0 = 3 \cdot 10^{-16}$ s, $a = 6 \cdot 10^{8}$ s^{-2} we can fit both, the B dependence for T = 150 K (Fig. 2) and the T dependence in range I (Fig. 3). In the intermediate range (II) $T_1^{-1}(B)$ for T = 100 K was analysed in terms of equation (1), too, yielding $E_A = 0.02$ eV, $\tau_0 = 1.1 \cdot 10^{-8}$ s, $a = 2.1 \cdot 10^{7}$ s^{-2}. The solid line in Fig. 3 presents the sum of these two T_1^{-1} contributions. For a correct description of the high-temperature wing of the $T_1^{-1}(T)$ peak in region I the contribution being dominant in region II has to be cut off at about the temperature where the peak maximum occurs, i.e. for $T \gtrsim 170$ K.

We are going to discuss these results in terms of the local environment of the Li^+ ion in the glass and the motional processes in the liquid state as they are established in the literature. BAIANU et al. /9/ investigated the structure of $LiCl \cdot n\,H_2O$ (n = 2...10) at 100 K by ^{1}H-NMR and found that the random network consists of $(Li^+4H_2O)Cl^-$ clusters with the water molecules surrounding Li^+ tetrahedrally and the excess water incorporated interstititally. We thus have a

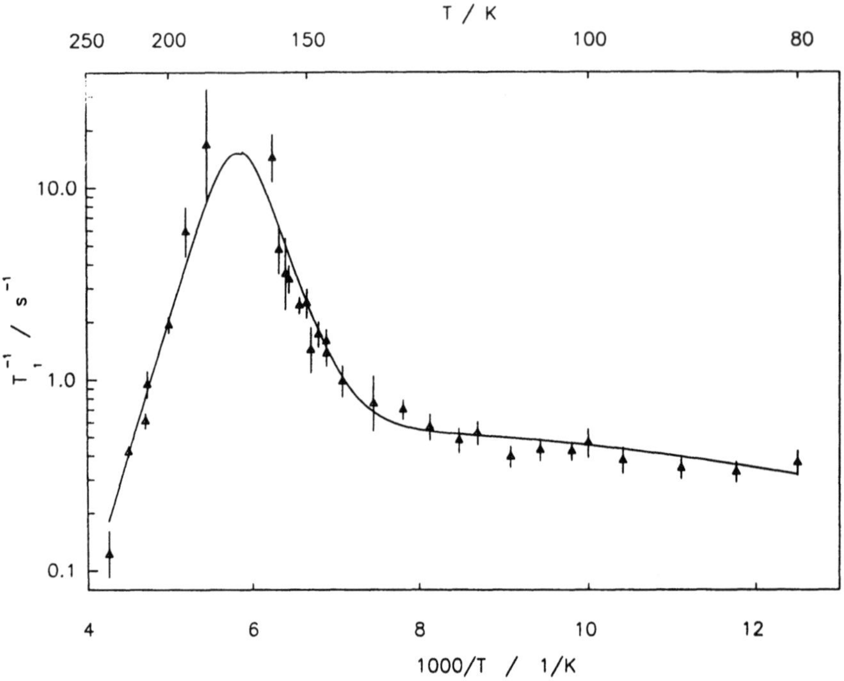

Figure 3: Log T_1^{-1} vs 1/T for 80 K $\lesssim$ T $\lesssim$ 240 K and B = 500 mT. The line through the data represents the sum of the two relaxation rates described in the text.

well defined hydration shell for Li^+. GEIGER and HERTZ /4/ concluded from ^{7}Li-NMR studies that the relaxation mechanism is quadrupolar. A comparison of ^{7}Li- and ^{1}H-NMR showed that motions of solvent molecules produce the fluctuating field gradients. LANG et al. /5,6/ developed the motional model of orientational fluctuations in the hydration shell of the cation: (I) the water molecules diffuse around the ion and chemical exchange occurs, and (II) they execute small amplitude librations superimposed onto anisotropic orientational fluctuations about the vector connecting the ion with the oxygen atom.

In agreement with the evidence /9/ that the local structure of the electrolyte glass is closely related to that of the liquid we suppose that the two dynamic processes mentioned above both exist in the glass as well. Here, in range II, process II governs the relaxation rate whereas above T_g process I is dominating. There is no obvious discontinuity of T_1^{-1} at the caloric glass transition temperature. One should note that recent ^{2}H-NMR line width measurements /8/ showed an effect of motional narrowing only above 160 K. To further elucidate the motional processes additional measurements by inelastic neutron scattering and β-NMR on $LiCl \cdot 4D_2O$ are in progress.

Acknowledgements

We are grateful to B. Bader for his help in starting the new β-NMR spectrometer and to H. Ackermann for his helpful advice and constant support. The hospitality of the Institut für Festkörperforschung, KFA Jülich GmbH, is gratefully acknowledged. This work was sponsored by the Bundesministerium für Forschung und Technologie.

References

1. A. Elarby-Aouizerat, P. Chieux, P. Claudy, J. Dupuy, J.F. Jal and J.M. Letoffe: J. de Physique 46, C8-629 (1985)
2. U. Reichert, M. Schmidt and S. Hunklinger: Solid State Comm. 57, 315 (1986)
3. G. Kaspar and V. Röhring: Proc. 17th Int. Conf. Low Temp. Phys., eds. U. Eckern, A. Schmid, W. Weber and H. Wühl (North-Holland, Amsterdam 1984) p. 385
4. A. Geiger and G. Hertz: Adv. Mol. Relax. Proc. 9, 293 (1976)
5. E.W. Lang and H.D. Lüdemann: Ber. Bunsenges. Phys. Chem. 89, 508 (1985)
6. E.W. Lang and F.X. Prielmeier: Ber. Bunsenges. Phys. Chem. 92, 717 (1988)
7. G. Balzer-Jöllenbeck, O. Kanert, J. Steinert: Solid State Comm. 65, 303 (1988)
8. A. Hagemeyer, G. Balzer-Jöllenbeck, O. Kanert: Verhandl. DPG (VI), 23 (1988) TT - 18.5
9. I.C. Baianu, N. Boden, D. Lightowlers, M. Mortimer: Chem. Phys. Lett. 54, 169 (1978)
10. P. Heitjans: Solid State Ionics 18&19, 50 (1986)
11. H. Kanno, I. Shirotani, S. Minomura: Bull. Chem. Soc. Jpn. 53, 2079 (1980)
12. P. Heitjans, B. Bader, H.J. Stöckmann, K. Dörr, G. Kiese, H. Ackermann, P. Freiländer, W. Müller-Warmuth and K. Meise-Gresch: Hyperfine Interactions 15/16, 597 (1983)
13. A. Schirmer, P. Heitjans, H. Ackermann, B. Bader, P. Freiländer, H.J. Stöckmann: Solid State Ionics 28-30, 717 (1988)

Dynamics of Two-Component Systems Near the Glass Transition

J. Bosse, U. Krieger, and J.S. Thakur

Freie Universität Berlin, Fachbereich Physik, Arnimallee 14, D-1000 Berlin 33, Germany

1. Introduction

It is an established fact that the structural relaxation time τ increases enormously as the temperature approaches the glass transition temperature T_g. The size of τ ranges from picoseconds in the liquid to hundreds of seconds at T_g. It seems to be generally accepted /1/ that it is this extraordinary slowing down of relaxation, which is accompanied by a corresponding increase of the shear viscosity η and a decrease of the diffusivity D, which causes the glass transition. For many experiments the time scale τ of relaxation processes will exceed by far the relevant experimental time scale τ_{exp} as the glassy state, i.e. the structural arrest to an amorphous solid is approched.

Theoretically, many of the subtle dynamical phenomena (such as *scaling* and *stretching* of relaxation functions, the temperature-smeared *jumps* in compressibility and specific heat, etc.) can be understood qualitatively, since in 1984 Leutheußer /2/ and Bengtzelius, Götze, and Sjölander /3/ in their pioneering work showed that the mode-coupling theory (MCT) of simple classical liquids /4/ develops a zero-frequency singularity. (For a review the reader is referred to the paper by W. Götze and L. Sjögren in this volume.) A thorough comparison between theoretical predictions and experimental results lends itself very difficult, however, for the following reasons

i. Long relaxation times require special care in analyzing experimental data. As will be demonstrated in the next section, often macroscopically measured "relaxation functions" are no longer equal to Kubo's relaxation functions, which the theoreticians calculate.

ii. MCT does *not* predict *universal* exponents as in second order phase transitions (in agreement with experimental findings). Therefore scaling functions describing the dynamics in the vicinity of the glass transition will be different for various liquids.

The latter point makes the study of binary liquid mixtures most attractive. Not only are conventional glass formers made up of several particle species, but there can occur *additional* dynamical phenomena absent in one-component liquids. The delocalization of small particles in a glassy matrix, presented in the third section, gives an example of a dynamical phase transition within the glass.

In the fourth section we will present results for mass- and charge-relaxation functions of a symmetric molten salt. The calculations on a schematic model with two coupled relaxation functions give a hint, why it was *not* possible to demonstrate the expected power-law behavior in recent neutron scattering data /5/ in the β-relaxation regime of $Ca_{0.4}K_{0.6}NO_3$, although earlier spin echo measurements on the same system /6/ in the α-relaxation regime were found in good agreement with theoretical predictions /7,8/.

Springer Proceedings in Physics, Vol. 37 Dynamics of Disordered Materials
Editors: D. Richter · A.J. Dianoux · W. Petry · J. Teixeira

2. Relaxation functions and measurements

In view of the difficulties mentioned above let us recall the *precise* definition of Kubo's relaxation function before turning to calculate density relaxation functions of binary liquids. A system with Hamiltonian H is assumed to be in a ***non-equilibrium*** state described by the statistical operator $\rho^{pert} = \rho(H + \delta H^{pert})$ at some initial time $t = 0$. Here $\rho(H)$ describes a grand canonical ensemble of particles and the perturbation takes the form

$$\delta\, H^{pert} = -\sum_{\nu} B_\nu\, \delta b_\nu^* \equiv -B \cdot \delta b^* \tag{1}$$

with a set of dynamical variables $B_\nu = B_{-\nu}^*$ and their conjugate fields $\delta b_\nu = \delta b_{-\nu}^*$. Then we expect the average value of a variable A at times $t \geq 0$ to deviate from its equilibrium value $\langle A \rangle$ calculated with $\rho(H)$ by an amount $\langle \delta A \rangle_t = \langle A \rangle_t - \langle A \rangle$ which in *linear* approximation with respect to the fields δb_ν is given by the formula

$$\langle \delta A^* \rangle_t = \Phi_{AB}(t) \cdot \delta b^* = \Phi_{AB}(t) \cdot \Phi_{BB}^{-1}(0) \cdot \langle \delta B^* \rangle_{t=0} \ , \tag{2}$$

where the second equality anticipates "independence" of the variables $\{B_\nu\}$ in the sense that the matrix $\Phi_{BB}(0)$ is non-singular. For classical systems Kubo's relaxation function is given by

$$\Phi_{AB}(t) = \langle \delta A^*(t) \delta B \rangle / (k_B T) \tag{3}$$

with $\delta A = A - \langle A \rangle$ and T denoting temperature. $\Phi_{AB}(t)$ describes the time evolution of a fluctuation $\langle \delta A^* \rangle_t$ for given initial fluctuations $\langle \delta B^* \rangle_{t=0}$. The latter define the non-equilibrium state at $t = 0$ and - depending on our experimental possibilities - stand for local fluctuations of density, temperature, electric field, etc.

Note that ergodicity implies $\langle \delta A \rangle_{t=\infty} = 0$ for *all* dynamical variales A (excluding constants of motion, for which $\langle \delta A \rangle_t = const. = \langle \delta A \rangle_{t=\infty}$). If there exists a non-conserved variable A with $\langle \delta A \rangle_t \neq 0$ for some very large time $t = t_\infty >\!\!> \tau_{exp}$, the system is said to be in a *non-ergodic* state, fluctuations of A being arrested (on experimental time scale).

In general, Kubo's relaxation function cannot be measured directly, because one must prepare the initial non-equilibrium state by applying a ***time-dependent*** field $\delta b_\nu(t) = \delta b_\nu\, h_\nu(t)$ to the equilibrium state. Linear response theory tells us:

$$\begin{aligned} \langle \delta A^* \rangle_t &= i \int_{-\infty}^{t} dt' \chi_{AB}(t-t') \cdot \delta b^*(t') \\ &= \Phi_{AB}^{exp}(t) \cdot \Phi_{BB}^{exp}(0)^{-1} \cdot \langle \delta B^* \rangle_{t=0} \end{aligned} \tag{4}$$

with the response function $\chi(t) = i\, \dot{\Phi}(t)$ and the "experimental relaxation function":

$$\begin{aligned} \Phi_{AB}^{exp}(t) &= -\int_{-\infty}^{t} dt' \dot{\Phi}_{AB}(t-t')\, h(t') \\ &\simeq \Phi_{AB}(t) - \Phi_{AB}(t + \tau_{exp}) \end{aligned} \tag{5}$$

where in the last line we inserted $h(t) \simeq \Theta(t + \tau_{exp})\Theta(-t)$ describing a field $\delta b_\nu(t)$ that was turned off at $t = 0$ after it had been turned on at a constant value for a length of time τ_{exp}. Eqs. (4,5) show:

i. only for sufficiently short relaxation times $\tau \ll \tau_{exp}$ can one measure the Kubo's relaxation function in a relaxation experiment. For macroscopically large τ near to the glass transition $\Phi^{exp}_{AB}(t)$ will depend onthe history of our probe treatment via $h(t)$. This phenomenon is well known experimentally /1/. A meaningful comparison between theoretical predictions on $\Phi_{AB}(t)$ and the measured $\Phi^{exp}_{AB}(t)$ therefore requires independent knowledge of $h(t)$, which is often not available.

ii. All $\Phi^{exp}_{AB}(t)$ will decay to zero even for a non-ergodic system, since according to Eq. (5) the experiment actually probes the *response function*. So, on the glass side away from the transition the relaxation experiment will render $[\Phi_{AB}(t) - \Phi_{AB}(\infty)]$, assuming its decay to be sufficiently fast compard to τ_{exp}.

In contrast to the macroscopic relaxation experiments discussed above the relation between the measured cross-section and the Kubo relaxation function of density fluctuations in a *scattering experiment* (neutron, X-ray, etc.) is quite well established ($\hbar\omega$ and $\hbar\vec{q}$ denote energy- and momentum transfer to the target, which contains particles of species $s = 1, \cdots, S$):

$$\frac{d^2\sigma}{d\Omega d\omega} \propto Re \int_0^{\tau_{exp}} dt\ e^{it\omega} \{\sum_{s,s'} (b_s^{coh})^* I(\vec{q},t)_{ss'}\ b_{s'}^{coh} + \sum_s N_s |b_s^{inc}|^2 \psi^{(s)}(\vec{q},t)\ I_s(\vec{q},t)\}. \tag{6}$$

Here the coherent intermediate scattering functions $I(\vec{q},t)_{ss'}$ are related to the density fluctuation spectra via

$$\begin{aligned} I(\vec{q},t)_{ss'} &= \langle N_s(-\vec{q},t)^*\ N_{s'}(-\vec{q})\rangle \\ &= \langle N_s(-\vec{q})\rangle^* \langle N_{s'}(-\vec{q})\rangle + \int_{-\infty}^{+\infty} \frac{d\omega}{\pi} e^{-it\omega} \Phi''(q,\omega)_{ss'} \hbar\omega / \left[1 - e^{-\frac{\hbar\omega}{k_B T}}\right]. \end{aligned} \tag{7}$$

where $\Phi''(\omega) = \int_{-\infty}^{+\infty} \frac{dt}{2} e^{it\omega} \Phi(t)$ and, for classical systems,

$$\Phi(q,t)_{ss'} = \langle \delta N_s(-\vec{q};t)^* \delta N_{s'}(-\vec{q})\rangle / (k_B T) \tag{8}$$

with the Fourier transformed particle density of species s:

$$N_s(\vec{q},t) = \sum_{1 \le j \le N_s} exp\left[-i\ \vec{q} \cdot \vec{r}_j^{\,s}(t)\right]. \tag{9}$$

Relations similar to Eqs. (7,8) hold for the incoherent intermediate scattering function with $N_s(\vec{q})$ replaced by the density of a tagged particle of species s:

$$N_s^0(\vec{q};t) = exp\left[-i\ \vec{q} \cdot \vec{r}_0^{\,(s)}(t)\right]. \tag{10}$$

$\psi^s(\vec{q},t)$ describes internal target particle dynamics and may be set equal 1 for neutron scattering.

For thermal neutrons the time of interaction with the target is of the order $\tau_{exp} \simeq 10^{-6}s$ and energy transfers correspond to $\omega^{-1} \simeq 10^{-12}s$, so we may put $\tau_{exp} = \infty$. Therefore, the scattering cross section Eq. (6) will produce an increasingly pronounced low-frequency peak as the relaxation time τ begins to exceed microscopic values.

Taking the derivative of Eq. (6) with respect to the average velocity v of the projectile, one finds

$$-v\frac{\partial}{\partial v}\left(\frac{d^2\sigma}{d\Omega d\omega}\right)^{coh} \propto \tau_{exp} Re\{\sum_{ss'}(b_s^{coh})^* \left[I(\vec{q},\tau_{exp})_{ss'} - \int_0^{\tau_{exp}} \frac{dt}{\tau_{exp}} q \frac{\partial}{\partial q} I(q,t)_{ss'}\right] b_{s'}^{coh} + o(\Delta v)\} \quad (11)$$

with $v = (v_0 + v_1)/2$, $\Delta v = (v_0 - v_1)/2$; $\tau_{exp} = 2L/v(1 + (\Delta v/v)^2 + \cdots)$; $q = (2m_n v/\hbar)\sin^2\theta/2 + o(\Delta v^2)$ and $\hbar\omega = 2m_n v\Delta v$, where v_0 and v_1 denote incoming and outgoing projectile velocities and L is the radius of the interaction sphere. For sufficiently small Δv, $v \approx v_0 = \sqrt{2E_0/m_n}$ is known and the left hand side of Eq. (11) can be measured. This is done in an elegant way by the neutron spin-echo method /6/ collecting information of the density relaxation function up to times of the order $\tau_{exp} \simeq 10^{-7}s$. (For sufficiently small scattering angle θ the second term containing the derivative with respect to q is assumed to be small.)

If density fluctuations get arrested at a certain temperature T_c, there will be an *elastic coherent* scattering intensity. (Note that the product of average densities in Eq. (7) which is responsible for Bragg-scattering in crystals does not contribute to elastic scattering in homogeneous liquids or glasses.) On the other hand, a finite long-time limit of tagged particle relaxation functions leads to an *incoherent recoilfree* scattering contribution which for neutrons will also be elastic. From Eq. (6) we find:

$$\begin{aligned}
\frac{d^2\sigma}{d\Omega d\omega}\Big|^{coh}_{elastic} &\propto \sum_{ss'}(b_s^{coh})^*\Phi(q;\infty)_{ss'}\; k_B T\; b_{s'}^{coh}\delta(\omega) \\
\frac{d^2\sigma}{d\Omega d\omega}\Big|^{incoh.}_{recoilless} &\propto \sum_s N_s |b_s^{inc}|^2 \psi^{(s)}(\vec{q},\omega) k_B T \Phi_s(q;\infty),
\end{aligned} \quad (12)$$

In view of Eq. (12) we define non-ergodicity parameters

$$\begin{aligned}
f(q)_{ss'} &= \Phi(q,\infty)_{ss'}/\sqrt{\Phi(q,0)_{ss}\,\Phi(q,0)_{s's'}} \\
f_s(q) &= \Phi_s(q,\infty)/\Phi_s(q,0)
\end{aligned} \quad (13)$$

noting that $f(q)_{ss}$ is the Debye-Waller factor (DWF) of species s, while $f_s(q)$ is its Lamb-Mößbauer factor (LMF).

3. Non-ergodicity parameters of a binary mixture of hard spheres

Starting from the exact equations of motion for the set of normalized density relaxation functions $\phi(q;t)_{ss'} = \Phi(q,t)_{ss'}/[\Phi(q,0)_{ss}\Phi(q,o)_{s's'}]^{1/2}$, which have the form of coupled oscillator equations with generalized friction forces /9,10/, one finds for the DWF, $f(q)_{ss'} = \lim_{t\to\infty}\phi(q,t)_{ss'}$,

$$\sum_{\bar{s}}\{\Omega^2(q)_{s\bar{s}} f(q)_{\bar{s}s'} + K(q;\infty)_{s\bar{s}}[f(q)_{\bar{s}s'} - \phi(q,0)_{\bar{s}s'}]\} = 0, \quad (14)$$

and a similar exact expression results for the LMF, $f_s(q) = \lim_{t\to\infty} \phi_s(q,t) = \lim_{t\to\infty} \Phi_s(q;t)/\Phi_s(q;0)$:

$$v_s^2 q^2 f_s(q) + K_s(q;\infty)[f_s(q) - 1] = 0. \tag{15}$$

Here $v_s^2 = k_B T/m_s$, $\Omega^2(q)_{ss'}$ and $\phi(q,0)_{ss'}$ are determined by the partial static structure factors $S(q)_{ss'}$. $K(q,\infty)_{ss'}$ and $K_s(q,\infty)$ are, respectively, the long-time limits of the friction kernels entering the equations of motion for coherent and tagged-particle density relaxation functions. Physically, Eqs. (14),(15) express the fact that under stationary conditions restoring forces will be compensated by friction forces. In a liquid the stationary friction forces are essentially built up by feed-back processes. This results in the friction kernels $K(q;t)$ being functionals of the relaxation functions $\phi(t)$, which they determine. Mode-coupling theory provides approximate expressions for these functionals /9/:

$$\begin{aligned} K(q;\infty)_{ss'} &= \frac{v_s^2}{q^2 V} \cdot \sum_{ij\mu\sigma} \sum_{\vec{k}} \{C_{si}(k) C_{sj}(k) (\vec{k}\cdot\vec{q})^2 (S_{ss}(\vec{q}-\vec{k}) S_{s's'}(\vec{q}-\vec{k}))^{1/2} \\ &* \ (S_{ii}(k) S_{jj}(k))^{1/2} \ \delta_{\mu s'} \delta_{\sigma j} \\ &+ \ C_{si}(k) C_{s'j}(\vec{q}-\vec{k}) (\vec{k}\cdot\vec{q})(q^2 - \vec{k}\cdot\vec{q}) \\ &* \ (S_{ss}(\vec{q}-\vec{k}) S_{jj}(\vec{q}-\vec{k}))^{1/2} (S_{ii}(k) S_{s's'}(k))^{1/2} \delta_{\mu j} \delta_{\sigma s'}\} f_{s\mu}(\vec{q}-\vec{k}) f_{i\sigma}(k) \end{aligned} \tag{16}$$

and

$$K_s(q) = \frac{v_s^2}{q^2 V} \cdot \sum_{ij} \sum_{\vec{k}} f_s(\vec{q}-\vec{k}) (\vec{k}\cdot\vec{q})^2 C_{si}(k) C_{sj}(k) (S_{ii}(k) S_{jj}(k))^{1/2} f_{ij}(k) \ , \tag{17}$$

where direct pair correlation function $C_{ss'}(k)$ are related to $S_{ss'}(k)$ by $S_{ss'}(k) = [I - C(k)]^{-1}_{ss'}$. Note that Eqs. (14),(16) form a closed set of equations for $f(q)_{ss'}$, requiring $S(q)_{ss'}$ as input information. After finding the DWF these can be used together with $S(q)_{ss'}$ as input for calculation the LMF from Eqs. (15),(17).

We solved Eqs. (14-17) numerically for a binary mixture of hard-spheres with input information derived from Percus-Yevick-Equations /11/. The mixture is described by three parameters: size ratio $\delta = \sigma_1/\sigma_2 \leq 1$, concentration of small particles $c = N_1/(N_1 + N_2)$, and total packing fraction $\eta = (\pi/6)(N_1\sigma_1^3 + N_2\sigma_2^3)/V$. We fixed $\delta = 0.2$ and $c = 0.9$ and varied the packing fraction η. Figs. 1,2 show the LMF of large and small particles, respectively, for $0.51 \leq \eta \leq 0.57$. At high densities ($\eta \geq 0.53$) both species are localized (non-zero LMF) making up an amorphous solid ($f_{1,2}(q)$ vary smoothly with q and $S(q)_{ss'}$ have values similar to liquids). The diffusion constants $D_{1,2}$ vanish in the approximation Eq. (17). The width of the bell-shaped curves is a direct measure for the inverse localization length $l_{1,2}$ of the particles. While the large particles are strongly localized ($30 \gtrsim \sigma_2/l_2 \gtrsim 10$) for $\eta \geq \eta_B \simeq 0.515$, the localization length of the small spheres is larger by a factor 10 at $\eta \simeq 0.57$ and it grows continuously as η is decreased - diverging at $\eta = \eta_A \simeq 0.525$; i.e. $f_1(q) \equiv 0$ for $\eta \leq \eta_A$. Physically this means that the small spheres get delocalized at η_A, they are diffusing within the glassy matrix produced by the localized large spheres. Further decrease of η below η_B will melt the glass. At $\eta = 0.51$ both particle species are delocalized, we have a liquid mixture.

Note that the large particle's LMF can be approximated by a Gaussian very accurately (Fig. 1b), while this is not possible for the small particle's LMF (Fig. 2b). It is remarkable that Eqs. (15),(17) produce essentially Gaussian q-dependence for the strongly localized large spheres.

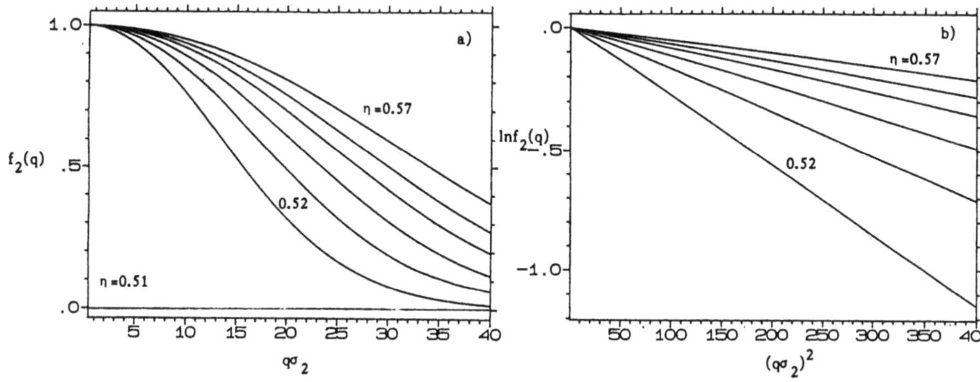

Fig. 1 LMF of large particles (mixed with $N_1 = 9\ N_2$ small spheres of diameter $\sigma_1 = \sigma_2/5$) for total packing fraction $\eta = 0.57 \cdots 0.51$ in steps of 0.01. (a) NOTE that $f_2(q) \equiv 0$ for $\eta < \eta_B \simeq 0.515$ (η_B glass melting point). (b) Gauss-function test: *straight* lines expected for *harmonically* bound particles.

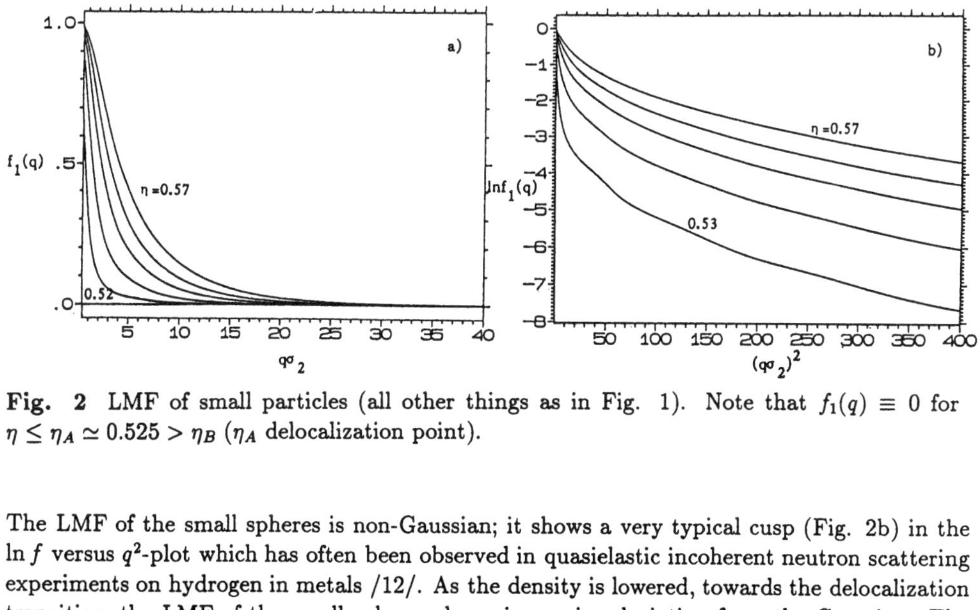

Fig. 2 LMF of small particles (all other things as in Fig. 1). Note that $f_1(q) \equiv 0$ for $\eta \leq \eta_A \simeq 0.525 > \eta_B$ (η_A delocalization point).

The LMF of the small spheres is non-Gaussian; it shows a very typical cusp (Fig. 2b) in the $\ln f$ versus q^2-plot which has often been observed in quasielastic incoherent neutron scattering experiments on hydrogen in metals /12/. As the density is lowered, towards the delocalization transition, the LMF of the small spheres shows increasing deviation from the Gaussian. The large slope at $q = 0$ reflects the size of l_1. One may use the above results to derive an approximate expression for the quasielastic incoherent scattering law of the small particles in the glass /13/

$$S_{inc}^{qe}(q;\omega) \equiv \frac{1}{\pi}\phi_1''(q;\omega) \simeq \frac{1}{\pi}\frac{\gamma(q)}{\omega^2 + \gamma^2(q)} f_1(q) \tag{18}$$

with $\gamma(q) = q^2 D_h f_1(q)$. Here the hopping diffusion constant D_h was introduced ad hoc to achieve an approximation for amorphous systems on the level of the Chudley-Elliot-model /14/. Note the dependence of $\gamma(q)$ on the LMF. The Lorentzian ω-dependence is, of course, an artefact of our simple approximation. Near the delocalization transition non-analytic frequency dependence is expected. To account for them properly the full mode-coupling equations must be solved for all times (not just for $t = \infty$). This has not been done yet. The DWFs $f_{ss'}(q)$ which stay finite for both species as η is decreased to the melting point, will be discussed elsewhere /10/.

4. Schematic model of a symmetric molten salt (SMS)

We have studied the mode-coupling equation for a simple one-wavenumber model of a SMS /15/. Due to the symmetry of the system cross correlations between charge- and mass-density fluctuations vanish. Eqs. (14),(15) then simplify if one transforms to mass- ($M(\vec{q}) = N_1(\vec{q}) + N_2(\vec{q})$) and charge- ($Q(\vec{q}) = N_1(\vec{q}) - N_2(\vec{q})$) density variables. The oscillator equations of motion for the normalized relaxation functions of $M(\vec{q})$ and $Q(\vec{q})$ (denoted by $\phi_+(t)$ and $\phi_-(t)$, respectively) read:

$$\ddot{\phi}_\pm(t) + \Omega_\pm^2 \phi_\pm(t) + \int_0^t dt'\; K_\pm(t-t')\dot{\phi}_\pm(t') = 0 \quad , \tag{19}$$

with initial conditions $\phi(t = 0) = 1$, $\dot{\phi}(t = 0) = 0$. Now the full MCT-expression for the friction kernels, Eqs. (16,17), is drastically reduced to a schematic model with three interaction parameters conserving, however, the structure of terms implied by the symmetry of the SMS:

$$K_+(t) = \Omega_+^2 \lambda[\alpha\phi_+^2(t) + (1-\alpha)\phi_-^2(t)] + 2\Gamma_+\delta(t) \tag{20}$$
$$K_-(t) = \Omega_-^2 \lambda r \phi_+(t)\phi_-(t) + 2\Gamma_-\delta(t) \quad . \tag{21}$$

This model shows a transition from a liquid to a glass phase (type B-transition) if the coupling constants $\lambda\alpha$, $\lambda(1-\alpha)$, and λr reach certain critical values. An elegant discussion of the full phase diagram of the model has recently been given /16/. We will in the following fix α and r, put $\Gamma_+ = \Gamma_- = 5\Omega_+, \Omega_-^2 = \Omega_+^2/r$, and treat λ as our transition driving parameter. Taking the $t = \infty$-limit of Eq. (19)-(21) leads to the analog of Eq. (14) for our model; solving for $f_\pm = \lim_{t\to\infty}\phi_\pm(t)$ we find for the two (α, r)-choices of Fig. 3 and Fig. 5 critical values λ_c and $f_\pm^c$ listed in Table I.

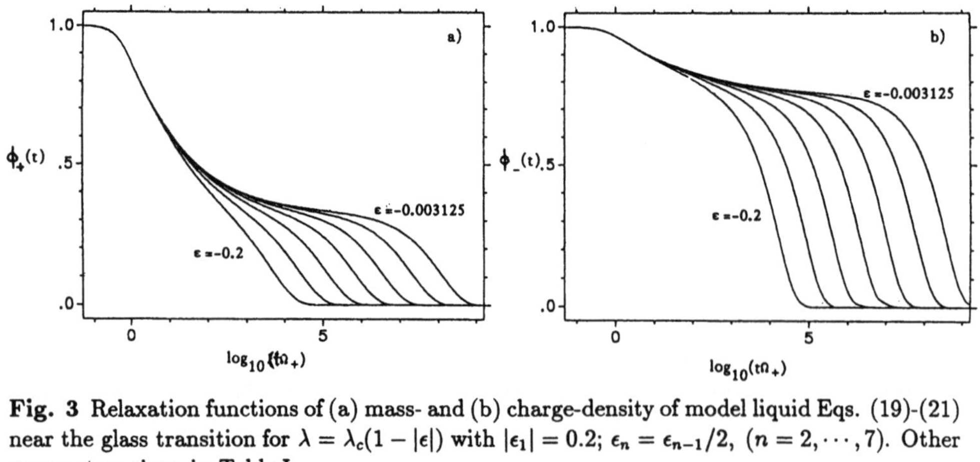

Fig. 3 Relaxation functions of (a) mass- and (b) charge-density of model liquid Eqs. (19)-(21) near the glass transition for $\lambda = \lambda_c(1-|\epsilon|)$ with $|\epsilon_1| = 0.2$; $\epsilon_n = \epsilon_{n-1}/2$, $(n = 2, \cdots, 7)$. Other parameters given in Table I.

Table I

	Fig. 3,4	Fig. 5,6
α	0.900	0.932
r	4.00	6.35
λ_c	3.16093	3.25186
f_+^c	0.32964	0.27175
f_-^c	0.76007	0.82179
$C(\alpha,r)$	0.710	0.907
a	0.3228	0.206
b	0.6244	0.289
γ	2.3497	4.193

Figs. 3a and 3b display mass- and charge-relaxation functions, respectively, in a supercooled melt with $\lambda = \lambda_c(1-|\epsilon|)$ close to its glass transition. The decay time is larger than the microscopic relaxation time ($\approx \Omega_+^{-1}$) by more than 4 orders of magnitude at $|\epsilon| = 0.2$ and it increases further as the transition is approached ($|\epsilon| \to 0$). As was noted before /15/,/8/, Eqs. (19)-(21) may also be interpreted as a model for a *one*-component liquid which takes account of coupling between two wavenumbers $q_\pm$. Although the special choice of coupling constants in Eqs. (20),(21) prevents us from drawing physical conclusions from this interpretation, the general mathematical properties derived for one-component systems with q-dependent relaxation functions /17/ will apply here. Therefore, $\phi_\pm(t)$ in Fig. 2 are expected to show decay with characteristic scaling behavior on two scales defined by $\omega_\epsilon = \Omega_+|\epsilon|^{1/2a}$ and $\omega_\epsilon' = \Omega_+|\epsilon|^\gamma = \omega_\epsilon|\epsilon|^{1/2b}$ with $\gamma = (a^{-1}+b^{-1})/2$. On the shorter time scale ω_ϵ^{-1} (called β-relaxation regime) one finds:

$$\phi_\pm(t) = \begin{cases} f_\pm^c + |\epsilon|^{\frac{1}{2}} A_\pm/(\omega_\epsilon t)^a & , \quad \omega_\epsilon t \ll 1 \\ f_\pm^c - B_\pm(\omega_\epsilon' t)^b & , \quad \omega_\epsilon t \gg 1. \end{cases} \tag{22}$$

Here $A_\pm$ and $B_\pm$ are positive constants obeying $A_+/A_- = B_+/B_- = f_+^c/(1-f_-^c)$ /15/. On the larger time scale $(\omega_\epsilon')^{-1}$ (called α-relaxation regime) one finds:

$$\phi_\pm(t) = g_\pm(\omega_\epsilon' t) = f_\pm^c - B_\pm(\omega_\epsilon' t)^b + \cdots \tag{23}$$

with scaling functions often resembling closely the Kohlrausch stretched-exponential form $g_\pm(\tau) \simeq \overline{f}_\pm \exp(-b_\pm \tau^\beta) \beta \leq 1$, over several decades /7/. The exponents $0 \leq a \leq 1/2, 0 \leq b \leq 1$ in Eqs. (22)-(23) are determined (see Fig. 2 of Ref. 8) by the exponent-parameter $C(\alpha, r)$ of our model

$$C(\alpha,r) = [\lambda_c \alpha f^3 + 3(1-\alpha)/(\lambda_c r^2 f)]/[2(1-\alpha)/r + f^3/(1-f)^3] \tag{24}$$

with $f = f_+^c$.

Our numerical solutions of Eqs. (19)-(21) for the melt shown in Fig. 3 clearly confirm the asymptotic behaviour described in Eqs. (20)-(22):

i. The α-relaxation scaling implies a shift of $\gamma log_{10} 2 \simeq 0.707$ between the curves belonging to adjacent ϵ-values, (because we chose $|\epsilon_{n+1}/\epsilon_n| = 1/2$) which is well fulfilled.

ii. The functional form of the relaxation functions for $|\epsilon| = 0.003125$ is in very good agreement with solutions of the corresponding scaling equation /7/, /8/.

iii. Looking at plots of $log_{10}\{(\phi_\pm(t) - f_\pm^c)/\sqrt{\epsilon}\}$ versus $log_{10}(\omega_\epsilon t)$ we convinced ourselves that the curves show β-relaxation according to Eq. (22). The t^{-a}-power law holds for the smallest $|\epsilon|$ between $10\ \Omega_\pm^{-1} \lesssim t \lesssim 10^4 \Omega_+^{-1}$ for mass relaxation, while for charge relaxation the interval is smaller $10^2 \Omega_+^{-1} \lesssim t \lesssim 10^4 \Omega_+^{-1}$ (it will increase for smaller $|\epsilon|$).

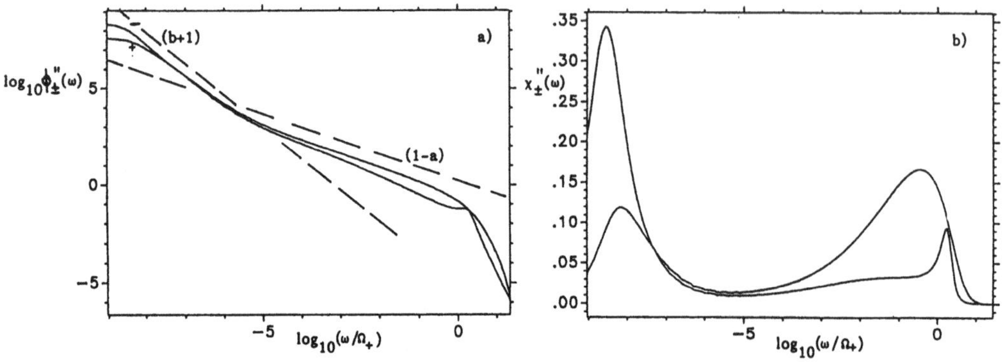

Fig. 4 a) Mass- and charge density relaxation spectra $\phi''_{\pm}(\omega)$, for $|\epsilon_7| = 0.003125$. Dashed lines: slopes corresponding to $1/\omega^{1-a}$ (large ω)- and $1/\omega^{b+1}$ (small ω)-behavior.
b) Susceptibility spectra $\chi''_{\pm}(\omega) = \omega\phi''_{\pm}(\omega)$ corresponding to (a). In both plots the mass spectra have the lower values for small ω and the higher values for large $\omega \gtrsim 5 * 10^{-8}\Omega_+$.

Fig. 4a presents the spectral functions $\phi''_{\pm}(\omega)$ for the smallest $|\epsilon|$, where both the $(\omega/\omega_\epsilon)^{-(1-a)}$ power law (corresponding to Eq. (22)) and the von Schweidler law $(\omega/\omega_\epsilon)^{-(b+1)}$ (corresponding to Eq. (22)) can be verified. Fig. 4b contains the very same information dislplayed in the susceptibility $\chi''_{\pm}(\omega)$. Note the spectral weight for the charge response around $\omega \simeq 10^{-2}\Omega_+$, which results from the slow β-relaxation in combination with the high plateau value f^c_-. This β-relaxation maximum is very often buried under a broad microscopic peak. For experimental detection of β-relaxation it seems more reasonable to observe the minimum position ω_{min} between α-relaxation and microscopic peak which scales with ω_ϵ /18/.

In Fig. 5 we plotted the solutions of Eq. (19)-(21) for parameter values (α, r) describing $Ca_{0.4}K_{0.6}(NO_3)_{1.4}$. The exponent parameter was fixed by fitting a power-law to the T-dependence of the viscosity ($\eta \propto 1/\omega'_\epsilon$). The curves show the expected α-relaxation scaling (see points (i), (ii) above). The master curves can be fitted with stretched exponentials with $\beta \simeq 0.6$ /8/. All this is in good agreement with neutron scattering data /6/. (Note that the MCT prediction $\eta \propto 1/\omega'_\epsilon$ is beautifully confirmed by the data. While a power-law fit to the measured viscosity $\eta(T)$ gives $T_c \simeq 90°C$ and $\gamma \simeq 4.2$, one finds from the shift $S_{n\to n+1}$ between adjacent curves in Fig. 1 of Ref. /6/ $\gamma = S_{n\to n+1}/log_{10}\,(T_n - T_c)/(T_{n+1} - T_c) \simeq 4.13$).

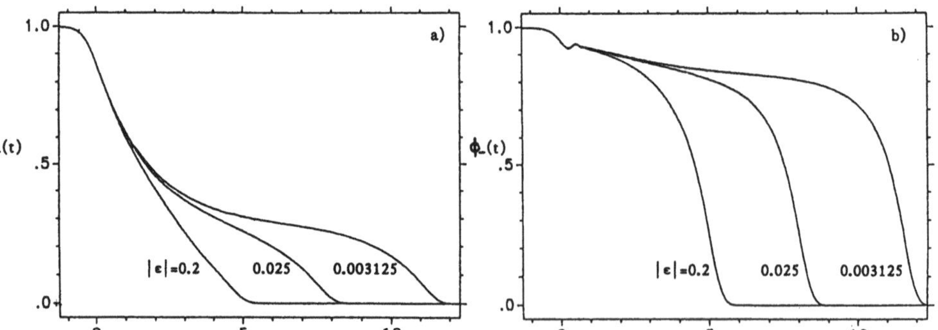

Fig. 5 Relaxation functions as in Fig. 3 for $Ca_{0.6}K_{0.4}(NO_3)_{1.4}$-model (parameters listed in last column, Table I.) for $|\epsilon_1| = 0.2$, $|\epsilon_4| = 0.025$ and $|\epsilon_7| = 0.003125$.

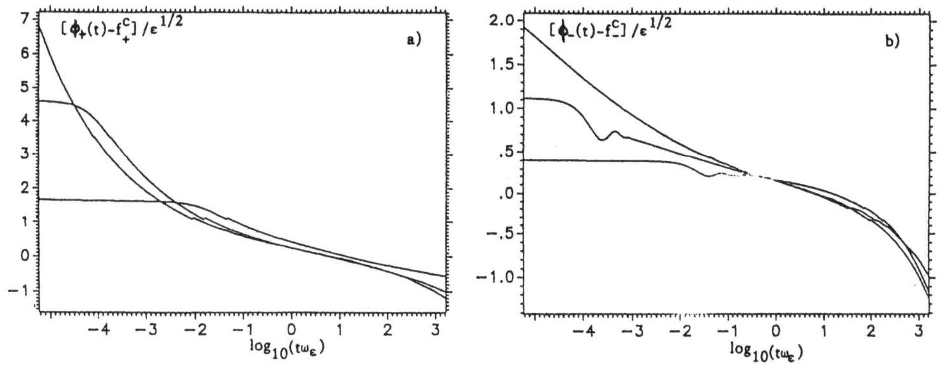

Fig. 6 Test of *first* scaling law (β-relaxation) for (a) mass- and (b) charge-density relaxation for $Ca_{0.6}K_{0.4}(N)_3)_{1.4}$-model for ϵ as in Fig. 5. The scaling times t_ϵ with $log(t_\epsilon\omega_\epsilon) = 0$ are indicated by arrows in Fig. 5.

Now, Figs. 6a and 6b, however, show that the scaling properties Eq. (22) of the first regime are destroyed. This again agrees with recent experiments: Analysis of neutron scattering data on $Ca_{0.4}K_{0.6}(NO_3)_{1.4}$ /5/ forced to the conclusion that Eq. (22) was not applicable in this case. The reason for the break-down of the simple behavior Eq. (22) is easily seen from Fig. 2 of Ref. /8/. The strong temperature variation of viscosity ($\gamma \simeq 4$) implies a large exponent parameter $C(\alpha, r)$ (close to 1) which in turn leads to small values of a, b. In the limit $C(\alpha, r) \to 1, a$ and b will vanish, and it is tempting to assume that a *new* time-dependence-slower than any power law - will take over at this singular point $C(\alpha, r) = 1$. Indeed Figs. 6a,b show for $|\epsilon| \lesssim 0.025$ a dependence

$$\phi_\pm(t) = f_\pm^c - \sqrt{\epsilon} D_\pm \ln(t\, \tilde{\omega}_\epsilon). \tag{25}$$

over about 2.5 decades around $t \simeq \omega_\epsilon^{-1}; (\tilde{\omega}_\epsilon \simeq \omega_\epsilon)$.

A logarithmic time dependence near endpoint singularities was predicted recently /16/. The parameters of the $Ca_{0.4}K_{0.6}(NO_3)_{1.4}$ bring us close to the swallowtail singularity in the phase diagram of our model shown in Fig. 8 of Ref. /16/. So we conclude, that the destroyed β-relaxation scaling properties are due to a crossover to a *new* behavior showing *logarithmic* time-dependence in the vicinity of the point ($\lambda_c r \simeq 25, \lambda_c\alpha \simeq 3.5, \lambda_c(1-\alpha) \simeq 0.2$).

5. Conclusions

In chapter 3 MCT has been applied to a mixture of large and small hard spheres in order to calculate the glass phase LMF of both particle species. While the LMF of the large spheres shows almost Gaussian q-dependence close to the glass melting point, the LMF of the small spheres exhibits large characteristic deviations from the Gaussian, which have been observed as typical for the quasielastic incoherent neutron scattering intensity of hydrogen in metals. With decreasing packing fraction η the localization length of the large spheres increases to about 10% of their diameter at η_B, and jumps to infinity for $\eta < \eta_B$ (melting of the glass in accordance with Lindemann's criterium). The localization length of the small particles behaves quite differently: it increases continuously with decreasing η diverging at $\eta_A > \eta_B$. For $\eta_A > \eta > \eta_B$ the small particles diffuse through the glassy matrix of large spheres.

In chapter 4 relaxation properties near the glass transition were studied for a schematic model of an ionic melt. Especially, it was demonstrated that for small exponent parameter $1/2 \leq C(\alpha, r) \lesssim 3/4$ the relaxation functions approach their plateau values $f^c_\pm$ via power laws in time, while for large $C(\alpha, r) \rightarrow 1$ there is a crossover to a logarithmic time dependence. Moreover, the relaxation functions slowly decay from their plateau values $f^c_\pm$ to zero in a way that can typically be fitted by streched exponentials.

Since hopping diffusion processes are neglected in this idealized model, it leads to a divergence of relaxation times (and viscosity) at a temperature $T_c > T_g$, where T_c is defined by $\lambda(T_c) = \lambda_c$, i.e. $\epsilon = (\lambda - \lambda_c)/\lambda_c = const.\ (T - T_c)$ assuming $\lambda = \lambda(T)$ a regular function of T. Therefore, the above results are expected to have physical relevance for $T > T_c$ only. $T_c \simeq 90°C$ (while $T_g \simeq 60°C$) for $Ca_{0.4}K_{0.6}(NO_3)_{1.4}$. The predictions of the above schematic model have been confirmed in recent neutron scattering results on $Ca_{0.4}K_{0.6}(NO_3)_{1.4}$.

Acknowledgment

This work was supported by DFG grant Bo 873/1-1.

References

1. C.A. Angell; J. Non Crys. Sol. (1988)
2. E. Leutheußer; Phys. Rev. *A29*, 2765 (1984)
3. U. Bengtzelius, W. Götze, A. Sjölander; Phys. Rev. *C17*, 5915 (1984)
4. J. Bosse, W. Götze, M. Lücke; Phys Rev. *A17*, 447 (1978)
   J. Bosse, W. Götze, A. Zippelius; Phys. Rev. *A18*, 1214 (1978)
   G. Wahnström, L. Sjögren; J. Phys. *C15*, 401 (1982)
5. W. Knaak, F. Mezei, B. Farago; Europhys. Lett. (1988)
6. F. Mezei, W. Knaak, B. Farago; Phys. Rev. Lett. *58*, 571 (1987)
7. J. Bosse, U. Krieger; J. Phys. *C10*, L609 (1986)
8. U. Krieger, J. Bosse; Phys. Rev. Lett. *59*, 1601 (1987)
9. J. Bosse; Proceedings of Int. Workshop on Static and Dynamic Properties of Liquids, Dubrovnik (Yugoslavia) June 1988; to be published in Springer Proc. in Physics
10. J.S. Thakur, J. Bosse; to be published
11. J.L. Lebowitz; Phys. Rev. *133*, A895 (1964)
12. N. Wakabayashi, B. Alefeld, K.W. Kehr, T. Springer: Sol. State Commun. *15*, 503 (1974); see also K. Sköld; in: Hydrogen in Metals I, edited by G. Alefeld and J. Völk (Springer Verlag, Berlin, 1978), p. 267

13. J. Bosse, J.S. Thakur; Phys. Rev. Lett. *59*, 998 (1987)

14. C.T. Chudley, R.J. Elliot; Proc. Phys. Soc. (London) *77*, 353 (1961)

15. U. Krieger and J. Bosse, in: Proceedings of the NATO Advanced Study Institute of Amorphous and Liquid Materials, Passo del Mendola, Italy, 1985, eds. E. Lüscher, G. Fritsch, G. Jacucci, NATO Advanced Study Institute, Series E, Vol. 118 (Nijhoff, Boston, MA, 1987)

16. W. Götze, R. Haussmann; Z. Phys. B, Condensed Matter *72* 403 (1988)

17. W. Götze; Z. Phys. *B60*, 195 (1985)

18. W. Götze, L. Sjögren; J. Phys. *C21*, 3407 (1988)

Relaxation of the Incoherent Density-Density Correlation Functions of Binary Soft-Sphere Glasses and Jump Motion of Atoms Near the Glass Transition

G. Pastore[1], *B. Bernu*[2], *J.P. Hansen*[3], *H. Miyagawa*[4], *and Y. Hiwatari*[4]

[1]International School for Advanced Studies, Strada Costiera 11, I-34014 Trieste, Italy

[2]Laboratoire de Physique Théorique des Liquides, Université Pierre et Marie Curie, F-75252 Paris Cedex 05, France

[3]Laboratoire de Physique, Ecole Normale Superieure de Lyon, F-69364 Lyon Cedex 07, France

[4]Department of Physics, Faculty of Science, Kanazawa University, Marunouchi 1-1, Kanazawa, Ishikawa 920, Japan

1. Introduction

Molecular Dynamics (MD) techniques have played an important role in the development of our understanding of the physics of the condensed phases. In recent years the same numerical techniques have been applied to the study of the liquid-glass transition (GT). The existence of a recent new theoretical description of the GT (e.g. see GOETZE /1/) and the introduction of new experimental techniques (MEZEI et al. /2/) have increased the interest inor a model study which could usefully supplement the real experiments.

Unfortunately the slowing down of all the relaxation processes close to GT is a serious problem for this kind of approach and very long runs on large systems are required to reduce the statistical error. Moreover, one-component systems very often show nucleation of the crystalline phase. A two component system has to be used to avoid this of problem (BERNU et al /3/). The situation is reminiscent of that found in real metallic glasses where alloys are easily quenched in an amorphous solid.

We have studied in detail soft-sphere mixtures composed by N_1 atoms of mass m_1 and diameter σ_1 and N_2 atoms of mass m_2 and diameter σ_2. The pair potentials are

$$\phi_{\alpha\beta} = \varepsilon(\sigma_{\alpha\beta}/r)^{12} \qquad , \sigma_{\alpha\beta} = \frac{(\sigma_\alpha + \sigma_\beta)}{2} \tag{1}$$

with $\sigma_2/\sigma_1 = 1.4$ The phase diagram of mixtures of atoms interacting with the potentials (1) can, within an effective one-fluid approximation, be "projected" onto the phase diagram of a one component system characterized by the dimensionless interaction $\phi(r)/k_bT = \Gamma_{eff}^4/r^{12}$ where the coupling constant is $\Gamma_{eff} = (\sigma_{eff}/\sigma_1)^3 \rho^* (T^*)^{-1/4}$, r is measured in unit of σ_1 , $T^* = k_bT/\varepsilon$, $\rho^* = \rho\sigma_1^3$ and $\sigma_{eff}^3 = \sum_\alpha\sum_\beta x_\alpha x_\beta \sigma_{\alpha\beta}^3$ $(x_\alpha = N_\alpha/(N_1 + N_2))$. The natural unit of time is $\tau_0 = \sqrt{m_1\sigma_1^2/12\varepsilon}$.

Springer Proceedings in Physics, Vol. 37 **Dynamics of Disordered Materials**
Editors: D. Richter · A.J. Dianoux · W. Petry · J. Teixeira

Realistic values of the parameters would yield for τ_0 a number of the order of $10^{-12}/10^{-13}$ sec. Γ_{eff} is about 1.15 at the freezing point and about 1.6 at the glass transition.

Details of the model and extensive characterization of the system have been reported elsewhere (BERNU et al. /3/ and /4/). Here we wish to summarize and critically discuss the results recently obtained for the incoherent part of the density-density correlation functions (IDCF) (PASTORE et al. /5/, MIYAGAWA et al. /6/, KAMBAYASHI and HIWATARI /7/) focusing in particular on the scaling of the IDCF (section 2) and the jump motion at low temperature (section 3).

2. Scaling of the Density-density Correlation Functions

The IDCF is defined by and it is evaluated from the formula

$$F_s^{(\alpha)}(k,t) = 1/N_\alpha \sum_{i=1}^{N_\alpha} < \exp\{i\mathbf{k}[\mathbf{r}_i^\alpha(t) - \mathbf{r}_i^\alpha(0)]\} > \tag{2}$$

where $< \cdots >$ indicates an ensemble average.

We have systematically computed $F_s^{(\alpha)}(k,t)$ for 10 values of wavenumber k in an equimolar mixture of 500 particles over a time interval of $180\tau_0$. For these samples the equations of motion were integrated for 10^4 time steps ($k = k_0 n, k_0\sigma_1 = 0.642, n = 1{,}10$ and for $\Gamma_{eff} = 1.154$, 1.352, 1.425, 1.523, 1.584, 1.670). A structural arrest is evident in the MD data at the lowest studied temperatures. It is signalled by a non decaying component in the $F_s^{(\alpha)}(k,t)$ within the experimental time. Approaching the transition from the liquid and very close to it only a very slow relaxation is detected. This slow relaxation seems to be strongly related to the small residual diffusion originating from jump processes as we will discuss in the section 3.

A common feature of the relaxation processes close to the liquid-glass transition is the time scaling i.e. the possibility of drawing a single "master curve" with data belonging to different temperatures by taking as unit of time for each set of data its own relaxation time. In most cases the master curve can be well fitted by a "stretched exponential" form. In our case the natural time scale proportional to the measured inverse diffusion constant D_α^{-1} does not give an exact scaling. However it is possible to find empirically a sequence of scaling times $\tau_1^{(\alpha)}$ such that one curve can be built for each k. These empirical times are weakly k dependent and increase with Γ_{eff} more rapidly than the evaluated $1/D_\alpha$, diverging approximately as

$$|\Gamma_{eff} - \Gamma_{eff}^g|^{-3.7}$$

where $\Gamma_{eff}^g = 1.67$ is the estimated transition coupling. The difference between $\tau_1^{(\alpha)}$ and $1/D_\alpha$ is probably due to a progressive overestimation of D_α in our cooler runs.

We have found that a stretched exponential law

$$f^{(\alpha)}(k, t/\tau_\alpha) = \xi_\alpha \exp\left[-(t/\tau_\alpha)^{\beta_\alpha}\right] \tag{3}$$

or a power law

$$f^{(\alpha)}(k, t/\tau_\alpha) = f_\alpha - (t/\tau_\alpha)^{b_\alpha} \qquad (4)$$

both give a good fit of our empirical master curves with parameters given in tables 1 and 2. We do not put too much emphasise on the values of the parameters (better statistics could affect them) but we are confident that the main qualitative findings are correct: a dependence of β_α or b_α on the wavenumber is clearly displayed by our fit of the master curves. We notice that a fit of the individual curves for each Γ with the above functional forms would require a slight systematic change of the parameters but their variation with the coupling can not be unambiguously determined from our data.

Table 1. Optimum fitting parameters of a stretched exponential representation of the master curves (Eqn. 3). The relaxation times are in unit of τ_0 .

$n = k/k_0$	β_1	τ_1	ξ_1	β_2	τ_2	ξ_2
10	0.61	3.08	0.70	0.68	4.99	0.79
9	0.65	3.90	0.74	0.70	6.37	0.82
8	0.68	5.34	0.78	0.76	8.90	0.85
7	0.73	7.0	0.82	0.80	11.7	0.88
6	0.82	10.0	0.86	0.85	16.2	0.91
5	0.85	14.2	0.90	0.88	23.3	0.93
4	0.92	23.1	0.93	0.94	37.0	0.96
3	0.93	40.6	0.96	0.95	65.5	0.97
2	0.96	91.5	0.98	0.99	143.0	0.99
1	0.96	383.0	0.995	0.99	579.0	0.996

Table 2. Optimum fitting parameters of a power law representation of the master curves (Eqn. 4). The relaxation times are in units of τ_0 .

$n = k/k_0$	b_1	τ_1	f_1	b_2	τ_2	f_2
10	0.36	26.8	0.74	0.46	21.2	0.82
9	0.39	27.5	0.77	0.49	22.7	0.84
8	0.42	27.9	0.81	0.50	30.3	0.87
7	0.46	29.9	0.84	0.53	36.5	0.89
6	0.56	30.0	0.87	0.57	47.5	0.92
5	0.60	37.3	0.91	0.60	66.2	0.94
4	0.67	54.0	0.94	0.70	86.0	0.96
3	0.71	92.0	0.96	0.76	133.0	0.98
2	0.93	106.0	0.98	0.97	150.0	0.99
1	0.9	578.0	0.99	0.9	843.0	0.997

3. Jump Motion

In the runs at low temperature we have identified the presence of a subset of atoms which have a higher mobility. While the majority of the atoms vibrates around the equilibrium positions, some of them suddenly jump to a neighbour position. In some cases, coordinated motion of groups of atoms has been observed. The residual diffusion close and below the liquid-glass transition is responsible for the existence of this additional relaxation channel. We think that a more detailed investigation of this jump motion and its relation with the experimental β relaxation deserves more detailed investigations. As a first step towards the characterization of this jump motion we have tried a very crude empirical representation by using the phenomenological model of CHUDLEY and ELLIOT /8/.

The assumptions of the model are : i) the atoms perform many oscillations around their equilibrium positions between two successive jumps; ii) the jumps are uncorrelated; iii) in average the jump directions are isotropically distributed; iv) the jump lengths l_α are of the order of the first neighbour distances as measured by the position of the first peak of $g_{\alpha\alpha}(r)$; v) the flight time during a jump is much shorter that the time interval between jumps. Under these conditions the ICDF is given by

$$F_S^{(\alpha)}(k,t) = \exp(-a_\alpha k^2/2)\exp\left[t/\tau_{jump}^{(\alpha)}(-\sin(kl_\alpha)/kl_\alpha)\right]. \qquad (5)$$

The first exponential is the Debye-Waller factor describing the oscillatory motion, the second one might be refined by weakening some of the hypotheses i-v but at the price of introducing additional parameters like, for example, a distribution of jump lenghts. We have taken for l_α a characteristic nearest neighbour distance ($1.2\sigma_1$) and we have obtained a_α and τ_{jump}^{α} from a least square fit of our $F_S^{(\alpha)}(k,t)$. We consider (5) as a good fit of the data if the dependence of the two parameters on the wavenumber k is negligible.

In the stable and supercooled liquid regime (5) is not able to reproduce the MD data but, for temperatures close enough to the transition, we observe a remarkable improvement. In addition, for our coolest sample, in a run extending over a total interval of 350 τ_0 we get a mean time $\tau_{jump} \simeq 10^4 \tau_0$ between subsequent jumps of any atom. This implies that, in that sample of 500 particles, only a few tens of them will have jumped during the run, in agreement with the direct enumeration of the jump events.

4. Conclusions

Computer experiments on the liquid-glass transition are able to give a detailed atomic-level description of the transition when approaching it from the high temperature side. Accurate results can be obtained in the supercooled regime with systems of a few hundreds particles if the interest is on the one-particle properties. The accurate determination of collective properties like the coherent scattering function $F_{\alpha\beta}(k,t)$ or the shear viscosity could benefit from using larger systems. The calculation of these quantities by computer experiments seems to be one of the most interesting tasks for future numerical studies on this subject. Results for the frozen amorphous solid should be carefully checked because in that region some strong history dependence effect seems to appear. Also in this case larger systems are required to extract precise numerical values.

References

1. W. Goetze: Z. Phys. Chem. 156 , 83 (1988)
2. F. Mezei, M. Knaak, B. Farago: Phys. Rev. Lett. 58 , 571 (1987)
3. B. Bernu, Y. Hiwatari, J. P. Hansen: J. Phys. C 18 , L371 (1985)
4. B. Bernu, J. P. Hansen, Y. Hiwatari, G. Pastore: Phys. Rev. A 36 , 4891 (1987)
5. G. Pastore, B. Bernu, J. P. Hansen, Y. Hiwatari: Phys. Rev. A 38 , 454 (1988)
6. H. Miyagawa, Y. Hiwatari, B. Bernu, J. P. Hansen: J. Chem. Phys. 88 , 3879 (1988)
7. S. Kambayashi, Y. Hiwatari: J. Phys. Soc. Jpn. 56 , 2788 (1987)
8. C. T. Chudley, R. J. Elliott: Proc. Phys. Soc. 77 , 353 (1961)

Yukawa Systems: Model for Colloidal Glasses

N. Pistoor[1] *and K. Kremer*[2]

[1]Institut für Physik, Johannes-Gutenberg-Universität, Postfach 3980, D-6500 Mainz, Fed. Rep. of Germany
[2]Institut für Festkörperforschung, KFA Jülich, D-5170 Jülich, Fed. Rep. of Germany

1 The Model

The effective interaction between two charged macroions in a dilute aqueous solution can be described by means of a *Yukawa-Potential*

$$U(r) = U_0 \frac{a}{r} \exp(-\lambda r/a) \tag{1}$$

with

$$U_0 = \frac{Z_1^* Z_2^* e^2}{\epsilon} \left\{ \frac{\exp(2\lambda R/a)}{(1+\lambda R/a)^2} \right\} \tag{2}$$

where R is the radius of a macroion, Z_1^*, Z_2^* are effective charges, n is the number density of the macroions, from which $a = n^{1/3}$, a typical distance between near macroions, is derived, and λ is the screening parameter[1].

2 The Phase Diagram (One Component System)

The state of the system is specified by two dimensionless parameters, λ and $\bar{T} = k_B T/U_a$. The energy scale is given by $U_a = \langle U(a) \rangle = \langle U_0 \rangle \exp(-\lambda)$ where the average is performed over all pairs of particles.

Figure 1 shows the phase diagram as it was obtained by computer simulations[2,4] and recently has been verfied by experiments on latex spheres[5], displaying that the Yukawa potential provides a good representation of the interaction of the colloids. One finds three different stable phases: *fluid, bcc crystal,* and *fcc crystal.*

In a two component system one expects suppression of crystallization in favour of a *glassy state.*

3 The Simulation (Two Component System)

We integrated the equations of motion of the macroions

$$m \frac{d^2 \mathbf{r}_i}{dt^2} = -\frac{\partial}{\partial \mathbf{r}_i} \sum_{j \neq i} U(|\mathbf{r}_i - \mathbf{r}_j|) \tag{3}$$

for a system with the following parameters

charges	$Z_1^* = 350$	$Z_2^* = 500$
number of particles	$N_1 = 170$	$N_2 = 342$
screening parameter	$\lambda = 3.4$	

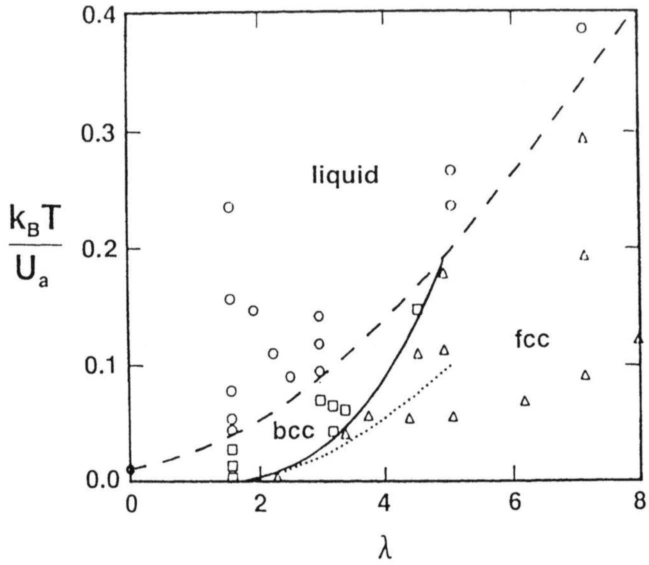

Figure 1: The phase diagram of one component Yukawa systems. Open circles indicate points where the liquid phase is stable. Points where the fcc or bcc phase are stable are indicated by triangles or squares, respectively. The dashed line separating liquid and solid phases indicates the value of $\bar{T}$ where the rms displacement reached $0.19a$. The dotted line is the harmonic result for the bcc to fcc phase transition, and the solid line interpolates the simulation results for this transition (from [4]).

Starting at $\bar{T} = 0.20$ we cooled the system down to $\bar{T} = 0.05$ in several steps by coupling it to a heat bath[3,4]. After the intermediate configurations at $\bar{T} = 0.20$, 0.15, 0.10 and 0.05 had relaxed further, the heat bath was switched off and the simulation went on in the microcanonical ensemble.

During the simulation we calculated the following quantities

- pair distribution functions

$$g_{\nu\mu}(\mathbf{r}) = \frac{1}{nN}\langle\sum_{n=1}^{N_\nu}\sum_{m=1}^{N_\mu}{}' \delta(\mathbf{r} - \mathbf{r}_n + \mathbf{r}_m)\rangle$$

- scattering functions

$$S_{\nu\mu}(\mathbf{k}) = \frac{1}{N}\langle\sum_{n=1}^{N_\nu}\sum_{m=1}^{N_\mu} \exp(-i\mathbf{k}\cdot(\mathbf{r}_n - \mathbf{r}_m))\rangle$$

Two examples at different temperatures, $\bar{T} = 0.15$ (liquid phase), and $\bar{T} = 0.05$ (solid or glassy phase) are shown in Fig. 2.

According to the well known Hansen-Verlet criterion[6] the system at $\bar{T} = 0.15$ is still in the liquid state (height of the first peak in $S(k)$ is below 2.8) whereas the system at $\bar{T} = 0.05$ appears to be frozen. But the shape of the pair correlation function and the absence of other peaks in the scattering function show that we do not have a crystal.

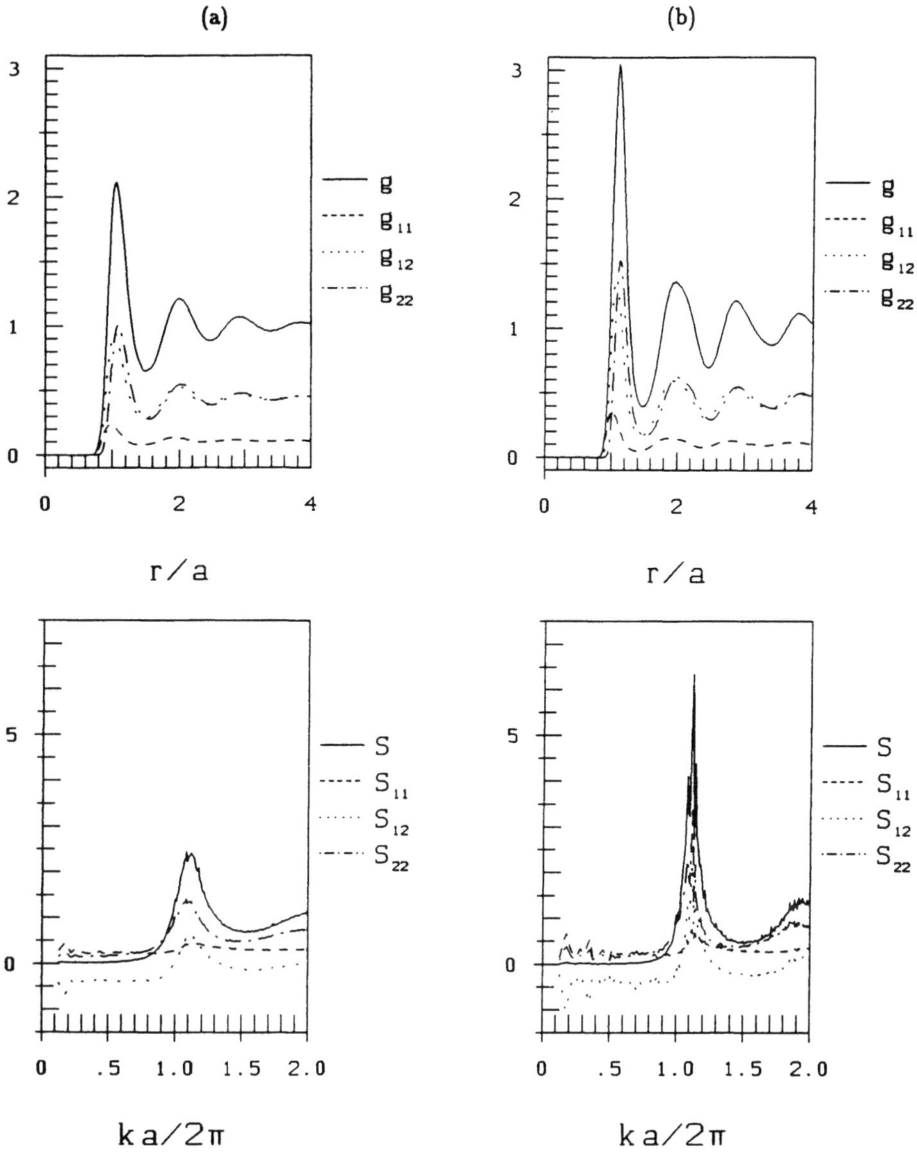

Figure 2: Pair distribution functions and scattering functions for a two component Yukawa system at (a) $\bar{T} = 0.15$ (liquid state) and (b) $\bar{T} = 0.05$ (solid or glassy state).

4 The "mean-field frequency"

From the pair distribution function $g(\mathbf{r})$ and the interaction potential $U(r)$ one can calculate a *mean effective potential* acting in the vicinity of the central particle. At places $\mathbf{r}$ that can only be reached by the central particle we have

$$U_{\text{eff}}(r) = U_0^{\text{eff}}(\lambda r/a)^{-1}\sinh(\lambda r/a) \tag{4}$$

where

$$U_0^{\text{eff}} = \frac{4\pi U_0}{a^2}\int_0^\infty g(r')r'e^{-\lambda r'/a}dr' \tag{5}$$

A Taylor expansion around the position of the central particle,

$$U_{\text{eff}}(r) \rightarrow U_0^{\text{eff}}\left[1 + \frac{1}{6}(\lambda r/a)^2 + \ldots\right] \qquad \text{for } r \rightarrow 0, \tag{6}$$

gives a characteristic *"mean-field frequency"*

$$\omega^2 = \frac{1}{3}\left(\frac{\lambda^2 U_0^{\text{eff}}}{ma^2}\right). \tag{7}$$

This frequency is temperature dependent, if $\tilde{T} \rightarrow 0$ we recover the Einstein frequency.
An estimation of the mean-field frequency is possible by doing a self-consistent harmonic approximation which replaces the particles by harmonic oscillators that are arranged in shells of distance r_n around the central particle. For this arrangement one can calculate the pair distribution function and from that the corresponding mean-field frequency. Setting this frequency equal to the frequency of the harmonic oscillators and assuming that

$$\frac{kT\lambda^2}{m\omega^2 a^2} \ll 1, \tag{8}$$

which was always the case for our systems, we finally get

$$\frac{m\omega^2 a^2}{U_0} \approx \lambda^2 C\left(1 - \tilde{T}/C\right)^{-1} \tag{9}$$

where

$$\tilde{T} = \frac{kT}{U_0} \quad \text{and} \quad C = \frac{1}{3}\sum_n N_n \frac{a}{r_n} e^{-\lambda r_n/a} \tag{10}$$

Figure 3 shows that the values of ω^2 calculated from $g(\mathbf{r})$ can be well described by the above harmonic approximation.

5 Summary

The first results of our investigation of Yukawa systems as a model for fluid and glassy colloids are the following

- The pair distribution functions describe the local surroundings of a particle. The difference in the repulsive forces between different types of pairs of particles can be observed clearly.
- At lower temperatures the local structure becomes more pronounced. This can also be seen from the growing peak of the scattering function, which indicates a frozen structure for the system at $\tilde{T} = 0.05$, but there are *no signs of crystallization* visible.
- The pair distribution function of the system under investigation shows at $\tilde{T} = 0.05$ a slight splitting of the second peak as can also be observed in *glassy states*.

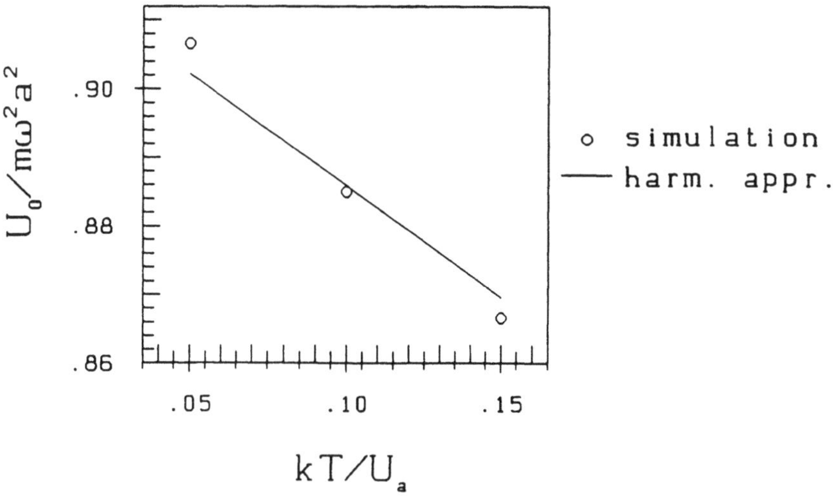

Figure 3: Temperature dependence of the "mean-field frequency" ω

- Generalizing the concept of the Einstein frequency we defined a *mean-field frequency* characteristic for the given system. Its temperature dependence can be well estimated by a self-consistent harmonic approximation.

Acknowledgement

This research was supported in part by Deutsche Forschungsgemeinschaft, Sonderforschungsbereich 262.

References

[1] S. Alexander, P. M. Chaikin, P. Grant, G. J. Morales, P. Pincus, D. Hone, J. Chem. Phys. **80**, 5776 (1984)

[2] K. Kremer, M. O. Robbins, G. S. Grest, Phys. Rev. Lett. **57**, 2694 (1986)

[3] K. Kremer, G. S. Grest, M. O. Robbins, J. Phys. A **20**, L181 (1987)

[4] M. O. Robbins, K. Kremer, G. S. Grest, J. Chem. Phys. **88**, 3286 (1988)

[5] Y. Monovoukas, A. P. Gast, J. Coll. Int. Sci. (in press)

[6] J. P. Hansen, L. Verlet, Phys. Rev. **184**, 150 (1969)

Non-exponential Decay of Density Correlation Function in Dense Microemulsions

Sow-Hsin Chen[1], *E.Y. Sheu*[2], *and J.S. Huang*[2]

[1]Department of Nuclear Engineering and
Center for Materials Science and Engineering,
Massachusetts Institute of Technology, Cambridge, MA 02139, USA
[2]Exxon Research and Engineering Company, Annandale, NJ 08801, USA

Extensive quasi-elastic light scattering (QELS), small angle neutron scattering (SANS), and neutron spin echo spectroscopic (NSE) measurements have been made on a three-component microemulsion system containing AOT (surfactant), D_2O and decane (oil) at two temperatures in the one-phase region. By fixing the molar ratio of water to AOT at w = 40.61, the average radius of spherical, surfactant coated water droplets can be maintained at 58 Å with a polydispersity index of 22%. The volume fraction ϕ of the droplets can be continuously varied from 0.1 to 0.75 simulating the entired range from a dilute gas to liquids to a close-packed solid by adjusting the oil content of the microemulision. The static structure factors and the time-dependent density correlation functions of these densely packed spherical droplets were systematically measured as a function of the volume fraction. As ϕ increases from 0.1 to 0.75, the static structure factor gradually evolves from a smooth function of the order unity into a highly oscillatory function with the height of the first diffraction peak reaching about two, similar to that of a dense liquid. At the same time the time-dependent density correlations function exhibits pronounced non-exponential relaxation. The Kohlrausch-Williams-Watts (KWW) stretched exponential function, $\exp[-(t/T)^{\beta}]$, has been shown to fit the density correlation function over most of the time range of the measrements, which covers nearly three decade in time. At Q values of about 10^{-3} Å^{-1}, typical in a light scattering experiment, the exponential β, starting from an initial value of unity at ϕ = 0.1, gradually decreases to about 0.64 at ϕ_c = 0.65, and increase thereafter. The relaxtion time T shows a trend to diverge at $\phi \sim \phi_c$ approximately following the well known Vogel-Tammann-Fulcher law. These features bear stiking similarity to a glass-like dynamic transition predicted by recent mode-mode coupling theories of dense liquids. The KWW function also fits very well the normalized intermediate scattering function at ϕ = 0.6, measured by NSE at much higher Q (Q ~ 0.2 Å^{-1}). Thus, the KWW function is shown to describe the density

correlation function well for Q range spanning from the hydronamic region to the region of first and second diffraction peaks of the structure factor. The relaxation time T scales to a good approximation, like $Q^{-2/\beta}$.

Springer Proceedings in Physics, Vol. 37 **Dynamics of Disordered Materials**
Editors: D. Richter · A.J. Dianoux · W. Petry · J. Teixeira

1. Introduction

In general a "microemulsion" is a thermodynamicaly stable fluid mixture consisting of five components: brine (water with a small amount of salt added), oil, surfactant and co-surfactant (a short chain alcohol). It is an isotropic, optically clear liquid solution which contains colloidal particles of sizes of the order of 100 Å. However, the simplest microemulsion system can be realized with only three components: water, oil and surfactant. AOT (sodium di-2-ethylhexylsulfosuccinate)/water/decane is one such system. It is, moreover, a model microemulsion system in the sense that a single phase microemulsion is obtainable over a substantial portion of its ternary phase diagram at room temperature.

Previous light scattering [1] and small angle neutron scattering [2] experiments have established that near the oil corner of the ternary phase diagram, AOT/water/decane, microemulsions consist of surfactant coated spherical water droplets dispersed in the oil continuous medium. The size of the water core of the droplet is a function of a parameter w which is the molar ratio of water to surfactant; i.e., the number of water molecules per surfactant molecule in the solution. Kotlarchyk, Chen and Huang [3] showed by a SANS experiment that these droplets were polydisperse having a polydispersity index $p \simeq 0.3$ (ratio of the standard deviation over the mean size). The average radius of the water core, $\bar{R}$, can be accurately calculated by a formula [4]

$$(1+2p^2)\bar{R} = \frac{3v_w}{a_H} W + \frac{3v_H}{a_H} \tag{1}$$

where v_W is the volume of a water molecule, v_H the volume of a head group of an AOT molecule, and a_H the average area occupied by a head group of an AOT. In practice $\bar{R}$ is a function of temperature because the parameters a_H and p are temperature dependent, but dependence of $\bar{R}$ and p on volume fraction of the droplet, ϕ, is expected to be weak [4]. It has also been shown by Kotlarchyk, Chen, Huang and Kim [5] that near the center of the phase triangle where the volume fractions of oil and water are equal, the microemulsion can be regarded as a system of densely packed spheres of average radius $\bar{R}+\Delta$, Δ being the length of the surfactant molecule. Further evidence of the droplet structure of AOT/D_2O/decane microemulsion was provided by an observation of phase separation and critical phenomena in systems where the w value was fixed at 40.61 [6,7]. Under this constraint, the average radius of the water core of the microemulsion droplets has a constant value of 50 Å (see equation 1) and one can freely vary the volume fraction of the droplets by varying the volume fraction of oil in the solution according to: $\phi = 1-\phi_{oil}$. Thus if one regards the microemulsion droplets as giant molecules packed into a liquid like structure in an oil medium, one can increase the number density of these giant molecules at will simply by adding more AOT and water into decane, keeping the former molar ratio fixed. Figure 1 is a T-ϕ phase diagram for microemulsions made with $w = [D_2O]/[AOT] = 40.61$. This condition can in practice be realized by maintaining a constant AOT to water ratio at 3 grams of AOT per 5ml of water, and variable amount of decane. We call this system a 3/5 system. For $\phi = \phi_{AOT} + \phi_{D_2O}$ between 0.03 and 0.40, if one heats up the microemulsion it will phase separate at a certain temperature called the cloud point. Below the cloud point a microemulsion is in a single phase . Above the cloud point the microemulsion phase separates into two microemulsion phases containing different volume fractions of the droplets [4]. But the average size and size distribution of the microemulsion droplets in the two phases are the same within experimental error [4]. The cloud point curve can be regarded, in this sense, as a coexistence curve of a pseudo-binary fluid consisting of two components, the droplets and the oil. At a temperature of

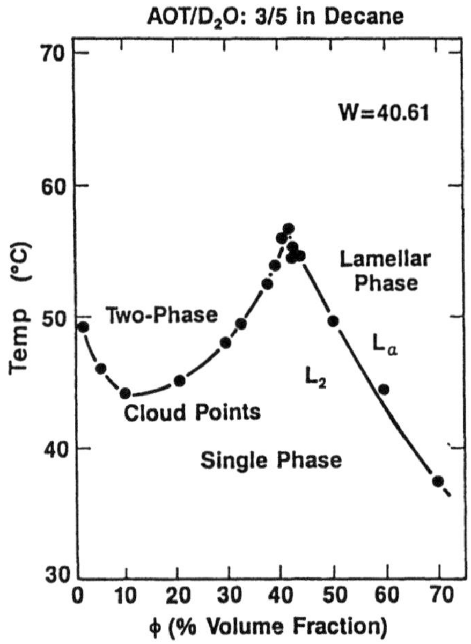

Fig. 1 : Boundaries between the one-phase microemulsion, denoted by L_2, and the two-phase microemulsions, called a cloud point curve (coexistence curve), and between L_2 and the lamellar liquid crystalline phase, denoted by L_α, for a system of three-component microemulsion, AOT/D_2O/Decane, at a constant water to AOT ratio w=40.611. The ordinate is the temperature and the coordinate is the volume fraction of water plus AOT. Experimental paths used in this paper correspond to horizontal lines with temperatures equal to 17.9°C and 22.6°C.

$40^{\circ}C$, if we increase the volume fraction of the droplets above 0.6, the single phase microemulsion will undergo a first order phase transition to a lamellar phase. In the lamellar phase the droplets no longer exist. The single phase microemulsion we will discuss in this paper is usually denoted as the L_2 phase in the literature [8]. For temperatures below $30^{\circ}C$ and above $15^{\circ}C$, the L_2 microemulsions can be well represented as liquid like systems having a collection of surfactant coated water spheres of an average radius 58 Å dispersed in an oil continuous medium. The volume fraction of the spherical droplets can be varied continuously from 0.03 to 0.75 without being interrupted by the lamellar phase.

In this paper we shall exploit the fact that L_2 microemulsions can be regarded as dense liquids with continuously variable particle packing fractions. We are thus able to investigate the dynamics of density fluctuations of the spherical droplets as functions of volume fraction and temperature by quasi-elastic light scattering and neutron spin echo spectroscopy. An exciting possibility is to test the predictions of recently developed self-consistent mode-mode coupling theories of dense liquids [9,10,11]. Especially, we would like to observe the dynamic singularity of the glass-like transition predicted by these theories [9,10,12]. Our starting observation is that the shear viscosity , η, of the microemulsions at $22.6^{\circ}C$ appears to increase rapidly (one hundred fold) as a function of the volume fraction ϕ, and seemingly becomes divergent toward the critical volume fraction $\phi_c = 0.65$, as shown in figure 2a. In figure 2b we test the applicability of a form of power law divergence

$$\eta = \eta_s\left(1-\frac{\phi}{\phi_c}\right)^{-\alpha} \tag{2}$$

where η_s is the shear viscosity of the solvent (decane). We plot $(\eta_s/\eta)^{1/\alpha}$ as a function of ϕ/ϕ_c. Equation (2) corresponds to the solid line. Experimental data are plotted by choosing $\alpha = 1.8$ and 2.5, respectively. We can see that the $\alpha = 2.5$ case is closer to the theoretical formula. The mode-mode coupling theories predict $\alpha = 1.8$ but Angell made a survey of a number of glass forming simple liquids and concluded that the range of α is between 1.5 and 2.5 [13].

In the following sections we shall discuss our recent SANS, QELS and NSE experimental results. Some preliminary QELS results were previously published by Chen and Huang [14]. The full details of these experiments and their analyses will be published elsewhere [14b]. This paper describes only some highlights of these new measurements.

Figure 2a

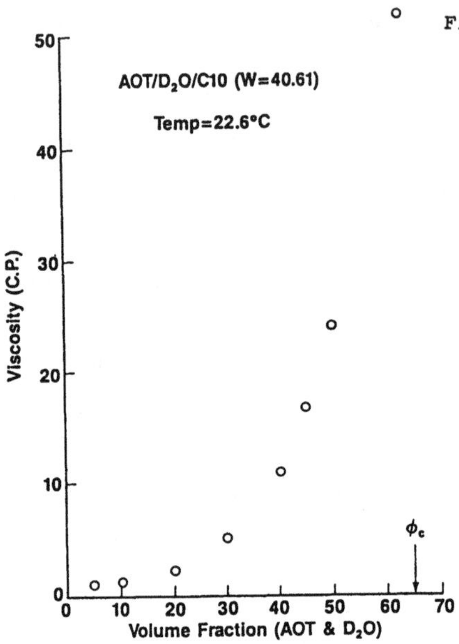

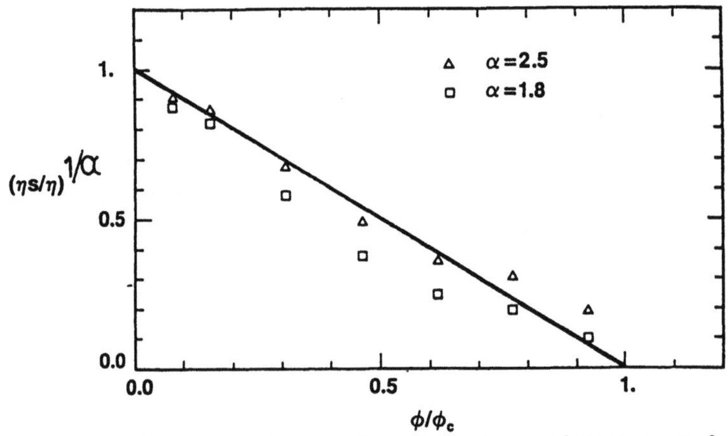

Figure 2b

Fig. 2. : Shear viscosity η of w=40.61 microemulsions as a function of volume fraction $\phi = \phi_{AOT} + \phi_{D_2O}$ at a temperature 22.6°C. (2a) is the actual data and (2b) is a reduced plot of $(\eta_s/\eta)^{1/\alpha}$ vs (ϕ/ϕ_c) where η_s is the shear viscosity of decane and ϕ_c = 0.65 is the critical volume fraction defined in the text. Squares represent the data reduceed by α = 2.5 and triangles by α = 1.8. Solid line is the theoretical prediction described in section I.

2. Results

2.1 Small Angle Neutron Scattering

For a system of polydisperse spheres, having a Schultz distribution of sizes [15],

$$f(R) = \left(\frac{Z+1}{\bar{R}}\right)^{Z+1} \frac{1}{\Gamma(Z+1)} R^{Z} \exp\left[-\left(\frac{Z+1}{\bar{R}}\right)R\right] \tag{3}$$

where $\bar{R}$ is the average radius of the spheres and Z the width parameter, related to the polydispersity index p by $p = (1+Z)^{-1/2}$, the intensity distribution of small angle neutron scattering can be expressed concisely as:

$$I(Q) = (\Delta\rho)^2 \phi_w \left(\frac{4\pi}{3}\bar{R}^3\right) \frac{(Z+6)(Z+5)(Z+4)}{(Z+1)^3} \langle\tilde{P}(Q)\rangle \langle S(Q)\rangle \tag{4}$$

In Eq. (4), $\Delta\rho$ is the contrast between D_2O and the hydrocarbon in the solvent, ϕ_w the volume fraction of D_2O, $\bar{R}$ the mean radius of water cores in microemulsion droplets. The size averaged particle structure factor is defined as:

$$\langle\tilde{P}(Q)\rangle = \langle R^6\tilde{P}(QR)\rangle / \langle R^6\rangle \tag{5}$$

with $\tilde{P}(QR) = [3j_1(QR)/QR]^2$ and the brackets denoting an average over the distribution Eq. (3). $\langle S(Q)\rangle$ is the size averaged inter-droplet structure factor [16], which is the Fourier transform of the effective pair correlation function of the droplets.

Since the information on the interparticle interaction between droplets is not precise enough at the present time, $\langle S(Q)\rangle$ could not be computed reliably by the statistical mechanical theory. We therefore tried to extract $\langle S(Q)\rangle$ experimentally using knowledge of parameters obtained from a previous analysis of $\langle\tilde{P}(Q)\rangle$ [4]. We chose $\bar{R} = 50$ Å and Z = 20 for the Schultz distribution parameters to compute $\langle\tilde{P}(Q)\rangle$ according to Eq. (5). Then $\langle S(Q)\rangle$ was obtained by diving I(Q) by the known prefactor in Eq. (4) and the $\langle\tilde{P}(Q)\rangle$ factor. Fig. 3 shows the average structure factors thus extracted for $\phi = 0.75$, 0.65 and 0.55. Three features of these curves should be noted: (a) the values of $\langle S(Q = 0)\rangle$ are very low indicating a strong repulsive interaction resulting in low osmotic compressibilities, (b) the first diffraction peaks of heights around 2.0 occuring at $Q_1 = 0.079$ Å^{-1}, (c) the second diffraction peaks of heights about 1.2 occuring at $Q_2 = 0.14$ Å^{-1}. It should also be remarked that the size averaged structure factor for a hard sphere system generally has the first diffraction peak height lower than the structure factor of the corresponding mono-disperse system, but their peak positions are nearly the same [17]. It is therefore of interest to infer the average interparticle seperation distance D from Q_1. According to an empirical formula known to be valid in describing correlation peaks in dense micellar systems [18]

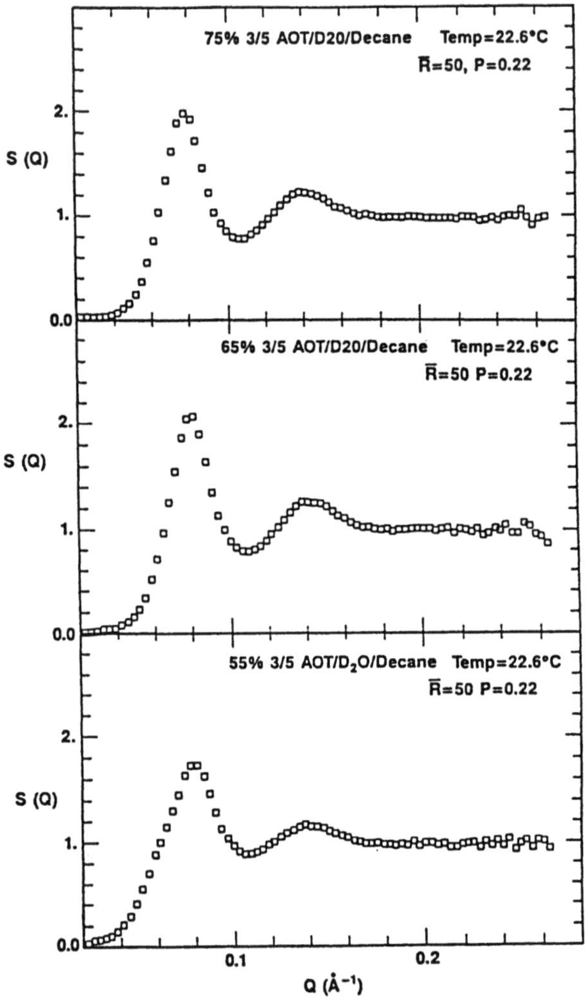

Fig. 3 : Interdroplet structure factors extracted from SANS data at ϕ=0.75,0.65 and 0.55 assuming an average droplet size of $\bar{R}$ = 50 Å and polydispersity index of 0.22, taken from a previous analysis [4]. A notable feature is the sharp first diffraction peak having height in the vicinity of 2.0 followed by a second diffraction peak ocurring at approximately twice the Q value of the first. This feature is characteristic of a dense liquid-like structure.

$$Q_1 D = 6.8559 + 0.0094D \qquad (6)$$

From this equation, using Q_1 = 0.079 $Å^{-1}$, we get D = 99 Å, which is about twice the $\bar{R}$ = 50 Å we used to extract these curves. Since the microemulsion droplets consists of a water core coated with a surfactant layer of about 8 Å thickness, this calculation indicates that at such high volume fraction, the surfactant layers actually overlap and the droplets are percolated. This seems to be supported by an independent experiment on AC electric conductivity for such systems by Kim and Huang [19].

2.2 Quasielastic Light Scattering

The photon correlation function measured in this experiment is denoted by C(t). It is usually normalized in such a way that C(0) = 1. In this latter case it is exactly equal to the droplet density correlation function, $\langle \rho_{-Q}(0)\rho_Q(t)\rangle / \langle |\rho_Q|^2 \rangle$ [20]. The sampling time used was 2.5 μsec per channel and a multiple sampling time mode was used to measure the correlation function up to a delay time of 1400 μsec. We now give some typical examples of the photon correlation function. Fig.4 shows ln[C(t)]+5 vs t (in msec) for a high temperature case, Temp = 22.6°C at three volume fraction ϕ = 0.1, 0.5 and 0.65. It is easily seen that, while for ϕ = 0.1 the decay is exponential, the decay is clearly non-exponential showing considerable long time tail for ϕ = 0.5 and 0.65. Fig. 5 displays three unnormalized C(t) for a low temperature case, Temp = 17.9°C. We see, as the volume fraction increases from 0.2 to 0.5 to 0.68, there is a much faster development of the long time tail, leading to nearly non-zero constant (D.C.) at 500 μsec.

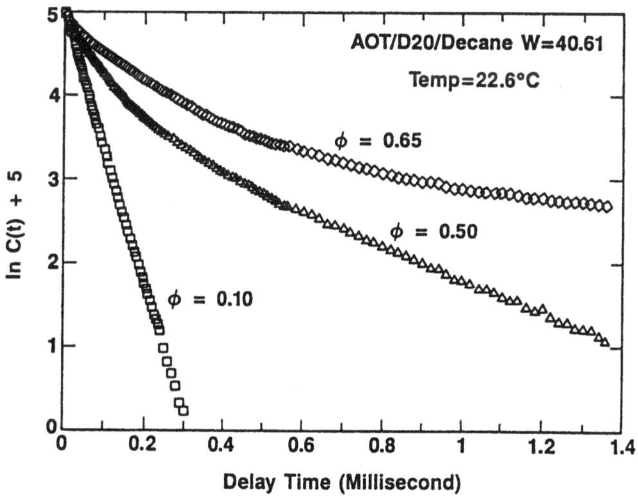

Fig. 4 : Normalized time dependent correlation functions for w=40.61 AOT/D_2O/Decane microemulsions at ϕ = 0.65, 0.50 and 0.10 at a temperature equal to 22.6°C. The ordinate is in a logarithmic scale and the abscissa is in a linear time scale of milliseconds. Note the increasingly pronounced deviation from exponential behavior as ϕ increases.

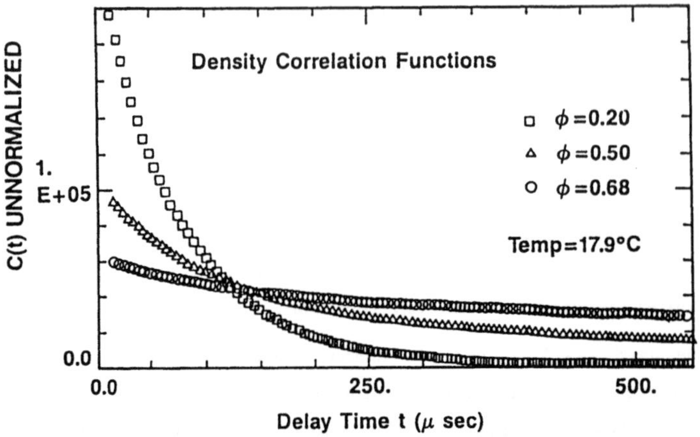

Fig. 5 : Unnormalized time dependent correlation functions for ϕ = 0.2,0.5 and 0.68 at 17.9°C. The feature to be noted is the sharp decrease of the amplitude of the initial decay of the correlation funcitons as ϕ increases. This is accompanied by development of long time tails which seem to contain a DC part.

These curves, however, can all be fitted to a good accuracy by a Kohlrausch-Williams-Watts (KWW) function of the form

$$C(t) = \exp[-(\frac{t}{T})^{\beta}] \ . \tag{7}$$

Fig. (6a) and Fig. (6b) illustrate the applicability of Eq.(7) vividly. In Fig. (6a) we display ln[C(t)] vs time in a linear scale. The curve spanning delay time between 2.5 μsec to 1200 μsec is clearly not a straight line. However, in Fig. (6b) the same data, when dispalyed in terms of a scaled time t^{β}, where β=0.68, shows a good straight line for the delay time longer than 100 μsec. The slope of the curve is just $T^{-\beta}$.

Since the non-exponential behavior seems to be the dominant feature of the decay of the time-dependent density correlation function, our strategy of analysing C(t) to extract useful physical quantities is as follows: we consider a multiple relaxation mechanism of some kind which leads to the non-exponential behavior [14,21]. We therefore assert

$$C(t) = \int_0^{\infty} d\lambda f(\lambda) e^{-\lambda t} \tag{8}$$

$$\doteq \exp[-(\frac{t}{T})^{\beta}] \tag{9}$$

Figure 6a

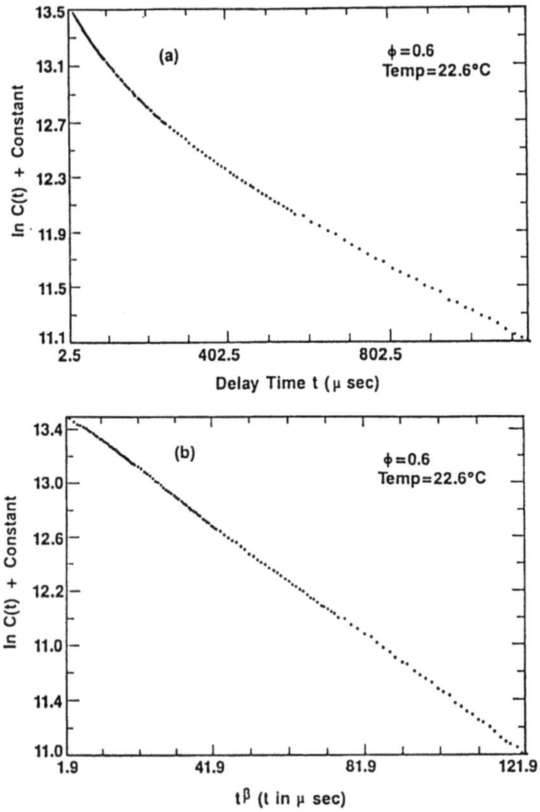

Figure 6b

Fig. 6 : Stretched exponential behavior of a time-dependent density correlation function C(t). (a) Logarithm of C(t) vs t plot for t from 2.5 μs to 1140 μs. (b) the same ln[C(t)] plotted as a function of t^{β} with β = 0.68. This figure illustrates vividly that the stretched exponential behavior is valid for delay times longer than 100 μs.

where $f(\lambda)$ is a distribution function of the relaxation rate λ. Eq. (9) is only an approximate equality over a time interval not too short and not too long. In the limits of short and long time C(t) has to be exponential. It follows then

$$\text{The first cummulant} \equiv \frac{dC}{dt}\bigg|_{t=0} = \int_0^{\infty} d\lambda\lambda f(\lambda) = \bar{\lambda} \qquad (10)$$

$$\text{Area} \equiv \int_0^{\infty} dtC(t) = \int_0^{\infty} d\lambda\left(\frac{1}{\lambda}\right)f(\lambda) = \langle\lambda^{-1}\rangle = \frac{1}{\beta}\Gamma\left(\frac{1}{\beta}\right)T \qquad (11)$$

Thus the two obvious quantities to extract are $\bar{\lambda}$ and $\langle\lambda^{-1}\rangle$ or β and T from the latter. If for example we assume that $f(\lambda)$ has a bimodal distribution with a band of fast relaxations centering around λ_f and a band of slow relaxations around λ_s, then $\bar{\lambda}$ would be dominated by λ_f and $\langle\lambda^{-1}\rangle$ or equivalently $\frac{1}{\beta}\Gamma(\frac{1}{\beta})T$ would be dominated by λ_s.

We report first the average relaxation rate $\bar{\lambda}$ extracted from the first cummulant of the correlation functions as a function of the volume fraction for various Q values (different scattering angles) at the high and low temperatures. Fig. 7 shows $\bar{\lambda}$ vs. Q plots for three Q values 1.4×10^{-3} Å^{-1}. 1.98×10^{-3} Å^{-1} and 2.54×10^{-3} Å^{-1} We see that at Temp = 22.6°C the average relaxation rate decreases as a function of ϕ, reaching a minimum at certain ϕ_{min}, depending on the Q value, and then increases again. On the other hand at Temp = 17.9°C, Fig. 8 shows that at Q = 1.98×10^{-3}Å^{-1} (90° scattering angle), $\bar{\lambda}$ continues to decrease up to the highest volume fraction ϕ = 0.75. At the present time, the Q-dependence of ϕ_{min} is not yet well understood. If, however, one views the ϕ_{min} as the volume fraction where the cage effect in a dense liquid begins to set in then one may argue that at higher Q values where the dynamics are probed at a more local level the cage effect is easily seen at lower volume fractions. The cage effect may have some connection with percolation of microemulsion droplets mentioned earlier. It is known [19] that at Temp = 22.6^{oC} the percolation phenomenon sets in at a volume fraction of about 0.5, but for Temp = 17.9°C, it

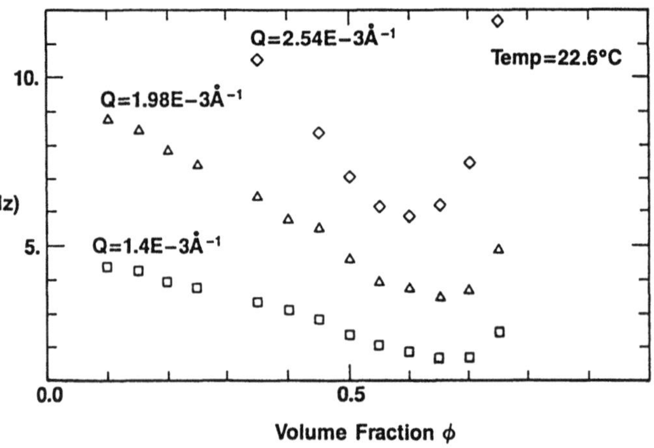

Figure 7

Fig. 7 : Average relaxation rate, $\bar{\lambda}$, of the density correlation funciton vs ϕ at different scattering angles for temperature equal to 22.6°C. Squares correspond to 60 degree scattering angle, triangles, 90 degree, and diamonds, 130 degree. $\bar{\lambda}$ is extracted from the first cummulant of the density correlation funciton at short time . The interesting feature of these curves is that a minimum appears between ϕ = 0.6 and 0.65, which we define as ϕ_c, depending on the Q values in the scattering.

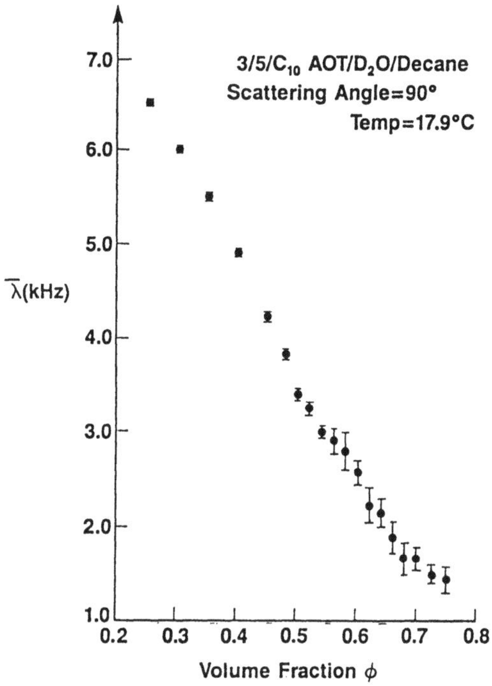

Fig. 8 : The same plot as Fig. 7 for temperature equal to 17.9°C and scattering angle equal to 90 degrees ($Q = 1.98 \times 10^{-3}$ Å^{-1}). Note that there is no minimum appearing in the vicinity of $\phi = 0.6$, signifying a significant structural difference of microemulsions at $\phi = 0.6$ between two temperatures, 22.6°C and 17.9°C.

would not set in until much higher volume fraction. This fact may explain why we do not observe a minimum in $\bar{\lambda}$ vs. ϕ curve in Fig. 8 up to $\phi = 0.75$. If we now conveniently define $\phi_c = \phi_{min}$ (at $Q = 1.98 \times 10^{-3}$Å^{-1}), then we conclude from Fig. 7 and Fig. 8 that $\phi_c = 0.65$ at Temp = 22.6°C and $\phi_c = 0.75$ at Temp = 17.9°C.

We turn next to the results of fittings of KWW function to the time-dependent correlation functions. The stretch exponent β is plotted as a function of ϕ for different Q values for the high temperature case in Fig. 9. We see clearly that β starts out with a value unity at the lowest volume fraction and decreases to a minimum in the vicinity of $\phi = 0.55 - 0.65$ depending on Q. Again we take $Q = 1.98 \times 10^{-3}$Å^{-1} as a reference point. We get $\beta_{min} = 0.70$ at $\phi_{min} = 0.60$. Fig. 10 shows a similar plot for the low temperature case. We get $\beta_{min} = 0.64$ at $\phi_{min} = 0.70$. The relaxation time T is an interesting quantity. Fig. 11 shows T vs. ϕ plots for the three Q values at high temperature. We observe maxima which are Q-dependent in these plots. Again take $Q = 1.98\times10^{-3}$ Å^{-1} case as a reference

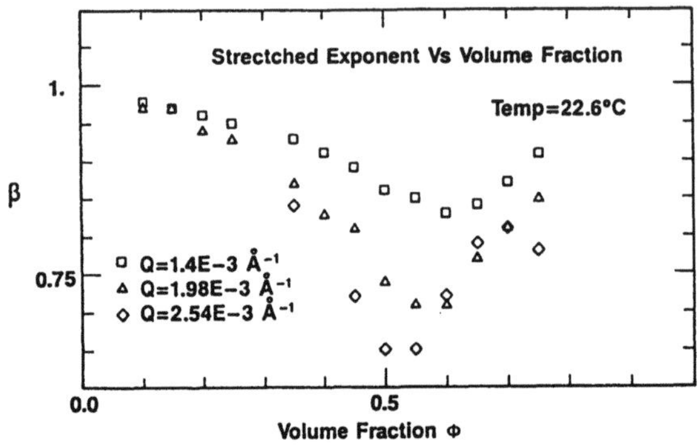

Fig. 9 : The stretch exponent, β, as a function of ϕ at three Q values. Squares corresponds to $Q = 1.4\times10^{-3}\ \text{Å}^{-1}$, triangle , $Q = 1.98 \times 10^{-3}\ \text{Å}^{-1}$, and diamond, $Q = 2.54 \times 10^{-3}\ \text{Å}^{-1}$. For these three scattering angles, a minimum β appeas at ϕ = 0.6, 0.58 and 0.52 respectively.

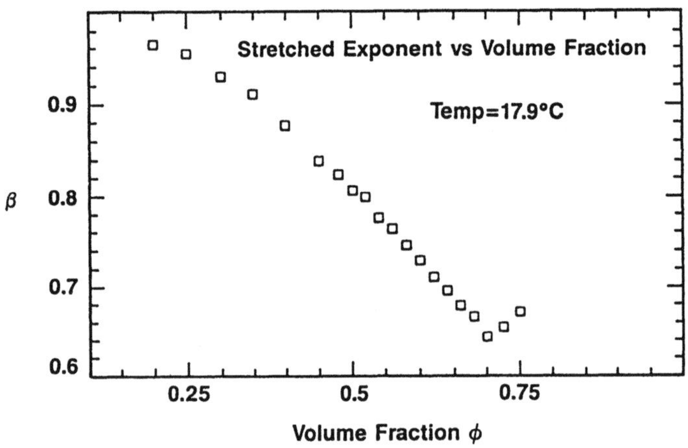

Fig. 10 : Same plot as Fig.9 for temperature equal to 17.9°C. The minium β occurs at $\phi = 0.70$.

point. We have T_{max} = 200 μsec occuring at $\phi_{max} = 0.65$. Fig. 12a give a similar plot for the low temperature case. We get T_{max} = 625 μsec at $\phi_{max} = 0.75$ for this case.

Since for the low temperature case T seems to be diverging toward $\phi_{max} = 0.75$, it is of interest to check whether T obeys the so-called Vogel-Tammann-Fulcher (VFT) formula [21,22]:

$$T = T_0 \exp[\frac{K}{\phi_c - \phi}] \qquad (12)$$

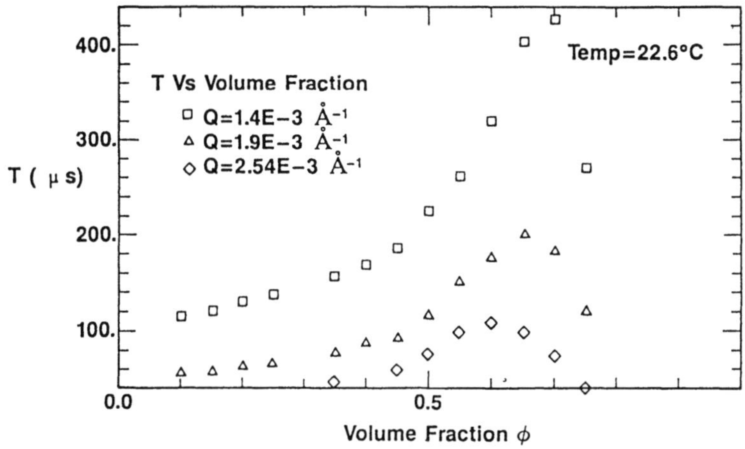

Fig. 11 : Relaxation time T vs ϕ plot for the high temperature case at the three Q values. The relaxation time appears to reach maximum at $\phi = 0.7$ for $Q = 1.4\times10^{-3}$ Å^{-1}, 0.65 for $Q = 1.98\times10^{-3}$ Å^{-1} and 0.6 for $Q = 2.54\times10^{-3}$ Å^{-1}.

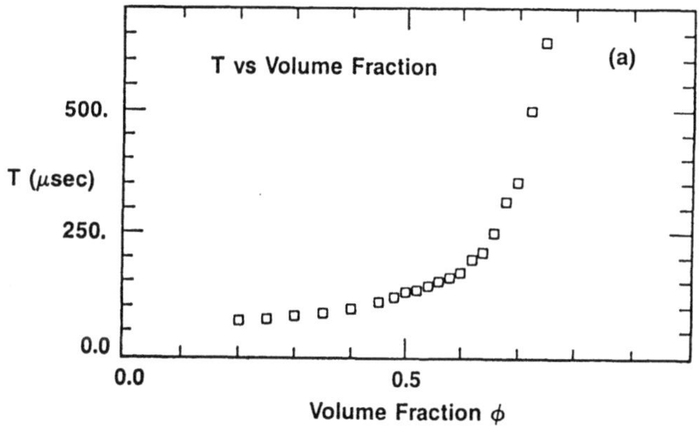

Figure 12a

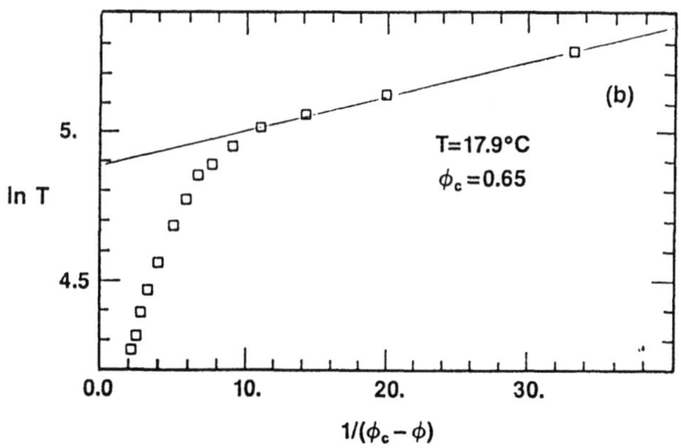

Figure 12b

Fig. 12 : (a) The relaxation time, T, vs ϕ plot for the low temperature case. (b) A Vogel-Fulcher plot exhibits linear behaivor which is indicative of a glass-like transition. ϕ_c is taken as 0.65 in the plot.

In Fig. 12b we plot lnT vs. $(\phi_c-\phi)^{-1}$ with ϕ = 0.65. It is seen that for ϕ sufficiently near ϕ_c the plot shows a linear behavior with the slope K = 0.012.

2.3 Neutron Spin-Echo Spectroscopy

NSE spectroscopy measures the normalized intermediate scattering function, I(Q,t)/S(Q), which is essentially the same as the normalized photon correlation function, C(t) [23]. There is, however, a major difference in time scale of the measurement. While photon correlation spectrosocopy measures C(t) at a time scale of milliseconds, NSE spectroscopy measures at a time scale of nanoseconds. This is because the Q-range of the two techniques differ by a factor of 10^3 and the relaxation rates normally scale like Q^{-2}. We have measured a sample with ϕ = 0.60 at Temp = 22.6°C. Fig. 13 shows the results of these measurements at four Q values, 0.086, 0.111, 0.136 and 0.185 $\AA^{-1}$. The points are experimental data and the solid lines are the KWW fits with a fixed β = 0.67 and a variable T. Fig. 14 summarizes all the photon correlation data and the NSE data in one graph. We plot a scaled relaxation time $T^{-\beta}$ for all the volume fractions as a function of Q^2 in a log-log scale. The solid line is a straight line going through all the data points having a slope equal to two.

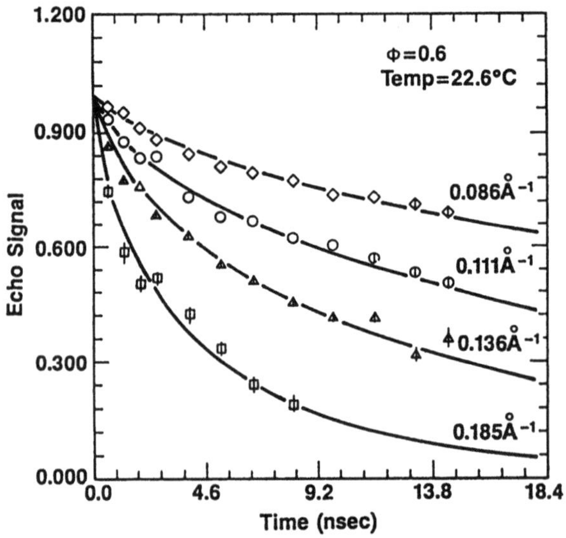

Fig. 13 : The normalized intermediate scattering function, I(Q,t)/S(Q) obtained from neutron spin-echo measurement for a AOT/D_2O/Decane microemulsion with w = 40.61 at ϕ = 0.6. The solid lines are the stretched exponential fits to the data at different Q ranging from 0.086 to 0.185 $\AA^{-1}$. The fits were made with β set equal to 0.67 but with various values of T.

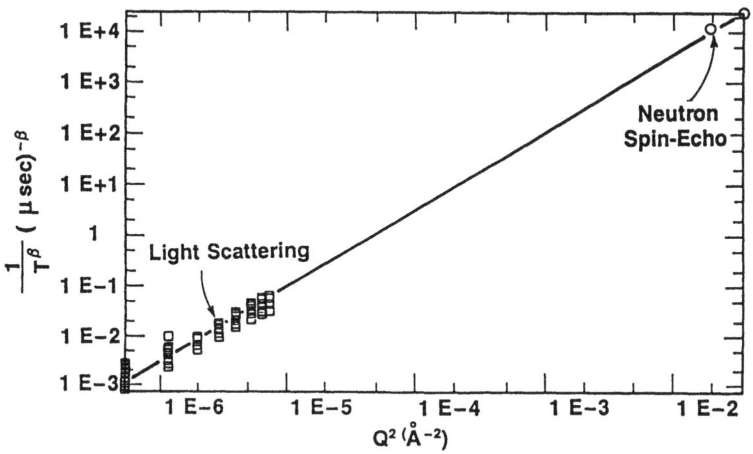

Fig. 14 : ($1/T^{\beta}$) vs Q^2 plot for all data taken in QELS measurements and NSE experiments. Within the statistical spread of the data it fits a straight line going through the origin over nearly six decade of Q^2 range.

3. Conclusions

It is clear from the experimental evidence presented in Fig. 2 through Fig. 14 that various transport and dynamical properties of AOT/D_2O/decane = 3/5 microemulsions around the droplet volume fraction ϕ_c = 0.65 - 0.75 exhibit singular behavior. The central theme of this paper is to show experimentally that the non-exponential decay of the time-dependent density correlation functions of microemulsions arises from dynamic correlations between the constituent particles of the surfactant coated water droplets packed in a dense liquid-like structure. This type of microemulsion system is an ideal model system for testing the predictions of the current self-consistent mode-mode coupling theory of dense liquids and dynamics of glass-like transition [9,10,11,12,24]. Specifically, we have used static shear viscosity measurements, small angle neutron scattering, photon correlation spectroscopy and neutron spin echo spectroscopy to study the transport coefficient, the static structure factor and the dynamic structure factor in the whole range of volume fractions, including ϕ_c = 0.65, a critical volume fraction where the glass-like dynamic transition occurs. We showed that (a) the shear viscosity diverges toward ϕ_c in a power law behavior with α = 2.5, (b) the static structure factor near ϕ_c resembles densely packed spheres, (c) the average relaxation rate of the density correlation function shows a minimum near ϕ_c, (d) the time-dependent density correlation function follows, to an excellent approximation, the KWW form over at least two decades of delay time, (e) the stretch exponent β, starting from unity at $\phi = 0.1$, gradually decreases to a minimum value near ϕ_c, depending on Q value of the scattering experiment, (f) the relaxation time T, on the other hand, increases toward ϕ_c and shows a diverging behavior similar to the Vogel-Tammann-Fulcher formula, and finally, (g) the KWW stretched exponential decay behavior is also observed in NSE measurements at high Q values, beyond the second diffraction peak of the structure factor. Thus, the anomalous dynamical behavior is similar all the way from the hydrodynamical regime to the kinetic regime and the relaxation time T scales like $Q^{-2/\beta}$ from $Q = 0.0014$ Å^{-1} to $Q = 0.185$ Å^{-1}.

Finally, a word about the reason why ϕ_c occurs between 0.65 - 0.75. We have been using the definition $\phi = \phi_W + \phi_{AOT}$ all along. But in microemulsions surfactant tails of AOT are well mingled with decane because of their chemical similarity. For W = 40.61 microemulsions we used $\phi_W = 0.665\ \phi$. Thus, if we regard the excluded volume effects to arise from interference of water cores of the droplets alone, then ϕ_c would be reduced to values between 0.43 to 0.50, closer to the theoretically predicted value [25].

The authors are grateful to Bruce Carvalho and Szu-Li Chang for typing the manuscript, and to Drs. Farago and Richter of the Institut Laue-Langevin for assistance in neutron spin-echo measurements. This reasearch is supported by an NSF grant administered through the Center for Materials Science and Engineering at MIT and by Exxon Research and Engineering Co.

References

1. M. Zulauf and H. F. Eicke, J. Phys. Chem. 83, 480 (1979).
2. C. Cabos and P. Delord, J. Appl. Cryst. 12, 502 (1979).
3. M. Kotlarchyk, S. H. Chen and J. S. Huang, J. Phys. Chem. 86, 3273 (1982).
4. M. Kotlarchyk, S. H. Chen, J. S. Huang and M. W. Kim, Phys. Rev. A29, 2054 (1984).
5. M. Kotlarchyk, S. H. Chen and J. S. Huang, Phys. Rev. Lett. 49, 941 (1984).
6. J. S. Huang and M. W. Kim, Phys. Rev. Lett. 47, 1452 (1981).
7. M. Kotlarchyk, S. H. Chen and J. S. Huang, Phys. Rev. A28, 508 (1983).
8. Per Ekwall, L. Mandell and K. Fontell, J. Coll. int Sci. 33, 215 (1970).
9. E. Leuthensser, Phys. Rev. A29, 2765 (1984).
10. V. Bengtzelius, W. Gotze and A. Sjolander, J. Phys. C.17, 5915 (1984).
11. T. R. Kirkpatrick, Phys. Rev. A31, 939 (1985).
12. S. P. Das and G. F. Mazenko, Phys Rev. A34, 2265 (1986).
13. C. A. Angell, " Perspective on the Glass Transition" an invited contribution to the Perspectives section of the J. Phys. Chem. Solids (1988).
14. S. H. Chen and J. S. Huang, Phys. Rev. Lett. 55, 1888 (1985).
14b. E. Y. Sheu, S. H. Chen, J. S. Huang and J. Sung (to be published).

15. M. Kotlarchyk and S. H. Chen, J. Chem. Phys. 79, 2461 (1983).
16. S. H. Chen, Annu. Rev. Phys. Chem. 37, 351 (1986).
17. P. Var Beurten and A. Vrij, J. Chem. Phys. 74, 2744 (1981).
18. S. H. Chen, E. Y. Sheu, J. Kalus and H. Hoffmann, J. Appl. Cryst. 21, Dec. (1988).
19. M. W. Kim and J. S. Huang, Phys. Rev. A34, 719 (1986).
20. S. H. Chen and A. V. Nurmikko, "Photon Correlation Spectroscopy in Biology", Chapt.12, Spectroscopy in Biology and Chemistry-Neutron X-Ray and Laser, Edited by S. H. Chen and S. Yip, Acadamic Press (1974).
21. J. Jäckle, Philosoph. Magazime B, 56, 113 (1987).
22. M. I. Klinger, Physics Reports, 165 275 (1988).
23. J. B. Hayter, "Quasielastic Neutron Spin-Echo Spectroscopy", in Scattering Techniques Applied to Supramolecular and Non-equilibrium Systems, p. 49, Editors S. H. Chen, B. Chu, R. Nossal, NATO Advanced Study Insstitute Series Vol. 73, PLenum Press (1981).
24. G. H. Fredrickson, "Recent Developments in Dynamical Theories of the Liquid-Glass Transition", Annu. Rev. Phys. Chem. 39, (1988).
25. V. Bengtzelius, Phys. Rev. A33, 3433 (1986).

Free-Volume Distribution in Organic Glasses Studied by 2H NMR of Deuterated Probes

E. Rössler, M. Taupitz, and H.-M. Vieth

Institut für Atom- und Festkörperphysik, Freie Universität Berlin, Arnimallee 14, D-1000 Berlin 33, Germany

Introduction

During recent years nuclear magnetic resonance has found wide application to solid state systems. Due to improved equipment and novel techniques using, for example, cross-polarization, ccmplex pulse sequences or multi-dimensional Fourier transformation a high degree of spectral and temporal resolution has been achieved giving detailed information on the local structure and molecular mobility. Still, the low intrinsic sensitivity of NMR makes careful studies for many materials a painstaking or even impossible task. Particularly for most diamagnetic organic compounds without large-angle motion inefficient spin polarization, caused by slow spin-lattice relaxation prohibits the direct investigation of their crystalline or glassy state.

An indirect NMR study of the glassy state is possible when an NMR active probe molecule with short relaxation time is doped into the system. For short relaxation times sufficient motion of the probe molecule is necessary. Molecules as benzene or hexamethylbenzene are known to have anisotropic motion in the neat crystal; they form rotor phases. In many cases the type of motion is conserved when the probe is doped into a glassy matrix. Only, the values of parameters such as correlation time and activation energy are changed. In contrast to the crystalline state, for disordered systems generally a distribution of correlation times and activation energies is expected. The changes of the probe parameters depend on the local environment, therefore, structural properties of the solid matrix can be studied. For surpressing probe-probe interactions the ccncentration of the guest molecules has to be small, therefore, high sensitivity of the probe is an important requirement [1].

We used 2H NMR for measuring the spin lattice relaxation and the line shape of deuterated probe molecules in protonated organic glasses. For 2H NMR the dominant interaction is given by the quadrupolar coupling with the intramolecular electric field gradient of the C-2H bond. In contrast to 1H NMR,spin diffusion plays a minor role. No averaging between subensembles with different correlation times, e. g. between different sites in the matrix, is observed. Thus, 2H NMR is a suitable method for studying motional heterogeneities of the guest molecules. It is expected that by investigating the correlation time distribution in a glass the activation energy distribution of the probe molecules is measured. From this one can obtain information on the heterogeneous packing, i. e. the free-volume distribution of amorphous systems can be revealed.

Results

Here, we report studies on the system 2.9 % hexamethylbenzene-d_{18} (HMB) in polystyrene and 2.2 % benzene-d_6 in phthalic acid n-butyl ester (PDB) in the temperature range between 40 and 310 K. The spin lattice relaxation behavior was measured at a frequency of 46.07 MHz (7.04 T) by using a saturation recovery pulse sequence. In all cases a non-exponential relaxation behavior is observed as shown in Fig. 1 for HMB in polystyrene.

Springer Proceedings in Physics, Vol. 37 Dynamics of Disordered Materials
Editors: D. Richter · A.J. Dianoux · W. Petry · J. Teixeira

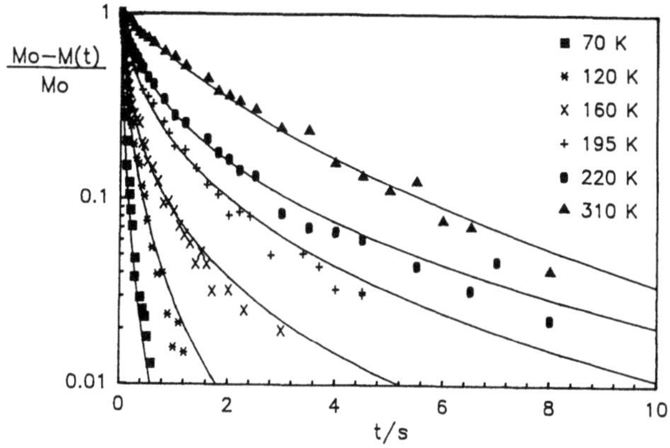

Fig.1: Spin lattice relaxation for HMB-d_{18} in polystyrene; solid line: log-Gauss fit

We described the relaxation behavior by assuming a distribution of relaxation times $F(T_1)$ due to a distribution of correlation times $p(\tau)$ of the probe motion. The relaxation of the magnetization $M(t)$ to its equilibrium value M_0 is given by:

$$\Phi(t) = (M_0-M)/M_0 = \int_0^\infty p(\tau)\exp(-t/T_1(\tau))d\tau \; . \qquad (1)$$

In good approximation the interdependence of T_1 and τ can be described by the well-known relation from Bloembergen, Pound, Purcell theory:

$$1/T_1 = C\left[\frac{\tau}{1+(\omega_0\tau)^2} + \frac{4\tau}{1+(2\omega_0\tau)^2}\right] \; . \qquad (2)$$

The coupling constant C is determined by the effective relaxation process. It is a specific property of the probe molecule and has been taken from the maximum relaxation rate in the neat crystalline probe. ω_0 stands for the Lamor frequency. When assuming a thermally activated motion, $\ln(\tau/\tau_0)=E/RT$, the distribution of activation energies transforms into a distribution $G(\ln\tau)$, where $G(\ln\tau)d\ln\tau=p(\tau)d\tau$.

As a first approach in our analysis we used a so-called log-Gauss distribution of correlation times

$$G(\ln\tau) = \beta/\sqrt{\pi}\exp(-\beta^2z^2) \text{ with } z = \ln(\tau/\tau_m) \; . \qquad (3)$$

The corresponding least square fits are also shown in Fig. 1, indicating a good agreement. The resulting width parameter β as a function of temperature is given in Fig. 3. The dependence of the mean correlation time τ_m on the reciprocal temperature is shown in Fig. 4. Included in the figures are the analogous data for benzene-d_6 in PDB For HMB-d_{18} two regimes have to be considered. At high temperatures the sixfold rotation of the molecule determines the relaxation, and the corresponding coupling constant C is used for the evaluation. For low temperatures the methyl group rotation with a different coupling constant dominates the relaxation. A broad intermediate regime of competing relaxation mechanisms cannot be evaluated in such a straightforward way. In the case of benzene only the sixfold rotation is possible. The data shown indicate a linear dependence of the width parameter β on temperature. This is in agreement with the assumption of a Gauss distribution of activation energy with a width of S_E, which leads to the relationship $\beta=RT/(\sqrt{2}\,S_E)$. The mean activation energy found for the sixfold rotation of HMB is 7.6 kJ/mol, and for the case benzene-d_6 it is 5.8 kJ/mol. For the HMB methyl group it is 1.6 kJ/mol. These values are only about one third as compared to the ones in the neat crystalline probes. In the glassy matrices the motion of the probe molecules is less hindered, indicating a loose packing of the matrix.

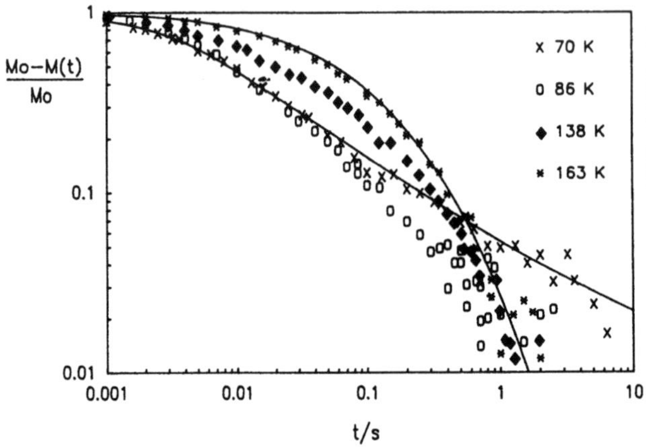

Fig.2: Spin lattice relaxation of benzene-d_6 in phthalic acid n-butyl ester; solid line: fit with a convolution of a log-Gauss with a Matsumoto-Higasi distribution

For the case of the benzene probe with its rather simple and well investigated motion more details about the motional heterogeneities in the glass can be unravelled from studies at lower temperatures. For the cases shown in Figs. 1, 3 and 4 the relaxation gets always faster for lower temperatures. However, for benzene-d_6 below 100 K the relaxation becomes slower, again. This is demonstrated in Fig. 2, where the spin lattice relaxation above and below 100 K is shown in a double logarithmic plot. Between 70 and 120 K significant changes in the shape of the relaxation function are observed. A fit with a log-Gauss distribution is no longer possible. The different form of the relaxation at high and low temperatures is an indication for a non-symmetric distribution function $G(\ln\tau)$. This relaxation behavior can be understood when looking on Fig. 5. In 5.I the relationship between the spin lattice relaxation time T_1 and the correlation time τ as given by Eq. (2), with its minimum around $\tau \approx 1/\omega_0$, is sketched. In 5.II two non-symmetric distributions $G(\ln\tau)$ of identical shape for the two asymptotic ranges $\tau \ll 1/\omega_0$ and $\tau \gg 1/\omega_0$ are shown. The resulting distribution of relaxation times $F(\ln T_1)$ is given in 5.III. For $\tau \gg 1/\omega_0$ the T_1 distribution has a shape comparable with the correlation time distribution $G(\ln\tau)$. However, for $\tau \ll 1/\omega_0$ a mirror image of the distribution $G(\ln\tau)$ is observed. The resulting relaxation curves $\Phi(t)$ can clearly be distinguished, as seen in 5.IV. Such differences are not given for a symmetric distribution [2].

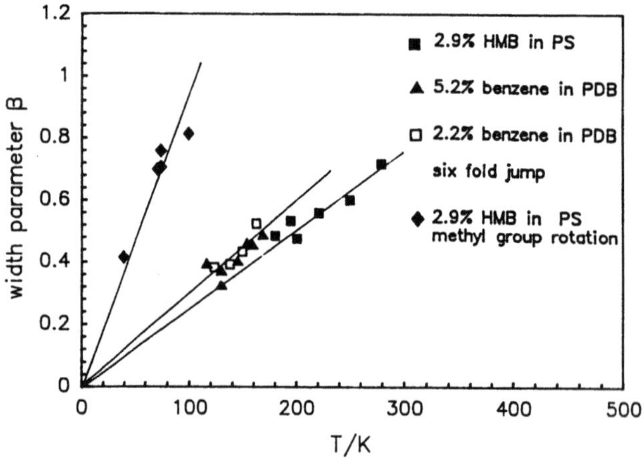

Fig.3: Width parameter β for the log-Gauss distribution as function of temperature

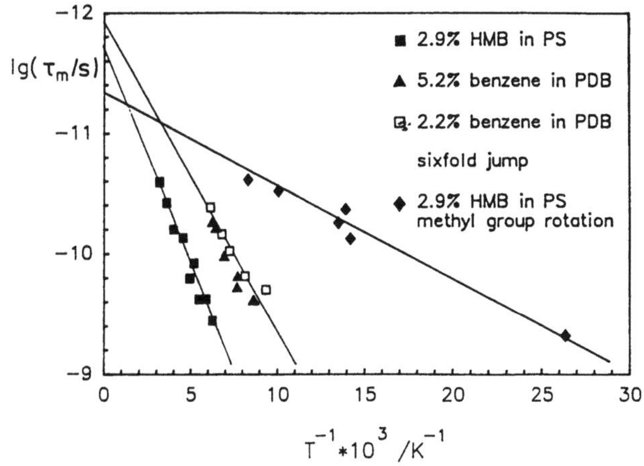

Fig.4: Mean correlation time τ_m (in seconds) of the log Gauss fit as function of reciprocal temperature

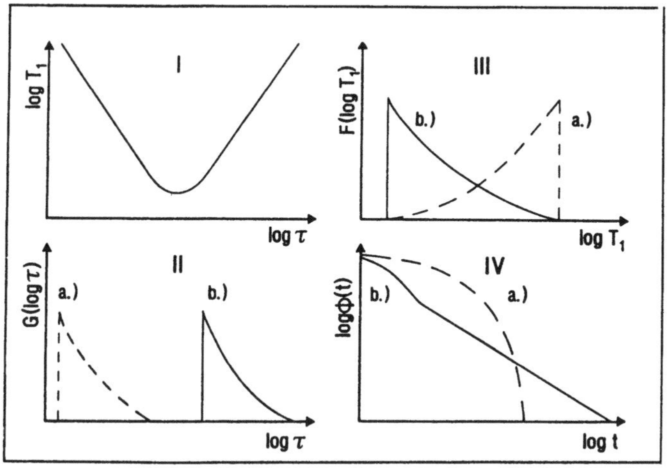

Fig.5: Schematic illustration for non-symmetric distributions of correlation times and the resulting relaxation above and below relaxation minimum (cf. text)

This can be stated quantitatively. Eq. (1) in connection with Eq. (2) reduces to simpler expressions when the two asymptotic cases are discussed

$$\Phi(t) = \int_0^\infty p(\tau)\exp(-tC_1\tau)d\tau = L(p(\tau)) \qquad \omega_o\tau << 1$$
$$\Phi(t) = \int_0^\infty p(\tau)\exp(-tC_2/\tau)d\tau = L(P(r)) \qquad \omega_o\tau >> 1 \quad . \tag{4}$$

C_1 and C_2 are appropriate coupling constants as given by Eq. (2).

The relaxation curve $\Phi(t)$ is given for $\tau << 1/\omega_0$ by a Laplace transformation L of the correlation time distribution $p(\tau)$ and for $\tau >> 1/\omega_0$ by the Laplace transformation of the rate distribution $P(r)$, where $r=1/\tau$. The well-known general problem, that the structure of a distribution function is obscured after a Laplace transformation [3] is less severe for NMR investigations, because a given distribution $p(\tau)$ can be checked by two different Laplace transformations. Inspecting Fig. 2, the most characteristic feature of the low temperature relaxation is a straight line in the double logarithmic plot, i. e. a power law. This relaxation behavior corresponds to $G(\ln\tau) \sim \tau^{-\beta}$ for large τ. This feature is reproduced by several well-known distribution functions, e. g. reverse Cole-Davidson, Maxwell-Boltzmann, and Matsumoto-Higasi [4]. The Matsumoto-Higasi distribution, $G(\ln\tau)=\beta\exp(-\beta z)=\beta(\tau/\tau_c)^{-\beta}$, is sketched in Fig. 5. The actual distribution function has to fit both, the high and the low temperature behavior. In Fig. 2 a convolution of a log-Gauss function with the Matsumoto-Higasi distribution has been used to approximate the data at high and low temperatures.

To obtain further support for such a distribution we used a different NMR method, namely the analysis of the 2H spectral lineshape in the temperature range between 5 and 130 K. The results are shown in Fig. 6 for HMB in polystyrene. At the highest temperature the lineshape is given by a powder spectrum for molecules rotating fast around their sixfold axis [5]. At the lowest temperature the splitting has increased, because the rotation is frozen in. For intermediate temperatures the lineshape can be well approximated by a superposition of these two asymptotic spectra corresponding to varying fractions of slow and fast rotating molecules. This is a direct consequence for the existence of a distribution of correlation times. It can be shown, that for an activated process the fraction of slowly rotating molecules as a function of the temperature is the integral V(T) over the distribution function:

$$V(T) \propto \int_T^{\infty} G(\ln\tau(T'))dT' \quad . \tag{5}$$

dV/dT renders the shape of the activation energy distribution. In Fig. 7 the result is shown. Again, the asymmetry of the distribution is clearly seen, its maximum lies at the low temperature side, i. e. the low activation energy side.

Summarizing, we think that an investigation of the non-exponential 2H NMR spin lattice relaxation is a well suited method to study in detail the activation ener-

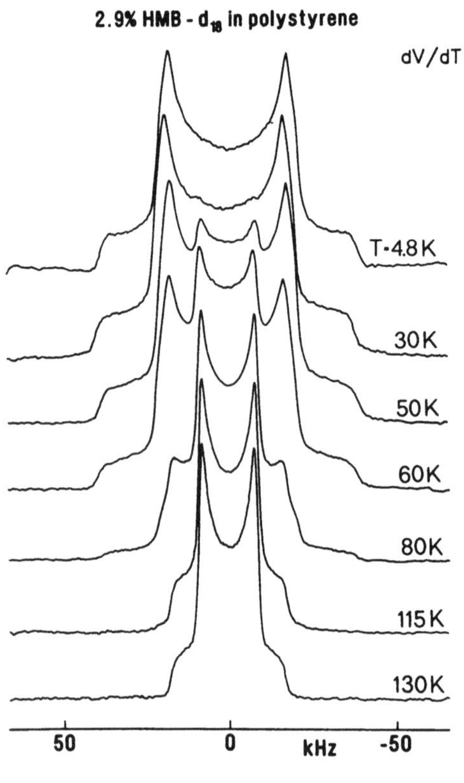

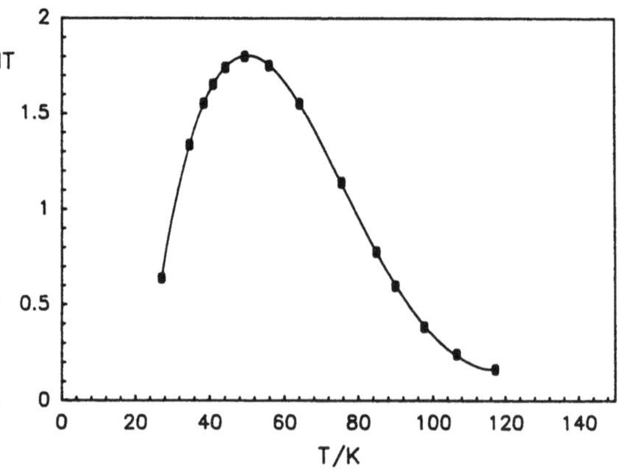

Fig.7 (above): dV/dT (in a. u.) as a function of temperature; this represents the shape of the activation energy distribution (cf. text)

Fig.6 (left): 2H lineshape for HMB-d18 in polystyrene as function of temperature gy distribution of probe molecules in disordered systems. Our experiments give indication for a non-symmetric distribution as expected for a non-symmetric packing of matrix molecules, i. e. for a non-symmetric free-volume distribution. These results may be related to the asymmetric dielectric loss peak for the β-relaxation [6] or to the asymmetric distribution of tunneling frequencies as discussed for two-level systems in glasses [7]. Also, the shape of the activation energy distribution in proteins is very similar to the one shown here [8].

Literature

1. B. Jansen-Glaw, E. Rössler, M. Taupitz, H.-M. Vieth, submitted to J. Chem. Phys.
2. E. Rössler, M. Taupitz, submitted to J. Chem. Phys.
3. C. P. Lindsay, G. D. Patterson, J. Chem. Phys. 73, 3348 (1980)
4. C. J. F. Böttcher, P. Borderwijk, Theory of Electric Polarization, Vol. II, Elsevier Scientific Publishing Company, Amsterdam 1978
5. A. Abragam, The Principles of Nuclear Magnetism, Oxford University Press, Oxford 1961
6. P. Johari, Journal de Chimie Physique 82, 283 (1985)
7. W. Köhler, J. Friedrich, Phys. Rev. Lett. 59, 2199 (1987)
8. H. Frauenfelder, Helvetica Physica Acta 57, 165 (1984)

Scaling Properties of Fast Motions in a Globular Protein

W. Doster[1], *S. Cusack*[2], *and W. Petry*[3]

[1]Physik-Department E13, Technische Universität München, D-8046 Garching, Fed. Rep. of Germany
[2]EMBL Grenoble, c/o Institut Laue-Langevin, 156X Av. des Martyrs, F-38042 Grenoble Cedex, France
[3]Institut Laue-Langevin, 156X Av. des Martyrs, F-38042 Grenoble Cedex, France

1. Introduction

Proteins have highly ordered, well defined structures. However, the absence of translational symmetry, a variety of local interactions produces dynamic properties resembling those of disorderd materials. Further the structure of globular proteins is close-packed. As a result fast motions are controlled by hard-core interactions /1/. The dynamic equations describing a hard core liquid may thus be a reasonable approximation to the average motion in a protein at short times. In this article we investigate this proposition experimentally using inelastic neutron scattering. We compare our results with the predictions of the mode coupling theory near the liquid-glass transition /2/.

Since the average position of each atom is preserved below the denaturation temperature a native protein is never fully on the liquid side and cannot peform a conventional liquid-glass transition. A scattering experiment will always produce an elastic peak. However, a transition from ergodic to non-ergodic dynamics is still possible. These motions contribute to the biological function of the molecule. In myoglobin, an oxygen storage protein, they enable the passage of dioxygen through the protein matrix and the binding to the heme iron inside the protein /3/.

A particular aspect of this system is the role of the hydration shell, a monolayer of bound water on the protein surface. The bound water displays a broad liquid-glass transition with $T_g \sim 200$ K /4/. The Lamb-Mössbauer factor of the heme iron in myoglobin shows a striking decrease above this temperature, suggesting a coupling of protein motions to the surface water /5,6/. We demonstrate below that a similar transition can be observed by using inelastic neutron scattering. We have measured both D_2O and H_2O hydrated sperm-whale myoglobin between 4 K and 350 K and at different levels of hydration using the instruments IN6 and IN13 at the ILL in Grenoble.

2. Incoherent Elastic Scattering

Fig. 1 shows an elastic scan of hydrated myoglobin (H_2O, 0.34 g/g) using a fixed window method on IN13 at 10 μeV resolution. Between 50 K and 200 K we find a linear decrease of $\ln I_{EL} \sim -Tq^2$ with temperature and the square of the momentum exchange q consistent with harmonic behaviour. Above 200K we observe an extra decrease in $\ln I_{EL}$ indicating the excitation of new degrees of freedom. The

Springer Proceedings in Physics, Vol. 37 **Dynamics of Disordered Materials**
Editors: D. Richter · A.J. Dianoux · W. Petry · J. Teixeira

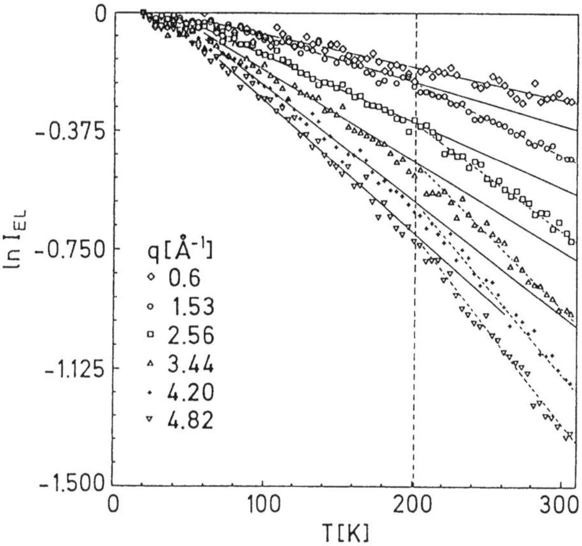

Fig. 1

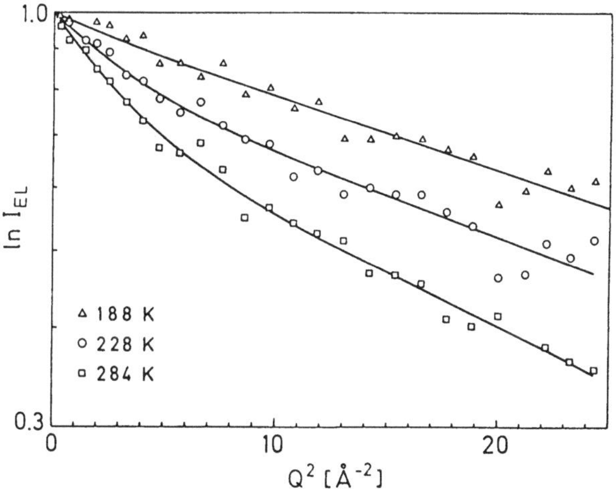

Fig. 2

geometry of these motions gives rise to a non-Gaussian scattering law, $I_{EL}(q)$, measured as a function of momentum exchange q (Fig. 2). The non-Gaussian behaviour grows with T suggesting transitions between states of different energy.

We can fit the data by assuming rotational jumps of protons between two sites of different energy. The powder averaged elastic intensity is then given by:

$$I_{EL}(q) \sim S(q,0) = \exp(-q^2\langle\Delta x^2_{vib}\rangle)\left[1-2p_1p_2\left(1-\frac{\sin qR}{qR}\right)\right] \quad (1)$$

The first term describes the vibrational contribution and the term in the brackets denotes the elastic incoherent structure factor of the two state model. p_1 and p_2 denote the probability to find a proton in the ground or excited state, respectively. The energy asymmetry defined by $p_2/p_1 \sim \exp(-\Delta H/RT)$ is found to be 10(±1) kJ/mol and the jump length R is 1.3(±0.1) Å.

3. Inelastic Scattering

The extra loss in elastic intensity is compensated by neutrons scattered into a broad quasielastic spectrum between 0 and 2 meV. To derive the actual shape of the spectrum, we have subtracted a phonon-like contribution by scaling a low-temperature spectrum (4 K - 180 K) with the Bose-occupation factor at each temperature. The shape of the residual spectrum is temperature independent between 250 K and 350 K.To demonstrate this property a rescaled susceptibility $\hat{\chi}''(\hat{E})$ is shown in Fig. 3.

$$\hat{\chi}''(\hat{E}) = \chi''(E/E_{\epsilon}^{min})/A_{\epsilon}^{min} \qquad (2)$$

This quantity is determined as follows: First the susceptibilities are derived from the dynamic structure factor S(E), (E=hω), according to $\chi''(E) = E \cdot S(E)$. The susceptibility has a minimum at E_{ϵ}^{min} with amplitude A_{ϵ}^{min} which depend on temperature, $\epsilon \sim T$. These values are used to construct $\hat{\chi}''(\hat{E})$ (solid dots, IN13, open symbols, IN6). E_{ϵ}^{min} and A_{ϵ}^{min} are shown in Fig. 4. $\hat{\chi}''(\hat{E})$ displays two scaling regions separated by a minimum. Mode coupling theory suggests the following interpolation formula:

$$\hat{\chi}''(\hat{E}) = A\,\hat{E}^{a} + B\,\hat{E}^{-b} \qquad (3)$$

The solid lines in Fig. 3 would indicate a = 0.25, b = 0.3.These numbers are approximately consistent with the mode coupling result /7/:

$$C = \Gamma^2(1-a))/\Gamma(1-2a) = \Gamma^2(1+b)/\Gamma(1+2b) \qquad (4)$$

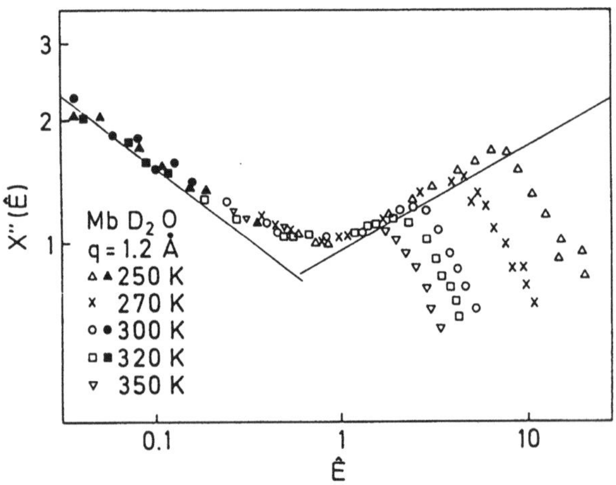

Fig. 3

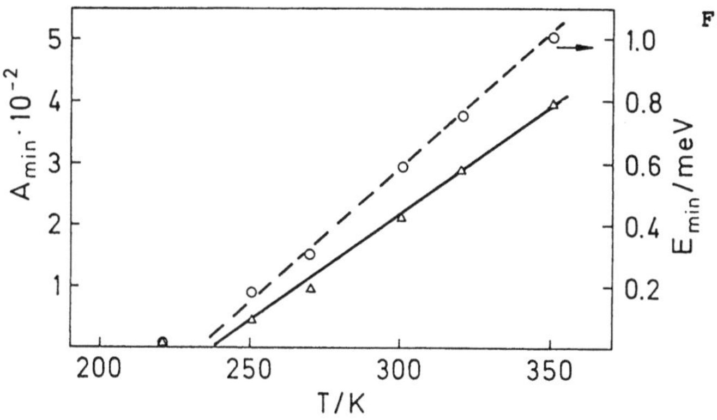

Fig. 4

The temperature dependence of the parameters A_ϵ^{min} and E_ϵ^{min} is shown in Fig. 4. Mode coupling theory predicts

$$A_\epsilon^{min} \sim (-\epsilon)^{0.5} \sim (T-T_c)^{0.5}, \; T>T_c. \tag{5}$$

The experimental exponent is however close to unity. The extrapolated $T_c \simeq 238$ K is about 40 K above T_g estimated from the liquid-glass transition of the hydration shell. E_ϵ^{min} should vary in proportion to $\epsilon^{1/2a} \sim (T-T_c)^2$. We find however approximately again a linear relationship consistent with $T_c = 238$ K. Starting with the value of E_ϵ^{min} at 250 K and using an exponent of 2.0, E_ϵ^{min} would rapidly exceed the microscopic time scale (0.33 ps or 2meV). This scaling law therefore can hold only close to T_c.The discrepancies could be related to activated processes. In conclusion the mode coupling approach decribes at least qualitatively the spectrum of myoglobin. This suggests that a more involved theory could lead to a more quantitative agreement with the experiment.

1. McCammon, J.A. Rep.Prog.Phys. 47, 1-46 (1984)
2. Götze, W., Sjögren, L., J.Phys. C21, 3407-3421 (1988)
3. Beece, D., Eisenstein, L., Frauenfelder, H., Good, D., Marden, M.C. Reinisch, L., Reynolds,A.H., Sorensen, L.B., Yue, K.T., Biochem. 19, 5147-5153 (1980)
4. Doster, W., Bachleitner, A., Dunau,R., Hiebl, M., Lüscher, E., Biophys.J. 50, 213-219 (1986)
5. Parak, F., Frolov, E.N., Mössbauer,R.L., Kononenko, R.L., Goldanskij, V.I., and Rubin, A.B., REBS Lett. 117, 368-372(1980)
6. Bauminger, E.R., Cohen, S.G., Nowik, I., Ofer, S., and Yariv, I., PNAS 80, 736-740 (1983)
7. Krieger, U.,Bosse, J., Phys.Rev.Lett., 59, 1601-1604 (1987)

The Structural and Dynamical Properties of the Sol-Gel Transition

R. Winter[1], *D.-W. Hua*[2], *X. Song*[2], *and J. Jonas*[2]

[1]Institute of Physical Chemistry, Philipps-University Marburg, D-3550 Marburg, Fed. Rep. of Germany
[2]Department of Chemistry, School of Chemical Sciences, University of Illinois, Urbana IL 61801, USA

Different experimental techniques have been employed to investigate the macroscopic and microscopic structural and dynamical properties of the sol-gel transition of tetramethoxysilicate. The experimental results are compared with recent theories for the gelation reaction.

1. Introduction

It has been established that the sol-gel process has a promising future as a unique process for preparing glasses, films and fibres /1,2/. In the sol-gel process hydrolyzed alkoxides, like e.g. $Si(OCH_3)_4$ (TMOS), polymerize via condensation reactions with the elimination of H_2O or alcohol ROH

$$\equiv Si\text{-}OH + HO\text{-}Si\equiv \longrightarrow \equiv Si\text{-}O\text{-}Si\equiv + H_2O \qquad (1)$$

$$\equiv Si\text{-}OH + RO\text{-}Si\equiv \longrightarrow \equiv Si\text{-}O\text{-}Si\equiv + ROH \qquad (2)$$

The viscosity of the solution increases with time until a gel is formed. The gel is subsequently dried and heat treated to yield the desired glass.

A variety of physical and chemical factors (e.g. temperature, pressure, pH, concentration of reactants and catalysts) influence the polymerization process and thus the properties of the final glass. In order to understand and control the structure of the product, a detailed understanding of the polymeric growth process and the sol-gel transition is essential. Also from a theoretical point of view, the nature of the sol-gel transition is a subject of extensive discussion /3,4/. A gelling solution represents a unique state of matter because it is neither liquid nor solid, but rather is in transition between these two states.

In this communication we report on recent viscosity, small angle neutron-scattering (SANS), NMR-, fluorescence polarization and dynamic light scattering (DLS) experiments on the polycondensation of TMOS, which lead to a detailed understanding of the macroscopic and microscopic structural and dynamical properties in course of the sol-gel transition. The experimental results are compared with growth models for the polymerization process and with recent theories for the gelation threshold.

2. Experimental

The samples were prepared by dropwise addition of a H_2O/CH_3OH mixture to the silicon alkoxide TMOS, which is diluted in CH_3OH /5/. The concentrations of the reagents were adjusted to yield a final mole ratio TMOS: $H_2O:CH_3OH$ = 1:4:4. The experiments have been conducted at neutral pH. The sol gelled at $t_{gel} = 32.45 \pm 0.1$ h at T=23°C.

Springer Proceedings in Physics, Vol. 37 Dynamics of Disordered Materials
Editors: D. Richter · A.J. Dianoux · W. Petry · J. Teixeira

In the following we define the gelation time by the absence of flow in the solution containing flask.
The viscosity measurements have been performed with a falling ball viscosimeter, the SANS experiments at the Intense Pulsed Neutron Source at Argonne National Laboratory and the NMR experiments on a 300 MHz pulsed NMR spectrometer GN300 (General Electric).
In all experiments the temperature of T=23°C was held constant with a water thermostated bath to within 0.2°C. For experimental details see /5,6/.

3. Results and Discussion

In order to follow the time dependence of the structural evolution of the silica network, Raman and NMR experiments have been performed /5/, which are able to detect individual species or groups of molecules during the polymerization process. As an example, we present the results from ^{29}Si-NMR experiments in Fig. 1, which shows the relative concentrations of monomers (A), dimeric and trimeric silica species (B) and the concentration of higher branched silica polymers (P) as a function of time t, normalized to the gelation time t_{gel}. As can be clearly seen, still about 10% of monomers are present at the gelation point and the amount of higher polymeric species P still increases considerably after gelation ($t/t_{gel}=1$).

Due to the large complexity and overlap of the signals of the higher branched polymer particles, the detailed structure of the polymer network cannot be obtained by these methods. However, in describing complex disordered structures the concept of fractal geometry provides a valuable means of quantitatively describing the average structure of these objects /7-9/.
We applied the SANS-technique to measure the fractal dimension D and to yield information about the underlying growth process of the polymeric structure. The intensity I(Q) of the scattered neutrons in the Porod region ($QR_G < 1$, R_G radius of gyration), can be analyzed in terms of a power law

$$I(Q) = Q^{-\mu} \qquad (3)$$

The exponent μ is related to the fractal dimension characteristics of the sample. For mass fractals of dimension $1 < D < 3$, $D = -\mu$.

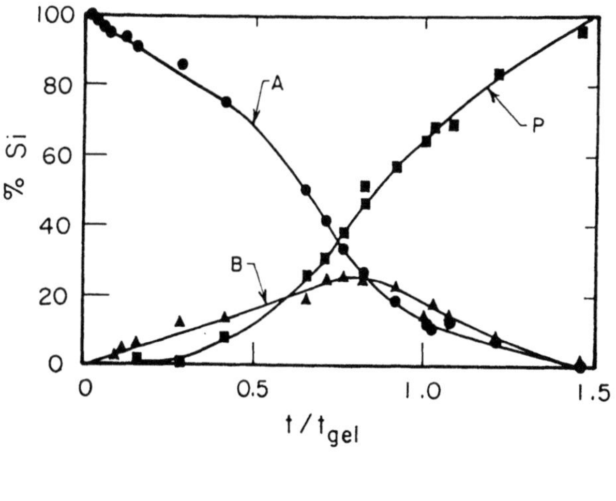

Fig. 1: The relative concentration (in % Si) of silicon monomers $Si(OR)_4$ A, dimeric and trimeric species B and of larger polymeric species P as a function of relative time t/t_{gel}

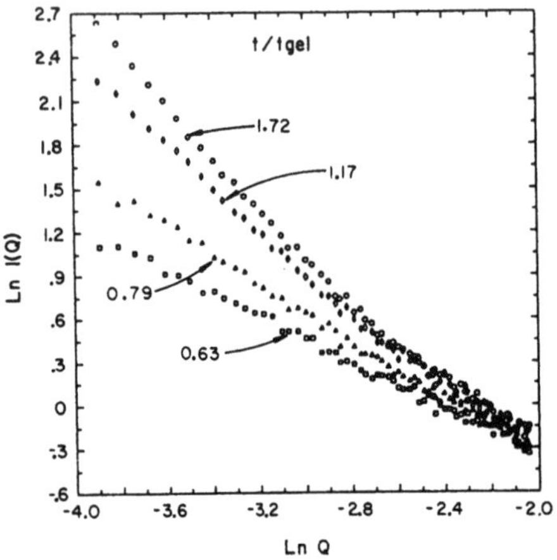

Fig. 2: The time dependence (in t/t_{gel}) of the scattering profiles lnI(Q) vs. lnQ of TMOS-sols

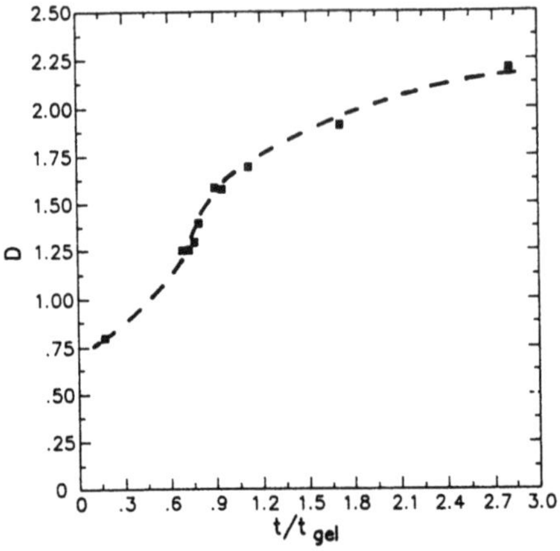

Fig. 3: The fractal dimension D of TMOS-sols as a function of relative time t/t_{gel}

Figure 2 shows a selection of scattering curves (plotted lnI(Q) vs. lnQ) in the Porod region as a function of relative time t/t_{gel}. The scattering intensity I(Q) and the slope of the curves progressively increase with time. Figure 3 depicts the time evolution of the fractal dimension D, which has been obtained from the slope of the scattering curves according to eq. 3. No significant change in the scattering profile or D-value is observed at the crossover from the sol to the gel. The D-value still increases with time after the gel-point and reaches a final value of $D=2.2 \pm 0.1$ not before $t/t_{gel}=3$. The time dependence of I(0) /8/ and D clearly demonstrates that the local structure of the polymer network still changes a long time after gelation. Taking into account the polydispersity of the polymer, a fractal dimension of 2 can be observed at $t/t_{gel}=1$. This value is only close to the value D=2.5 as obtained for 3d percolation at the gel-point /10/.
By computer simulation a variety of growth processes have been proposed for the silicon oxide systems. Our final value of about D=2.2 is close to that of the reaction-limited cluster-cluster growth model (D=2.1 /11/). The increasing Porod slope during the gelation process can be explained by the polymerization of incompletely hydrolyzed monomer spezies /12/.

The dynamical properties of the sol-gel transition can be described phenomenologically by changes in the viscoelastic properties, like e.g. the viscosity /4/. As depicted in Fig. 4 we measured the bulk viscosity η of the sol up to the gel in situ, i.e. in the reaction bath. In the first stage of the polymerization process, η does not significantly change with time, but increases rapidly after $t/t_{gel} \approx 0.8$ and it seems to diverge near t_{gel}, thus suggesting a power-law behaviour of the form

$$\eta = \left(\frac{t_{gel} - t}{t_{gel}}\right)^{-s} \qquad (4)$$

As can be seen in Fig. 5, a straight line is observed in the log-log plot for $t_{gel}=32.3$ h -which is within the experimental error that of the observed gelation time-, leading to an exponent of $s=0.75 \pm 0.05$.

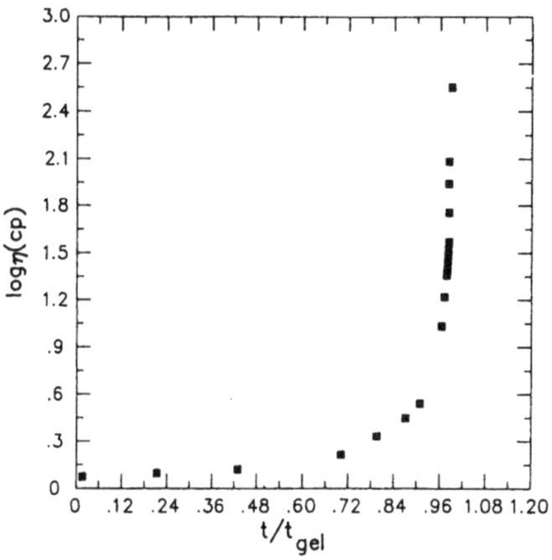

Fig. 4: The viscosity η of TMOS-sols as a function of relative time t/t_{gel}

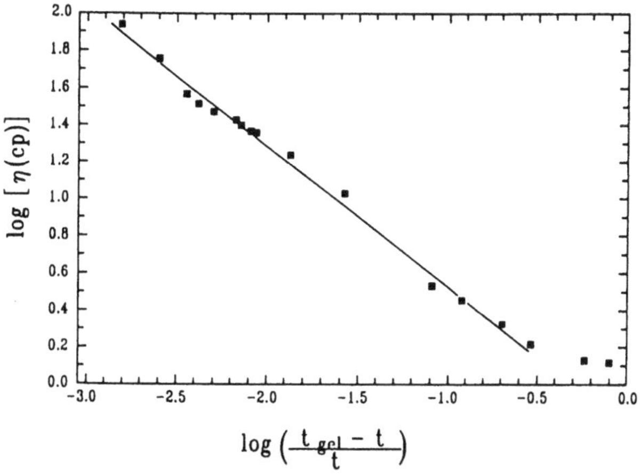

Fig. 5: The viscosity (in log η) of TMOS-sols as a function of reduced time $\log((t_{gel}-t)/t_{gel})$

This value is similar to that reported for the copolymerization of styrene-divinylbenzene /13/. The result is also in agreement with the percolation prediction of $s \approx 0.7$, which is based on an analogy between the viscosity divergence and the divergence of the conductivity in a superconductor-resistor network /14,15/. Although no exact value from theory is known for s until now, these results suggest that the gelation process can be described by the critical theory of percolation. In this approach, the gel-point is the statistical event which indicates the first appearance of a cluster that spans the whole sample region.

In the following we describe how the microscopic motional state of solvent and probe molecules is affected by the formation of the silica network and the sol-gel transition.
In replacing H_2O by D_2O in the sol, the deuteron spin-lattice relaxation time T_1 can be measured, which is directly correlated with the reorientational correlation time τ_c of the molecules ($1/T_1 \propto \tau_c$ /16/). Figure 6 shows the temporal change of T_1 of deuterons, which reflects the motional state of the solvent molecules (CH_3OD, D_2O) during the polymerization reaction. As can be clearly seen, T_1 decreases continuously up to $t' = t/t_{gel} \approx 0.8$, then it levels off and becomes nearly constant. This behaviour corresponds to a continuous increase in τ_c from about 8 ps to 11 ps around t', i.e., the overall change in motion is relatively small. According to the Debye-equation /see e.g. 17/

$$\tau_c = 4\pi r_H^3 \eta / 3kT \qquad (5)$$

(r_H hydrodynamic radius) τ_c can be related to the viscosity η. According to this equation the small increase in τ_c directly corresponds to an only 30% increase of η in course of the sol-gel transition, thus indicating that the fluidity around the solvent molecules is still very high after gelation. This hebaviour of η is completely different from that observed for the divergence of the bulk viscosity (see Fig. 4) and might be better characterized as microviscosity, which only reflects the local friction on the small solvent molecules. The

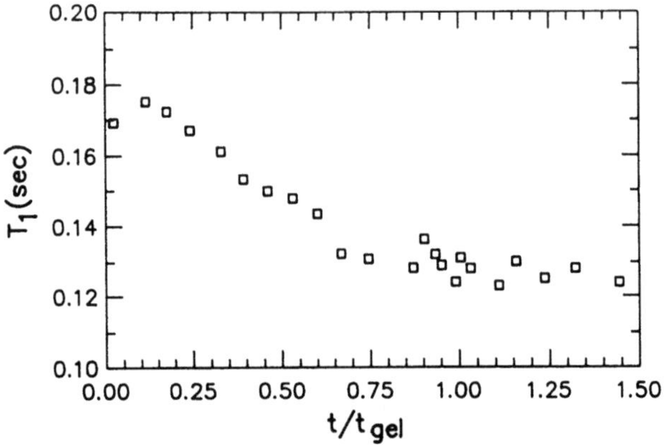

Fig. 6: The deuterium spin-lattice relaxation time T_1 of the solvent molecules D_2O, CH_3OD as a function of relative time t/t_{gel}

increasing τ_c before t' can be attributed to the continuously increasing restriction of motion during the developing connectivity of the silica network. Around t' the molecular mobility reaches its minimum probably due to the confinement of the solvent molecules in pores, which indicates the change from the sol to the gel state.

We additionally performed measurements of the fluorescence polarization of fluorophore probe molecules, which allow the evaluation of the rotational diffusion coefficient of the fluorophores during the network formation /6/. Similar to the NMR-experiments, these results indicate, that the microscopic motional state of probe molecules changes around t'=0.8-0.9 due to the formation of cages, i.e. about 10-20% earlier than macroscopic quantities like the bulk viscosity exhibit their largest changes.

The dynamical properties of larger entities, like the polymer clusters, in course of the gelation reaction can be probed by dynamic light scattering experiments. We measured the autocorrelation function of the scattered light intensity, which is determined by the motional state of the scattering particles, as a function of time /6/. A continuously slowing down of the motional correlation time with progressing reaction time is observed up to t/t_{gel}=0.9-1. At later times, additional scattering from the gel is seen, which probably arises from collective excitations of the whole silica network.

In conclusion, we note that no drastic changes in structure occur at the gelation threshold and the polymer network still changes considerably after gelation. The formed silica network exhibits a self-similar structure, its underlying growth process may be classified as reaction-limited cluster-cluster growth. In comparison, macroscopic dynamical properties like the bulk viscosity diverge at the gel point. Microscopic dynamical properties of molecules much smaller than the correlation length of the clusters exhibit only small motional changes in course of the sol-gel transition, which proceed the appearance of the macroscopic changes by about 10-20% in relative time t/t_{gel}. The gross features of the sol-gel transition of the silicon alkoxide TMOS can be described within the framework of percolation theory. However, the detailed chemical structure of the monomer, e.g. its time dependent functionality, and probably also solvent effects seem to play an important role in describing finer details and have to be taken into account for a more quantitative theoretical description of the gelation process.

Acknowledgements

This work was supported in part by the Air Force Office of Scientific Research under Grant AFOSR 85-0345 and the National Science Foundation under Grant NSF CHE 85-09870. R. Winter gratefully acknowledges a grant by the Deutsche Forschungs-Gemeinschaft. The SANS experiments were performed using the Intense Pulsed Neutron Source at Argonne National Laboratory. This facility is funded by the U.S. Department of Energy, BES-Materials Science, under Contract W-31-1o9-Eng.

References

1. L.L.Hench, D.R.Ulrich: Science of Ceramic Chemical Processing (Wiley 1986)
2. D.R.Ulrich: J.Non-Cryst.Solids 100,174-193(1988)
3. P.Meakin: In Time Dependent Effects in Disordered Materials, ed. by R.Pynn, T.Riste (Plenum Press 1987)
4. D.Stauffer, A.Coniglio, M.Adam: Advances in Polymer Science 44, 103(1982)
5. R.Winter,J.-B. Chan, R.Frattini, J.Jonas: J.Non-Cryst.Solids, in press
6. R.Winter, D.-W. Hua, X.Song, W.Mantulin, J.Jonas: to be published
7. D.W.Schaefer, K.D.Keefer: Phys.Rev.Lett. 53,1383(1984)
8. R.Winter, D.-W. Hua, P.Thiyagarajan, J.Jonas: J.Non-Cryst.Solids, in press
9. J.E.Martin, J.Wilcoxon, D.Adolf: Phys.Rev.A 36,1803(1987)
10. H.E.Stanley: J.Stat.Phys. 36,843(1984)
11. W.B.Brown, R.C.Ball: J.Phys.A 18,L517(1985)
12. C.J.Brinker: J.Non-Cryst.Solids 100,31(1988)
13. M.Adam, M.Delsanti, R.Okasha, G.Hild: J.de Phys.Lett. 40,L539(1979)
14. J.Bernasconi: Phys.Rev.B 18,2185(1978)
15. P.G.de Gennes: J.Phys.(Paris) 40,L197(1979)
16. T.C.Farrar, E.D.Becker: Pulse and Fourier Transform NMR (Academic Press 1971)
17. J.Jonas, T.De Fries, D.J. Wilbur: J.Chem.Phys. 65,582(1976)

Ordinary and Supercooled States in a Liquid Rubidium Model

U. Balucani[1], *T. Gaskell*[2], *and R. Vallauri*[1]

[1]Istituto di Elettronica Quantistica CNR, I-50127 Firenze, Italy
[2]Department of Physics of the University, Sheffield S37RH, UK

1. Introduction

In very recent years there has been an increasing number of works /1/ about the structural and dynamic properties of monatomic liquids in a supercooled regime. Since the quenching rates required for these studies ($> 10^{12}$ K/s) are still distinctly larger than those feasible in practice, computer simulation experiments are the only source of information for any microscopic theory in such simple systems. In this respect, an important step has been the recent development of theories /2/ based on mode-coupling concepts and eventually leading to a transition toward a glassy state. Although originally formulated for simple one-component systems, these theories yielded several results found to be in semiquantitative agreement with the actual situation occurring in some glassformers /3/.

In this contribution we consider a typical simple model system simulating liquid rubidium and report the results of several computer experiments starting from the ordinary liquid phase to deeply quenched supercooled states. In particular, accurate data are reported for the stress autocorrelation function, whose dynamical features control the value of the shear viscosity η . In the deeply quenched states, where the statistical accuracy of most dynamic correlations becomes poorer due to intervening nucleation events, bond-angle distribution functions are shown to provide an important probe of the increasing relevance of the configurational correlations at large angles.

2. The ordinary liquid phase

Our molecular dynamics (MD) studies deal with a system of N "rubidium" particles interacting through an effective pair potential which accounts for most of the static and dynamic properties of real liquid rubidium /4/. The system was always kept at constant density $n\sigma^3 = 0.9045$ ($\sigma = 4.405$ Å); in most of our studies N = 250, but situations with N = 108 and N = 500 were also considered. Initially, the system was brought in equilibrium by standard MD techniques at two temperatures, $T \sim 330$ K and $T \sim 318$ K, the latter being near the freezing point. In these ordinary liquid states several static and dynamic properties were investigated; for most of these data we refer to our previous works /5/. A less explored quantity which is of our special concern here is the Green-Kubo integrand for the shear viscosity, given by $\eta(t) = (k_B TV)^{-1} \langle \sigma^{\alpha\beta}(0)\, \sigma^{\alpha\beta}(t) \rangle$, where $\sigma^{\alpha\beta}$ is the α, β off-diagonal component of the stress tensor

$$\sigma^{\alpha\beta} = \sum_{i=1}^{N} \left(m\, v_i^{\alpha}\, v_i^{\beta} + r_i^{\alpha} F_i^{\beta} \right) \tag{1}$$

Springer Proceedings in Physics, Vol. 37 **Dynamics of Disordered Materials**
Editors: D. Richter · A.J. Dianoux · W. Petry · J. Teixeira

and $\vec{v}_i$, $\vec{r}_i$ and $\vec{F}_i$ are the velocity, the position and the total force acting on the i-th particle, respectively. At liquid densities $\eta(t)$ is dominated by its potential part, $\eta_{pp}(t)$, coming from the second term in (1) /6/. Broadly speaking, in the ordinary liquid $\eta_{pp}(t)$ exhibits a rapid decay at short times, followed by a slower tail at intermediate and long times. These features are common to other dynamical correlations in fluids; in general, the fast initial decay is due to highly localized collisional events, whereas the tail arises from collective effects due to couplings to slowly varying dynamical variables. Mode-coupling approaches appear to be the most convenient method to treat these long-lasting features; in particular, for the potential part of the stress autocorrelation function they predict /7/:

$$[\eta_{pp}(t)]_{M-C} = (n^2 k_B T/60\pi^2) \int_0^{q_c} dq \; q^4 \; [c'(q)]^2 \; F^2(q,t) \tag{2}$$

where c'(q) is the derivative of the direct correlation function (defined in terms of the structure factor S(q) by the relation $nc(q) = 1 - (1/S(q))$) and F(q,t) is the intermediate scattering function. The integration range $(0,q_c)$ comprises all the most slowly varying density fluctuation modes; in particular, the $q \to 0$ hydrodynamic modes (effective at very long times) as well as those with wavevectors near the main peak of S(q), which show a marked de-Gennes slowing down. In a true M-C regime, these modes should entirely control the decay of $\eta_{pp}(t)$ at intermediate and long times and the "irrelevant" cutoff wavevector q_c may be extended to infinity. Assuming S(q) to be known, in principle the intermediate scattering function in (2) can be deduced by other coupled M-C equations, and indeed it is just this self-consistent mechanism which may eventually give rise to a structural arrest in supercooled systems. However, in ordinary liquids a simpler procedure is often adopted by means of semiphenomenological ansätze: for example, in liquid rubidium a viscoelastic model is known to provide an overall good description of the dynamics of F(q,t) /8/, and can safely be used in (2) for all the wavevector range relevant at the times usually explored by MD. Finally, it must be noted that the result (2) is not valid at very short times, where there is a dependence on the cutoff, due to the local character of binary collisions which emphasizes the contributions of high wavevectors. One may of course exploit this circumstance by choosing for q_c a value which fits the t=0 result and reasonably describes even the short time regime, even if this is not strictly necessary for a proper check of the M-C result.

In Fig.1 we report the comparison between M-C theory and MD data at $T \sim 330$ K. In the theory, both the t=0 value and the initial decay of $\eta_{pp}(t)$ are accounted for by choosing $q_c\sigma = 11.1$, well above the value $q_m\sigma = 6.8$ where S(q) attains its main maximum. Beyond $t \sim 0.3$ ps, however, practically the same results are obtained by letting $q_c\sigma = 48$, the maximum value where S(q) is available. This is an indication that for $t > 0.3$ ps we are entering a true M-C regime. In this domain, the results of M-C theory are essentially parameter free, and appear to be excellent. It is important to remark that accurate simulation data such as those in Fig.1 typically require rather long MD runs: in this particular case, at $T \sim 330$ K , a N = 500 system has been followed for nearly 100,000 time steps Δt (the integration step Δt being 10^{-14} s).

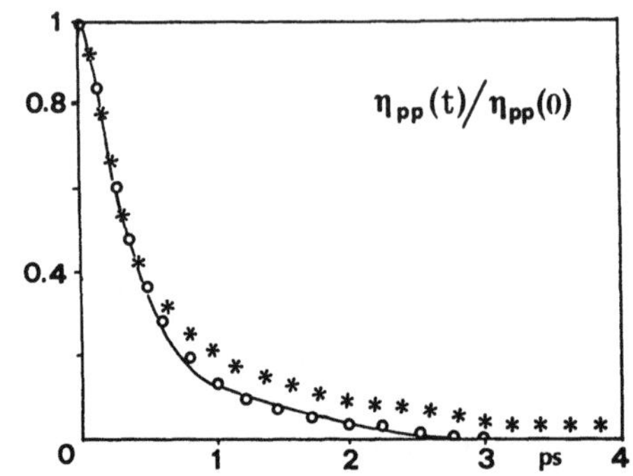

Fig. 1 - Normalized stress autocorrelation functions in ordinary and supercooled liquid rubidium. Circles: MD data at T~330 K; full line: mode-coupling theory at 330 K. Asterisks: MD data at T~270 K.

3. The supercooled phase

Starting from the ordinary liquid configurations at $n\sigma^3$ = 0.9045 and T~318 K, a rubidium system with N = 250 is progressively quenched by scaling the particles' velocities by a factor μ at each time step Δt. Until T~220 K two values of were used, $\mu \doteq 0.9999$ (effective quenching rate Q = 1.2 10^{12} K/s in this temperature range) and μ = 0.999 (Q = 1.2 10^{13} K/s). The decrease with T of the total energy E of the system is found to be nearly linear and independent of μ ; as found in previous studies /9/, the values of E(T) are always larger than those of the classical bcc solid rubidium at the same temperature. In this supercooled phase, several static and dynamic quantities were monitored by allowing suitable annealing periods at selected temperatures down to T~35 K (below 220 K, only μ = 0.999 was used). Most of these data will be reported elsewhere in detail; here it is worthwhile to discuss the important features of some of them, in connection with the situation found in the ordinary liquid.

In Fig.1 the normalized stress autocorrelation function at T~270 K is also reported. Whereas the initial decay is virtually the same as in the ordinary liquid, the tail at intermediate times is clearly enhanced. However, at this temperature the change with respect to the liquid is not as spectacular as the corresponding slowing down of F(q,t) for $q = q_m$ /10/. This is in qualitative agreement with the M-C result (2): the largest weight in the q-integral is not provided by $q=q_m$, where c'(q) = S'(q) = 0, but by values near q_m, for which the slowing down of F(q,t) with respect to the liquid is indeed found to be less marked.

The simulation data for $\eta_{pp}(t)$ at T~270 K were obtained after an annealing MD run of over 60,000 Δt; clearly, over this time the supercooled system should show no trace of nucleation (the latter being usually signalled by an abrupt rise in temperature). This is indeed the case at T~270 K, but at lower temperatures intervening nucleation events may severely limit the statistical accuracy of the MD averages requiring long runs. For example, such events were found to occur after a few thousand Δt at the starting temperature T~196 K; their evolution has been followed until the system reaches the stable state of a hot bcc crystal at T~260 K. We have not attempted a systematic analysis of these features by changing possibly important variables, as the size of the system /11/ and/or the quenching rates; even so, it appears clearly difficult to obtain in these strongly supercooled states reliable data for quantities requiring long runs.

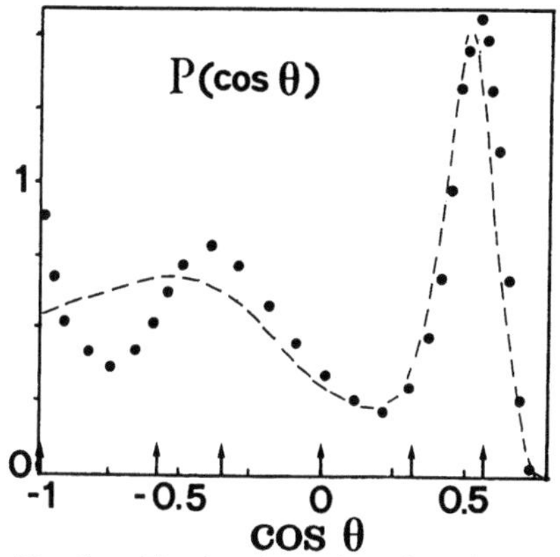

Fig.2 - The bond-angle distribution for "glassy" rubidium at T= 35 K. The dots are the MD data, and the dashed line is the theoretical result with the superposition approximation. All the values have been normalized to unit area in the interval (-1,1). The arrows indicate the locations of the δ-like spikes proper to a bcc crystal at very low temperatures.

The situation appears to be better for equal-time correlations, as well as for transport properties simpler than η, like the single-particle diffusion coefficient D. Often the latter can be reliably measured, before any sign of nucleation, from the mean-square atomic displacement; in our case, D is found to decrease linearly with T down to a sort of critical temperature $\sim$ 140 K, below which D is too small to be accurately determined.

As far as static quantities are concerned, a particularly important role is played by the bond-angle distribution P(cos θ), which measures the probability of occurrence of the angle θ between the directions joining a central atom with two nearest neighbours. In our MD simulations P(cos θ) has been obtained by considering all neighbours up to a distance R = 1.5 σ from the central atom. Clearly:

$$P(\cos\theta) = A \int_{(R)} d\vec{r} \int_{(R)} d\vec{r}\,' \; n^2 g^{(3)}(\vec{r},\vec{r}\,') \; \delta(\cos\theta - \cos\vartheta_{\vec{r}\vec{r}\,'}) \qquad (3)$$

where A is a normalization constant and $g^{(3)}$ is a triplet distribution function. In our MD simulations, upon quenching initially there is no essential difference between the ordinary and supercooled liquid. Only for deeply-quenched states one notes a marked sharpening of all the features (Fig.2). The bond-angle distribution of the "computer glass" at T$\sim$ 35 K is still quite different from the one proper of a bcc solid, and is not accounted for by a simple superposition approximation for $g^{(3)}$, except at small angles - a fact already noted in the crystal phase /12/. In particular, there is an increased weight of the collinear configurations with $\theta = \pi$; this feature is absent or hardly visible in the ordinary liquid, and in the supercooled phase its sudden increase is another indication of the onset of a nucleation process.

References

1. Among the more recent simulation works, see M.S.Watanabe, K.Tsumuraya: J.Chem. Phys. 88, 1991 (1988); J.Yang, H.Gould, W.Klein: Phys.Rev.Lett. 60, 25 (1988), and the references therein.
2. See W.Goetze: In Amorphous and Liquid Materials, ed. by E.Luescher, G.Fritsch, G.Jacucci, NATO ASI Series E, no.118 (Martinus Nijhoff, Dordrecht 1987) p.34.
3. F.Mezei, W.Knaak, B.Farago: Phys.Rev.Lett. 58, 571 (1987).
4. A.Rahman: Phys.Rev.Lett. 32, 52 (1974).
5. U.Balucani, R.Vallauri, T.Gaskell: Phys.Rev.A 35, 4263 (1987).
6. U.Balucani, R.Vallauri, T.Gaskell: Phys.Rev.A 37, 3386 (1988).
7. T.Munakata, A.Igarashi: Progr.Theor.Phys. 60, 45 (1978).
8. J.R.D.Copley, S.W.Lovesey: Rep.Progr.Phys. 38, 461 (1975).
9. R.D.Mountain: Phys.Rev.A 26, 2859 (1982).
10. U.Balucani, R.Vallauri, T.Gaskell: Phys.Scripta (in press).
11. J.D.Honeycutt, H.C.Andersen: J.Phys.Chem. 90, 1585 (1986).
12. A.D.J.Haymet: Chem.Phys.Lett. 107, 77 (1984).

Dynamics of o-Terphenyl in Its Glassy and Supercooled Liquid State as Observed by Incoherent Neutron Scattering

E. Bartsch [1], *M. Kiebel* [1], *F. Fujara* [1], *H. Sillescu* [1], *and W. Petry* [2]

[1]Institut für Physikalische Chemie, Universität Mainz, Postfach 3108, D-6500 Mainz, Fed. Rep. of Germany

[2]Institut Laue-Langevin, 156X Av. des Martyrs, F-38042 Grenoble Cedex, France

Introduction

Representatives of different classes of glass formers - ionic glasses, polymer glasses and van der Waals glasses - have been proposed as good model systems in order to study the glass transition by neutron scattering experiments and to test the predictions of mode-coupling theory [1,2], a microscopic theory describing this transition.

We have chosen van der Waals glasses as model systems because of the individuality of the molecules. The molecular glasses studied so far are 1,3,5-tri-α-naphthylbenzene (TNB) [3], 1,1'-di-(p-methoxyphenyl)-cyclohexane (DMPC) [3] and o-terphenyl (o-TP). Especially o-terphenyl (T_g = 243 K, T_m = 329 K) is a candidate of interest, since it comes closest to the requirements of theory (individual particles, short-ranged interactions, spherical shape). Moreover, o-TP is the most studied molecular glass [4,5].

Here we report on incoherent neutron scattering experiments on fully protonated o-TP and give a comparison of our results with the predictions of mode-coupling theory.

Experimental Data

We have performed elastic channel measurements ($S^{inc}(Q,\omega=0)$) on o-TP in a temperature range from 4 K to 340 K (Fig. 1) as well as measurements of the incoherent scattering function ($S^{inc}(Q,\omega)$) between 297 K and 327 K (Fig. 2) with the high resolution backscattering spectrometer IN13.

Whereas an anomalous decrease of the elastic intensity was observed already well below T_g for TNB [3], this effect sets in only near T_g for o-TP (Fig.1). At T = 297 K and

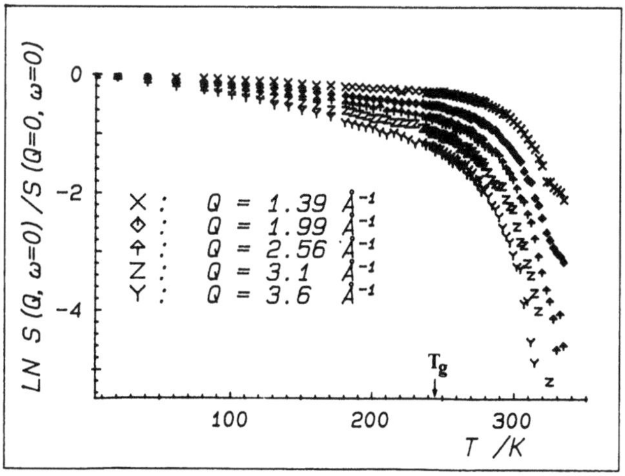

Fig. 1. Normalized elastic intensity of o-TP vs. temperature; plotted are selected Q-values, each containing the accumulated intensity of three detectors. Far below T_g the data are well described by the Debye-Waller factor $\ln(I^{DWF}) \sim T$

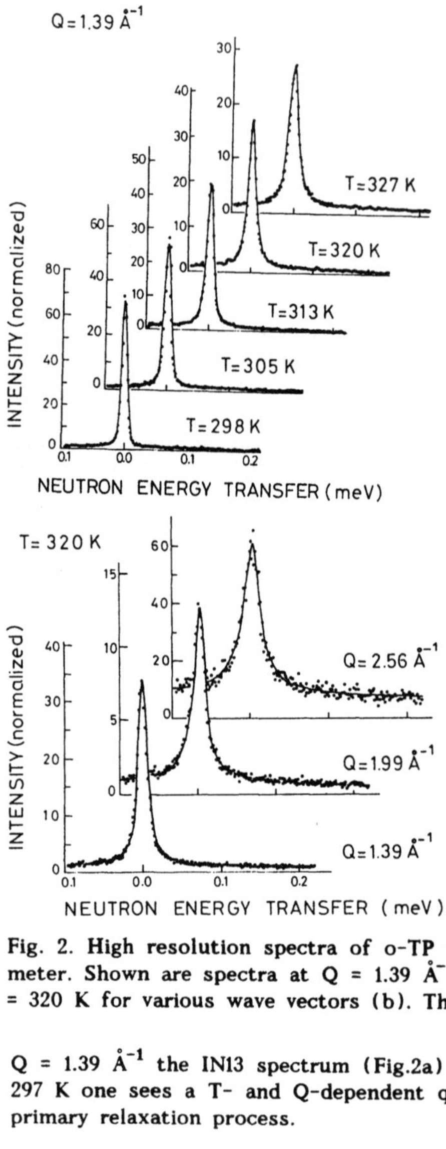

Fig. 2. High resolution spectra of o-TP measured using the IN13 backscattering spectrometer. Shown are spectra at Q = 1.39 Å^{-1} for various temperatures (a) and spectra at T = 320 K for various wave vectors (b). The solid lines indicate fits according to Eq. (1)

Q = 1.39 Å^{-1} the IN13 spectrum (Fig.2a) of o-TP shows no linewidth broadening. Above 297 K one sees a T- and Q-dependent quasielastic linewidth, indicating the onset of the primary relaxation process.

Comparison of Experiment and Theory

In order to extract $S^{inc}(Q,\omega)$ from the IN13 spectra we fitted the experimental data for each temperature and wave vector with theoretical lineshapes convoluted with the resolution function. We found that the intermediate scattering function $I^{inc}(Q,t)$ is well described by a stretched exponential law according to

$$I^{inc}(Q,t) = a\, e^{-(t/\tau)^{\beta}} \quad (1a)$$

$$S^{inc}(Q,\omega) = \int dt\, e^{-i\omega t}\, I^{inc}(Q,t) \quad (1b)$$

with numerical values of $0.42 \leq \beta \leq 0.7$ and $0.04 \leq \tau \leq 0.52$ ns. Use of a single Lorentzian for describing $S^{inc}(Q,\omega)$ failed to give satisfactory results. The β-values obtained agree reasonably well with those evaluated by photon correlation spectroscopy and dielectric relaxation [5] ($0.38 \leq \beta \leq 0.68$). Because of the small dynamic window of the backscattering spectrometer and the existence of a T- and Q-dependent background (secondary relaxation) the values obtained for the stretching exponent β do not show any systematic variation but a considerable scattering in the above mentioned range. On the basis of the available data it is therefore not possible to draw any conclusion concerning the temperature or wave vector dependence of β. Clearly an extension of the measurements towards higher frequencies is needed, calling for experiments on a time-of-flight instrument. The parameter τ, related to the mean relaxation time, however, shows the expected tendency to decrease with increasing temperature and wave vector.

Assuming that the glassy phase can be described by a harmonic solid we extracted the mean square displacement $\langle r^2 \rangle$ from the elastic intensity by exploiting the Q-dependence of the Debye-Waller factor according to

$$f_\varepsilon^{inc}(Q) = \frac{S^{inc}(Q,\omega=0)}{S^{inc}(Q=0,\omega=0)} = e^{-\frac{1}{3}\langle r^2 \rangle Q^2} \qquad (2)$$

for each temperature. We find that $\langle r^2 \rangle$ increases linearly with T - what would be expected for a Debye solid - up to a temperature slightly below T_g. Above that temperature a nonlinear variation is observed (Fig.3, left-hand side).

Such a nonlinear increase has been predicted by mode-coupling theory with $\langle r^2 \rangle^{1/2}$ varying according to [1]

$$\langle r^2 \rangle^{1/2} = r_o\left(1 - \frac{a}{2}\sqrt{T_c - T}\right) \qquad (3)$$

where T_c is the dynamic glass transition temperature [1,2] When trying to describe the nonlinear increase of $\langle r^2 \rangle^{1/2}$ using Eq. (3) we took care to consider only values of $\langle r^2 \rangle$ at temperatures where no quasielastic linewidth broadening was observed. We found that a range of values for T_c would yield satisfactory results. Obviously the neutron data are not sensitive enough towards variation of T_c to allow a reasonable estimation of T_c. Similar results have been obtained by FRICK et al. [6] who evaluated neutron scattering data on polybutadiene in terms of $\langle r^2 \rangle$. Following a suggestion of W. GÖTZE et al. [7] we used another experimental quantity, the shear viscosity η, to obtain a first estimate of T_c and then checked its compatibility with the neutron data. According to theory the viscosity near T_c can be described by [7]:

$$\eta = A(T - T_c)^{-\gamma} \qquad (4)$$

With T_c = 290 K and γ = 2.5 we were able to describe the high temperature behaviour of the viscosity data of CUKIERMAN et al. [8] (Fig. 3, right-hand side) as well as the nonlinear increase of the mean square displacement $\langle r^2 \rangle$.

However, one has to keep in mind that this value for T_c cannot be more than a first estimation, since it is derived from a single experimental quantity, namely the viscosity. Moreover, η is experimentally found to be continuous through T_c rather than showing a singularity. As GÖTZE et al. [7] pointed out, a T_c derived in this manner might be subject to a considerable uncertainty up to more than ten degrees.

The most direct determination of T_c (and at the same time a strong support of its existence) would be the observation of a linear increase of the mean square displacement above a certain temperature, corresponding to a Debye like decrease of the elastic neutron scattering intensity. T_c could then be just read off from the curves as demonstrated in Fig. 7 of reference [7]. As can be seen from Fig. 1 the neutron data do not yield such a behaviour. Instead, they show a monotonous decrease of the elastic intensity up to temperatures above the melting point. This is due to the onset of primary relaxation causing an additional loss of elastic energy (see Fig. 2a).

This effect could be corrected for in principle by using the information contained in the IN13 spectra and integrating $S^{inc}(Q,\omega)$. Considering the problems discussed already in connection with the determination of the β-values it seems advisable to look for a better way of checking for a linear variation of the elastic intensity above T_c. This has already been achieved for other systems - the ionic glass $Ca_{0.4}K_{0.6}(NO_3)_{1.4}$ by MEZEI

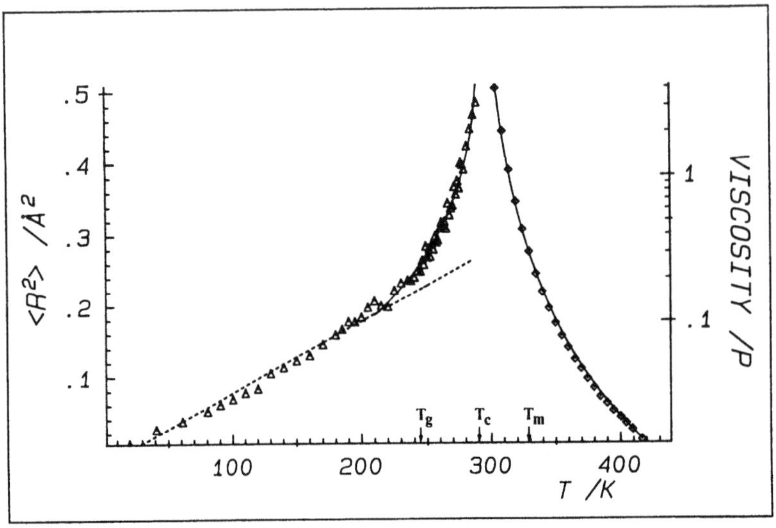

Fig. 3. Variation of the mean square displacement $<r^2>$ (left-hand curve and scale) and the viscosity η (right-hand curve and scale) of o-TP with temperature. The dotted line indicates the behaviour expected for a Debye solid. The solid lines have been calculated according to Eqs. (3) and (4) respectively, using T_c = 290 K and γ = 2.5.

et al. [9] and the polymer glass polybutadien by RICHTER et al. [10]. By performing neutron spin echo experiments with the instrument IN11, measuring directly the intermediate scattering function I(Q,t), they obtain the elastic intensity as the extrapolated value I(Q,t=0) and succeed in locating T_c.

Thus a logical extension of our study of o-terphenyl would be performing neutron spin echo experiments in order to locate T_c more precisely. Nevertheless we conclude that our neutron scattering data below a T_c of 290 K, monitoring the socalled β-process, and viscosity data above T_c, monitoring the α-process, can be interpreted consistently.

Acknowledgement

We gratefully acknowledge helpful discussions with Prof. W. Götze.

References

1. W. Götze, Phys. Scr. 34, 66 (1986)
2. W. Götze, in "Amorphous and Liquid Materials", ed. by E. Lüscher, G. Fritsch and G. Jacucci, Martinus Nijhoff Publ., Dordrecht 1987, p.34
3. F. Fujara, W. Petry, Europhys. Lett. 4, 921 (1987); W. Petry, F. Fujara, M. Kiebel, E. Bartsch, H. Sillescu, this conference
4. T. Dries, F. Fujara, M. Kiebel, E. Rössler, H. Sillescu, J. Chem. Phys. 88; 2139 (1986) and references therein
5. J. U. Hagenah, Ph. D. thesis, Mainz 1988
6. B. Frick, D. Richter, W. Petry, U. Buchenau, Z. Phys. B70, 73 (1988)
7. W. Götze, L. Sjögren, this conference
8. M. Cukierman, J. W. Lane, D. R. Uhlmann, J. Chem. Phys. 59, 3639 (1973)
9. F. Mezei, W. Knaak, B. Farrago, Phys. Scr. T19, 363 (1987)
10. D. Richter, B. Frick, B. Farago, Phys. Rev. Lett., preprint

Dielectric Relaxation in Microemulsions Near the Percolation Threshold

C. Cametti [1], P. Codastefano [1], A. Di Biasio [1], P. Tartaglia [1], and S.H. Chen [2]

[1]Dipartimento di Fisica, Università di Roma La Sapienza,
Piazzale A. Moro 2, I-00185 Roma, Italy
[2]Department of Nuclear Engineering, and
Center for Materials Science and Engineering,
Massachusetts Institute of Technology, Cambridge, MA 02139, USA

Recently the dielectric properties of microemulsion systems in the vicinity of a percolation point have attracted much interest both from the experimental [1-6] and the theoretical [7-10] point of view. Water in oil microemulsions are ideal systems to study the percolative behavior since they can be effectively considered as a heterogeneous two-component system made up of conducting spherical droplets of water, covered by a surfactant layer, immersed in a continuous nonconductive medium. It is also experimentally known [11] that the droplets maintain a constant radius if one keeps the molar ratio of surfactant to water constant. This fact gives as a consequence the possibility of describing the behavior of the system only in terms of the fractional volume ϕ of the dispersed phase (water and surfactant). The dielectric properties have been studied as a function of ϕ or temperature at various frequencies and show a remarkable universal behavior in the whole region of the percolative transition. The picture that emerges from the experiments is the following. As far as the dielectric constant ε is concerned, a power-law divergence has been observed [4] on approaching the percolation threshold T_p in temperature for $T< T_p$ at constant ϕ

$$\varepsilon \approx \left(T_p - T\right)^{-s} \tag{1}$$

with $s \approx 0.65$. The same behavior is expected when varying ϕ at constant temperature near the threshold ϕ_p. The low-frequency conductivity σ has a similar behavior both above and below the percolation limit

$$\sigma \approx \left(T_p - T\right)^{-s'} \quad \left(T<T_p\right) \qquad ; \qquad \sigma \approx \left(T - T_p\right)^{t} \quad \left(T>T_p\right) \tag{2}$$

with $1.17 \leq s' \leq 1.4$ and $1.6 \leq t \leq 1.94$ according to the systems investigated and the approach to the threshold in temperature [4,5] or fractional volume [1,2,5,6]. Both these kind of measurements give values independent from frequency, when the latter is kept below 100 KHz. On raising the frequency the dependence of ε and σ is given by [3,4]

$$\varepsilon \approx \omega^{-y} \qquad ; \qquad \sigma \approx \omega^{x} \tag{3}$$

with $x \approx 0.63$ and $y \approx 0.29$, and since ε and σ are related by a Kramers-Krönig relation one should have $x+y = 1$.

Editors: D. Richter · A.J. Dianoux · W. Petry · J. Teixeira

From the theoretical point of view percolating microemulsions have been treated using the effective medium theory applied to heterogeneous systems [8], scaling hypotheses close to the percolation threshold [7-9] and a dynamical percolation picture [10]. The former approaches are essentially based on a static percolation picture, i.e. the formation of an infinite cluster of conducting particles, a water channel in our case. The dynamical percolation theory takes into account the fact that clusters of conducting particles are continuously formed with a typical rearrangement time T_R and undergo Brownian diffusive motion in the solution and the overall conduction is dominated by interfacial processes. The two pictures give different results for the conductivity below the percolation threshold and for low frequencies. The main difference is the value of the index s' which has the values 0.6 and 1.2 for static and dynamic percolation respectively.

We studied the ternary system water - decane - sodium di - 2 - ethyl hexyl sulfosuccinate [AOT] along the line of constant molar ratio X = [water]/[AOT] = 40.8, where the microemulsion droplets are 50 Å in diameter. The fractional volume ϕ of the dispersed phase (water+AOT) was determined neglecting the oil penetration in the interfacial film, since these types of mixtures behave as ideal solutions, i.e. the volumes of the surface agents, oil and water can be added.

In order to measure the dielectric relaxation, two different experimental setups were employed, from 1 KHz to 10 MHz and from 1 MHz to 1 GHz. All the measurements were performed at (23.50 ± 0.05) °C, as a function of frequency for the following values of ϕ=0.20, 0.24, 0.28, 0.32, 0.36, 0.40, 0.44, 0.48, 0.52, 0.56, 0.60 and 0.64. We define the complex conductivity as

$$\tilde{\sigma}(\omega,\phi) = \sigma'(\omega,\phi) + i\,\omega'\,\varepsilon'(\omega,\phi) \tag{4}$$

where σ', ε' and ω' are the scaled conductivity, permittivity and frequency respectively

$$\sigma' = \frac{\varepsilon_d\,\sigma}{\varepsilon_0\sigma_w} \quad ; \quad \varepsilon' = \frac{\varepsilon}{\varepsilon_0} \quad ; \quad \omega' = \frac{\omega\,\varepsilon_d}{\sigma_w} \tag{5}$$

with ε_d the dielectric constant of decane, σ_w the conductivity of water and ε_0 the dielectric constant of free space.

We start our analysis from the low frequency conductivity (ν < 100 KHz) which is independent of frequency. On approaching the threshold ϕ_p the conductivity σ varies with ϕ according to the power-laws

$$\sigma = A\left(\phi_p - \phi\right)^{-s'} \quad \left(\phi < \phi_p\right) \quad ; \quad \sigma = B\left(\phi - \phi_p\right)^{t} \quad \left(\phi > \phi_p\right) \tag{6}$$

Figure 1 reports the scaled conductivities $(\sigma/A)^{-1/s'}$ and $(\sigma/B)^{1/t}$ as a function of ϕ for ϕ_p=0.31, with the exponents s' and t given by s'=1.2 and t = 1.94, and shows a very good agreement with the measured data. The value for t is the same that was recently obtain with a theoretical calculation [12] for static percolation, while the value of s' is in favor of a

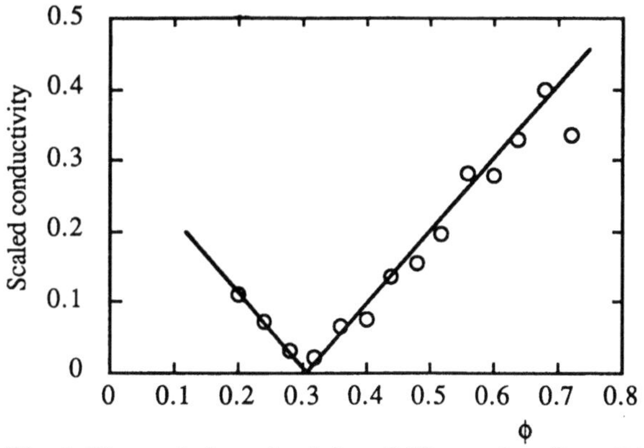

Fig. 1. The scaled conductivity of (6) as a function of ϕ for all the frequencies below 100 KHz.

dynamic percolation picture [10]. It is worth noting immediately that ϕ_p , s' and t are the only parameters we fitted, and using their values we can interpret the data in the whole range of composition and frequency in terms of dynamical scaling.

These results can be easily understood in terms of the dynamic percolation theory we referred to above. In fact this theory provides the following scaling expression of the conductivity

$$\sigma \approx \left(\omega + T_R^{-1} \right) \left| \phi - \phi_p \right|^{-s'} f\left(\omega \tau + T_R^{-1} \tau \right) \tag{7}$$

where $\omega = 2\pi\nu$ is the angular frequency, T_R the rearrangement time of the clusters and $\tau \approx \left| \phi - \phi_p \right|^{-s'-t'}$. The function f(x) is a scaling function of the scaled variable x with the asymptotic properties

$$f(x) \approx 1 \quad (x \ll 1) \quad ; \quad f(x) \approx x^{-\frac{s'}{s'+t}} \quad (x \gg 1) \tag{8}$$

The application of these properties to our low-frequency measurements, i.e. for $\omega T_R < 1$ assuming $1/T_R$ to be of the order of 100 KHz, gives the correct power-law below the threshold and a constant value when very close to the percolation point, as we observed in the experiment and as it is to be expected. At higher frequencies, when $\omega T_R > 1$, the results are of the same form as the scaling theories referred to above and give

$$\sigma \left| \phi - \phi_p \right|^{-t} \approx \omega \tau \quad (\omega \tau \ll 1) \quad ; \quad \sigma \left| \phi - \phi_p \right|^{-t} \approx \left(\omega \tau \right)^{u} \quad (\omega \tau > 1) \tag{9}$$

where the exponent u is given by $u = t / (s' + t) \approx 0.62$. We will see in the following that the scaled conductivity below the threshold is linear in the scaled variable $x = \omega\tau$ and then obeys a power law with exponent u for larger values of x.

The experimental results for the dielectric constant ε' below the threshold and both conductivity and dielectric constant above threshold, are very well accounted for by the scaling theory [7-9] which suggests a possible analogy with the scaling properties valid close to a critical point. These results are better summarized in terms of a complex scaling conductivity y which defines the scaled conductivity and dielectric constant

$$y \equiv \tilde{\sigma}(\omega, \phi)\left|\phi - \phi_p\right|^{-t} = \sigma'\left|\phi - \phi_p\right|^{-t} + i\,\omega'\varepsilon'\left|\phi - \phi_p\right|^{-t} \tag{10}$$

and the real scaled variable $x \equiv \omega'\left|\phi - \phi_p\right|^{-s'-t}$ which, apart from a constant, is the same x we already introduced. The scaling function which connects the two scaling variables, y=F(x), has now the properties [9]

$$F(x) \approx i\,B\,x \quad \left(\phi < \phi_p\right) \quad ; \quad F(x) \approx A + i\,B\,x \quad \left(\phi > \phi_p\right) \tag{11}$$

where A and B are constants, and the asymptotic behavior $F(x) \approx (ix)^u$ for $x \to \infty$. We report in Fig. 2 the real and the imaginary part of the complex variable y as a function of x, together with straight lines that in the log-log plot have slopes 1 and u. We notice that we only used the parameters s', t and ϕ_p, fitted from the low-frequency electrical conductivity below the threshold, in order to obtain the natural dynamical scaling of both conductivity and dielectric constant in the whole range of frequencies and volume fraction. All the predictions of the scaling theory of static percolation are therefore well verified, except for the conductivity below threshold which is accounted for by the dynamical percolation theory. In any case the index s' = 1.2 of the latter theory seems to give the correct answers even when applied to the scaling theory. This fact is in agreement with the measurement of ref.[3], but in contrast with the results of ref.[4], where the exponent s of static percolation is found to describe the behavior of the dielectric constant. A final comment regards the levelling off of the conductivity below threshold, a fact that is partially visible in the scaling plot, but of course is not a universal phenomenon. It could be partially taken into account with the method used in ref.[7] and considering the conductivity of decane, which was always neglected in our case.

The dielectric spectra for the microemulsion systems we investigated in the range from 1 KHz - 1 GHz show rather broad frequency dispersions covering three or more decades. Within

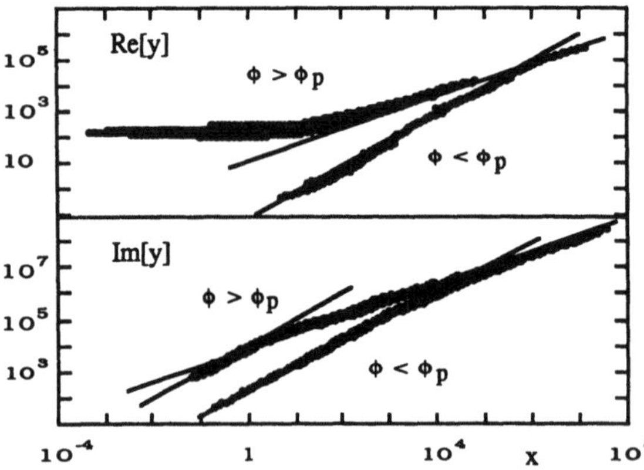

Fig. 2. Scaling plot for the dielectric relaxation. The lines refer to the limiting behavior of the relaxation function y=F(x) and have the slopes 1 and u ≈ 0.62.

our experimental accuracy an analysis has been possible, in all the samples we investigated, in terms of a dispersion of the Cole-Cole type for the normalized complex dielectric constant. The latter can be easily related to the stretched exponential time relaxation function, which is known to represent well the asymptotic behavior of nonexponential relaxation of various nonequilibrium parameters of macroscopic systems. The stretched exponential function has been used in the analysis of the experimental results of a dynamic light scattering experiment in the same microemulsion in a region of high values of ϕ [13]. The occurrence of this kind of relaxation in microemulsions is related to a mechanism of constrained diffusion, which is similar to the conduction by hopping of the surfactant molecules hypothesized by the dynamic percolation picture.

References

1. M. Lagues, R.Ober and C. Taupin, J. Physique 39,L-487(1978)
2. M. Lagues, J. Physique 40,L-331(1979)
3. M. A. van Dijk, Phys. Rev. Lett. 55,1003(1985)
4. S.Bhattacharya, J.P.Stokes, M.W. Kim and J.S.Huang, Phys. Rev. Lett. 55, 1884 (1985)
5. M.W. Kim and J.S.Huang, Phys. Rev. A 34, 719(1986)
6. M. Moha-Ouchane, J. Peyrelasse and C. Boned, Phys. Rev. A 35, 3027(1987)
7. A.L. Efros and B.I. Shklovskii, Phys. Stat. Sol. 76,475(1976)
8. D.J. Bergman, Y. Imry, Phys. Rev. Lett. 39,1222(1977)
9. J.M. Luck, J.Phys. A18,2061(1985)
10. G.S.Grest, I. Webman, S.A. Safran, and A.L.R.Bug, Phys. Rev. A33,2842(1986)
11. M. Kotlarchyk, S.H.Chen and J.S.Huang and M.W.Kim, Phys. Rev. A28,508(1983); Phys. Rev. A 29,2054(1984) and references therein
12. B.Derrida, D.Stauffer, H.J.Hermann, J.Vannimenus, J.Phys.(Paris)Lett. 44,L701(1983)
13. S.H.Chen and J.S.Huang, Phys.Rev.Lett.55,1888((1985)

Structure and Dynamics of Cyclohexane in Bulk and Constrained Geometry

J.C. Dore [1], M. Dunn [1], T. Hasebe [1], J.H. Strange [1], and M.C. Bellissent-Funel [2]

[1] Physics Laboratory, University of Kent, Canterbury, CT2 7NR, UK
[2] Lab. Leon Brillouin, CEN, Saclay, F-91191 Gif sur Yvette, Cedex, France

Differential thermal analysis (DTA), nuclear magnetic resonance (nmr) and neutron diffraction measurements have been made on cyclohexane in a porous silica (Spherisorb) with a characteristic pore size of 90 Å. The transition temperatures for phase changes from liquid to plastic crystal and plastic crystal to brittle crystal are substantially reduced. The molecular mobility is also affected by constrained geometry and is greatly increased in the plastic crystal phase over corresponding results for the bulk material. Structural studies confirm the formation of crystallites but suggest that defect properties resulting from restricted growth features are probably responsible for the unusual properties.

1 Introduction

The structural and dynamic properties of disordered materials are normally studied in the macroscopic (or bulk) phase, i.e. with characteristic dimensions substantially greater than 1μ. In recent years there has been a growing interest in the change of properties caused by restrictions in the spatial dimensions to values that are comparable with the atomic or molecular scale. This may be achieved by investigation of finely divided solid particles or alternatively within the geometrical constraints of a porous network. Recent work by neutron diffraction [1] has been centred on the properties of water and ice in microporous silicas with characteristic dimensions of 20, 90 and 150 Å. The effects of the geometrical constraints are found to depress the phase transition temperature and to alter the crystallographic growth characteristics. However, the silica surface is hydrophilic and therefore has a direct influence on the properties of the interfacial water layers affecting the hydrogen-bonding features of the system. It is therefore of interest to investigate alternative molecules which have a weaker van der Waals interaction with each other and with the confining boundary layer.

The present study concerns the properties of cyclohexane in a silica with 90 Å pore size characteristics. Cyclohexane is a globular molecule which is thought to exhibit unhindered rotation in both the liquid and plastic crystal phases [2]. The dynamic properties of the bulk phase have been investigated by many techniques and have shown that the plastic crystal contains a high-level of defects which influence its behaviour. DTA, nmr and neutron diffraction measurements have now been made for cyclohexane dispersed in silica and the results for this system represent the initial stage of a more extensive programme devoted to the structural and dynamic properties of various molecular systems in solid porous substrates of differing pore size and surface interactions.

2 Experimental Measurements

(a) DTA

The results of the thermal analysis for a sample containing 0.20 g/g of C_6H_{12} in Spherisorb S10W are shown in Fig.1; similar data for bulk C_6H_{12} are also given. It is clear that the transition temperatures for liquid to plastic crystal (280 to 237 K) and plastic crystal to brittle crystal (187 to 166 K) are depressed by approximately 43 K and 21 K respectively. The results were reproducible to

Springer Proceedings in Physics, Vol. 37 **Dynamics of Disordered Materials**
Editors: D. Richter · A.J. Dianoux · W. Petry · J. Teixeira

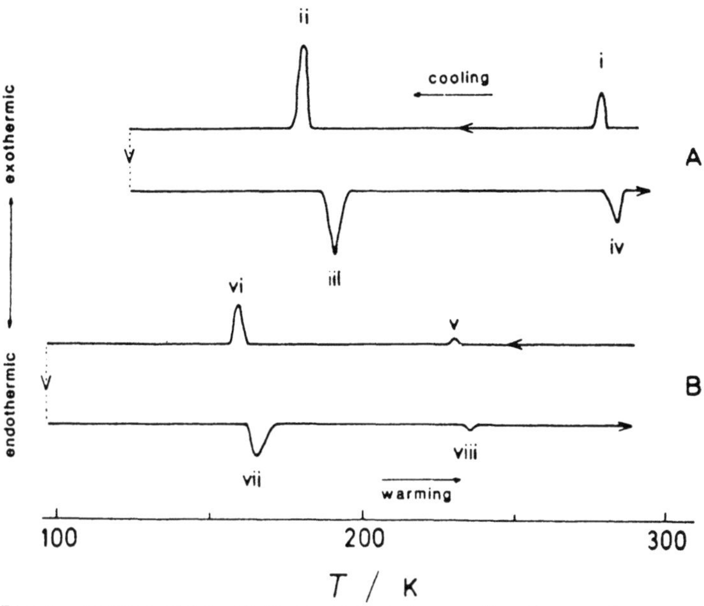

Fig.1 Differential thermal analysis results for cyclohexane in bulk and dispersed phase

better than ± 0·5 K but no attempt was made to investigate any dependence on cooling/heating rate which was left constant and approximately 1 $K.min^{-1}$. The data demonstrate the depression of the transition temperature but which is substantially larger than the change in the freezing point of water for the same material.

(b) NMR

Measurements of the proton relaxation times $T_{1\rho}$ and T_2 for 0·3 g/g C_6H_{12} in porous SiO_2 are shown in Fig.2. Above 235 K, corresponding to the super-cooled liquid phase, the relaxation times are equal and give an activation energy for molecular motion of 11·4 (± 1·0) kJ mol^{-1}. This is similar to that found in the bulk liquid. However, the relaxation times in the adsorbed material are two orders of magnitude lower, indicating a relative decrease in molecular mobility of the dispersed liquid. The behaviour is different for the plastic crystal phase (< 237 K) where there is a significant deviation in the $T_{1\rho}$ and T_2 values following the rapid fall at the onset of nucleation. The behaviour of T_2 suggests that the plastic crystal phase is behaving in a similar manner to that of the bulk phase but at a correspondingly lower temperature. It also implies that the self-diffusion rates and the characteristics of the molecular motion are similar for the corresponding states although it seems that the activation energy for the adsorbed phase (15 kJ mol^{-1}) is lower than that of the bulk phase (40 kJ mol^{-1}). The $T_{1\rho}$ results are particularly interesting for this temperature range. The existence of a broad minimum has recently been discussed by Tabony and Korb [3] who suggest that the anisotropy of the boundaries and the bounded nature of the diffusion have a direct effect on the data such that interpretation of the motion on the basis of conventional nmr theories may be inappropriate. Further work will be required to investigate these views but a treatment of the data on the basis of diffusional characteristics suggested by Villa and co-workers [4] can be used to study the effective dimensionality. For two-dimensional motion the spin-lattice relaxation has a logarithmic frequency dependence whereas for one-dimensional motion a dependence on $(frequency)^{1/2}$ is predicted. The present results seem to be in better agreement with the 1D case as shown in Fig.3 where the data show a linear relation with $\sqrt{B_1}^{1/2}$ in the higher temperature regime; B_1 is the applied r.f. field which is proportional to ω_1. Undoubtedly there are more complicated factors depending on the length and time characteristics for translational diffusion across the pores such that a full treatment involving fractal concepts may eventually be required for a complete interpretation.

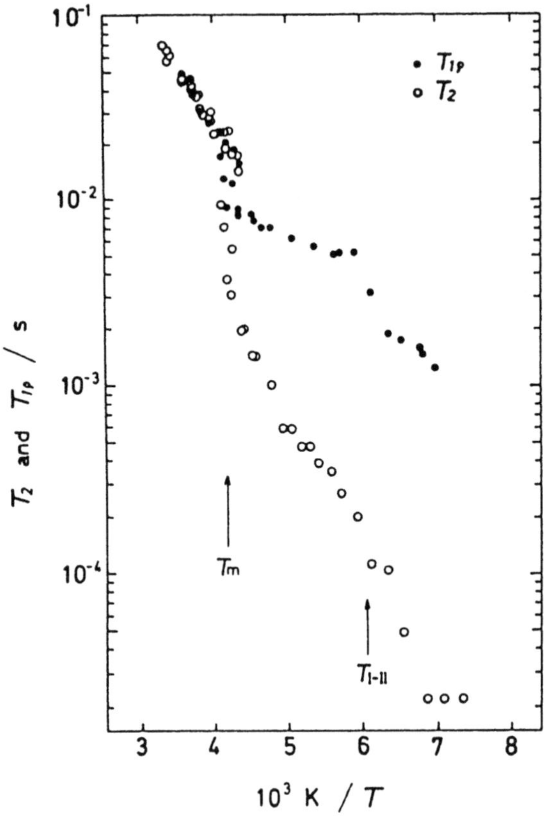

Fig.2 The proton relaxation times, $T_{1\rho}$ and T_2, for C_6H_{12} in porous silica as a function of temperature

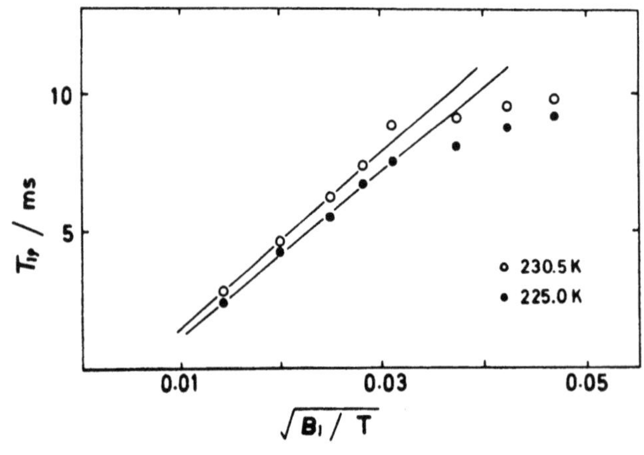

Fig.3 The plot of $T_{1\rho}$, against $B_1^{1/2}$ showing evidence of 'one-dimensional' diffusion in the high temperature range of the plastic crystal.

(c) Neutron diffraction
Structural studies of C_6D_{12} in the bulk liquid and plastic crystal phases have already been reported [4]. Similar measurements have now been made for C_6D_{12} in Spherisorb S10W and treated in the same manner as the earlier work on D_2O in silica [1]. The processed data are shown in Fig.4 as a function of the scattering vector, Q, at temperatures of 278, 263, 233 and 180 K. The three upper temperatures show little change and are typical of a disordered liquid phase. The data at 180 K are also similar in the high Q-value regime (> 3 $Å^{-1}$) but the main peak sharpens and there are two small peaks superimposed on the broad oscillatory structure. These features are characteristic of plastic crystal systems in which Bragg peaks are superimposed on the contribution from diffuse scattering arising from the orientational disorder. Since the mean lattice parameter is well defined there is a significant contribution from the crystal structure factor S(hkℓ) at low Q-values but local translational displacements of the molecular centres produces an additional disorder leading to a high value for the effective Debye-Waller factor and a rapid attenuation of the intensity at high Q-values. These phenomena are well know from other work on CBr_4 [5] and P_4 [6] and are apparently reproduced in C_6D_{12} for both bulk [4] and constrained phases. Since the isotopic variation (H/D) is not expected to change the normal transition temperatures by more than a few degrees, the results are in agreement with DTA and NMR data and confirm the presence of a plastic crystal phase for the middle temperature range of the present studies.

The real-space representation of the structure can be evaluated from the measured intensity by Fourier transformation [7]. The data for the 180 and 278 K measurements are shown superimposed in Fig.5. The characteristics of the plastic crystal phase are indicated by the regular oscillatory structure with a periodicity of 4·5 Å which is primarily due to the sharp features of the first Bragg peak. The relative displacement of the two curves in the intermediate range represents a change in the mean separation of molecular centres and corresponds to a microscopic density change of approximately 10%. The sharp peaks at low r-values are primarily due to intra-molecular correlations but

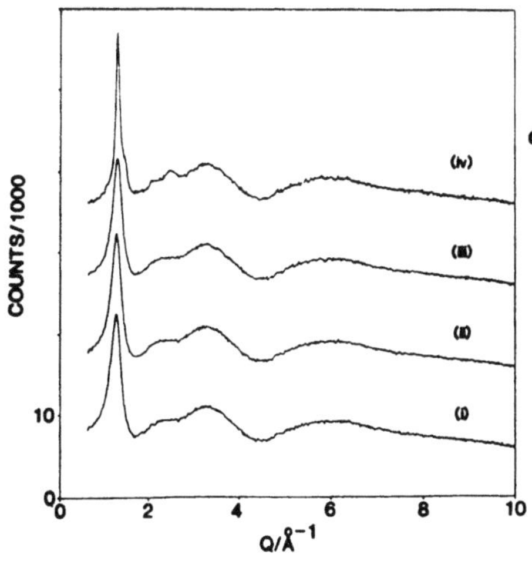

Fig 4 Neutron diffraction data at temperatures of i) 278K; ii) 263K; iii) 233K and iv) 180K

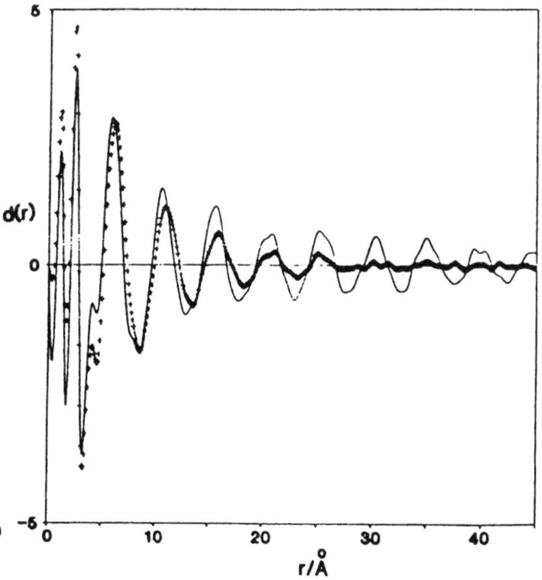

Fig 5 The real-space correlation function $d_L(r)$ derived from Fig.4 for 278 K and 180 K

there are also some smaller differences arising from correlations between adjacent molecules; a detailed analysis of these features has not yet been conducted but may possibly be related to positional changes caused by high defect concentrations.

3 Interpretation

It is clear from the results that the absorbed cyclohexane has properties that are related to the normal behaviour of the bulk material. However, the large temperature shift in the corresponding phases also indicates that there are significant changes in the microscopic structure and dynamics. The basic structural properties of the super-cooled liquid appear to follow the trends established by studies on the bulk phase and do not appear anomalous. The relaxation times are two orders of magnitude lower in the absorbed phase but the activation energy remains the same. This suggests that molecular mobility is reduced by the same factor and further nmr measurements on the self-diffusion are being made to distinguish the translational molecular motion from the molecular reorientation. The preliminary picture suggests that the liquid phase has the expected features but the intermediate and long-time diffusional behaviour is influenced by the volume constraints of the silica substrate.

The plastic crystal phase exhibits a number of interesting features. The present results provide the first evidence for the presence of defect contributions in a molecular crystal leading to higher mobility than for the corresponding bulk phase. This defect phenomenon appears to be parallel to that observed extensively in ionic systems [8] where there is enhanced ionic mobility in microscopically dispersed solids. The constraining geometry of the pore network appears to influence the defect structure of the cyclohexane crystallites but the way in which this occurs cannot be established from the present data. It would seem likely that the pore geometry has a direct effect on the nucleation and crystal growth processes [9] such that the observed phenomena are probably dependent on cooling rates and pore network characteristics. It is evident that further studies will be required to investigate these additional variables and that the present results represent only a preliminary observation of a uniquely disordered molecular assembly. Additional techniques can be utilised to study the properties and future work involving quasi-elastic neutron scattering is likely to provide an interesting insight into the mechanism of the molecular motion.

1. J.C.Dore, M.Dunn and P.Chieux, J.Cryst.Growth, in press.
2. J.N.Sherwood, In The Plastically Crystalline State (J.Wiley & Sons Ltd. 1979) p.71-80.
3. J.Tabony and J.P.Korb, Molec.Phys., 56, 1281 (1985).
4. H.Farman, J.C.Dore, J.C.Bellissent-Funel and D.G.Montague, Molec. Phys., 61, 583 (1987) and ibid, submitted to Molec.Phys.
5. M.More, J.Lefebvre, B.Hennion, B.M.Powell and C.M.E.Zeyen, J.Phys.C, 13, 2833 (1980).
6. J.R.Granada and J.C.Dore, Molec.Phys., 46, 757 (1982).
7. J.C.Dore, In Molecular Liquids: Dynamics and Interactions, A.J.Barnes, W.J.Orville-Thomas and J.Yarwood (Eds). D.Reidel (Pub) 1984.
8. C.C.Liang, J.Electrochem.Soc., 120, 1289 (1973).
9. M.Descamps, this meeting and private communication.

Dynamic Properties of an Inorganic System Near the Glass Transition

S. Jung[1], *P. Müller*[2], *and C. Morkel*[1]

[1]Fakultät für Physik, Technische Universität München, D-8046 Garching, Fed. Rep. of Germany
[2]Walther-Meißner-Institut für Tieftemperaturforschung, Bayerische Akademie der Wissenschaften, D-8046 Garching, Fed. Rep. of Germany

Quasielastic and inelastic neutron scattering experiments have been performed on supercooled, glassy and crystalline aqueous calcium nitrate. A fast diffusing component is revealed. The results on glass transition dynamics are discussed in the framework of recent mode coupling theories on glass transition dynamics.

In a series of neutron scattering experiments we studied the dynamics of the liquid-glass transition of aqueous calcium nitrate. The structural and transport properties of this system are well characterized /1-4/. On proper handling it stays in the supercooled state during a scattering experiment of several days without crystallization. We investigated both the protonated and the deuterated tetrahydrate $Ca(NO_3)_2 \cdot 4\,H_2O$ and $Ca(NO_3)_2 \cdot 4\,D_2O$, which form ionic liquids consisting of $(NO_3)^-$ ions and hydrated Ca^{++} ions with four water molecules in the first coordination shell. One therefore can consider this hydrate melt as a hard sphere system undergoing a liquid-glass transition upon cooling /2/. This point of view encourages the comparison of our results with predictions which mode coupling theory states for the dynamics of the liquid-glass transition /5-7/.

A prominent feature of this theory concerns the anomalous behaviour of the Debye - Waller factor near the glass transition temperature. Figures 1 and 2 show the temperature dependences of the elastic incoherent and coherent scattering intensities measured at IN13, ILL Grenoble. "Elastic" scattering in this context includes processes with energy transfers less than the experimental resolution of 9 μeV FWHM, corresponding to times longer than 10^{-9} s. In qualitative agreement with theory, intensity decreases with increasing temperature and momentum transfer, dropping at a fast rate at the approach of the phenomenological glass temperature $T_g \approx 217$ K and beyond in the supercooled state. The melting point is $T_m \approx 320$ K. A similar behaviour has also been observed in other inorganic glasses and in polymers /8-10/.

Interpreting the incoherent data ($Ca(NO_3)_2 \cdot 4\,H_2O$) in terms of a Debye - Waller factor $\exp(-1/2\ \langle r_0^2\rangle Q^2)$ yields the temperature dependence of the mean square displacement $\langle r_0^2\rangle$ (fig. 3). At low temperatures a harmonic oscillator like linear increase of the mean square displacement can be observed. In the vicinity of T_g, however, a significant deviation of the harmonic approximation occurs.

Mode coupling theory predicts a square root singularity of the elastic form factor at a critical temperature T_0 marking the transition from ergodicity (liquid) to nonergodicity (glass). Considering $\langle r_0^2\rangle$ as an order paramenter of the glass transition (fig. 3),

$$\langle r_0^2\rangle = \langle r_{0\,c}^2\rangle\ (1 - a\sqrt{\varepsilon}) \qquad \varepsilon = (T_c - T)/T_c,\ a > 0 \qquad (1)$$

yields $T_0 = (236.3 \pm 4.0)$ K, some 20 K above T_g, and $\langle r_{0\,c}^2\rangle^{1/2} = (0.18 \pm 0.04)$ Å. This length is about 5 % of the next neighbour distance, a feature which agrees with the Lindemann criterion.

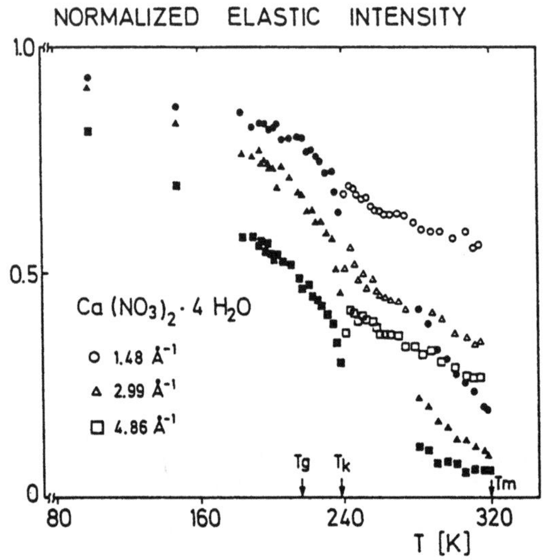

Fig. 1: Elastic intensities of $Ca(NO_3)_2 \cdot 4H_2O$ vs temperature, normalized to their 4.2K value. Full symbols denote data of the glassy and supercooled state, open symbols of the crystallized state. (The sample crystallized at T_k during heating.) T_g thermodynamic glass transition temperature, T_k crystallization temperature, T_m melting point.

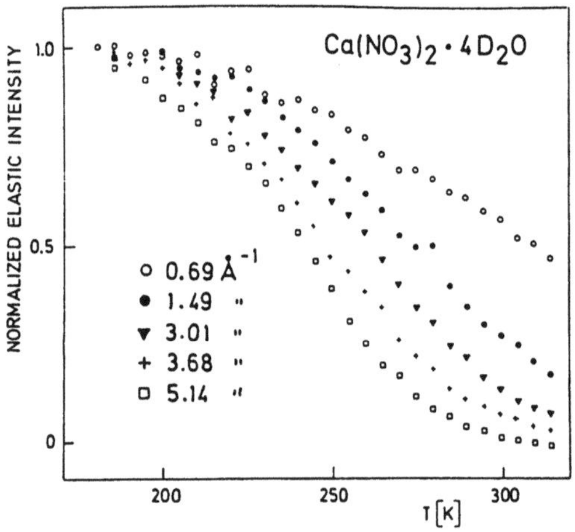

Fig. 2: Elastic intensities of $Ca(NO_3)_2 \cdot 4D_2O$ vs temperature in the glassy and supercooled region. The intensities are normalized to their 180 K value. No crystallization occured.

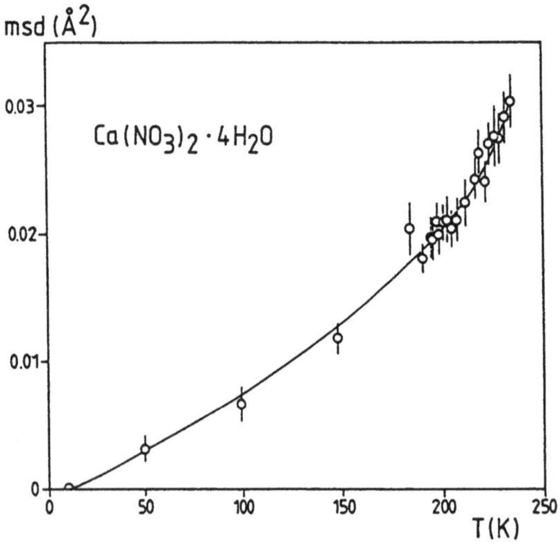

Fig. 3: Mean square displacements of $Ca(NO_3)_2 \cdot 4H_2O$ vs temperature, obtained by fitting the Q dependence of the elastic intensities as described in the text. The solid line fits the square root behaviour (equ. (1)).

The elastic intensity loss near T_g is connected with an intensity increase at finite energy transfers. Thus we investigated the temperature dependence of the dynamic structure factor $S(Q,\omega)$ of $Ca(NO_3)_2 \cdot 4\ D_2O$ with the time of flight spectrometer IN6, ILL Grenoble, in the temperature range between 0.95 Tg and 1.36 Tg. To elucidate the specific features of the glass transition, effects due to temperature dependent thermal occupation numbers are removed in harmonic approximation by applying the proper Bose factors. Figure 4 shows the resulting normalized dynamic structure factors for five temperatures. A drastic intensity increase with increasing temperature in an energy range below 5 meV is to be noted. Compare this effect with the network glass zinc chloride (fig. 5), measured on IN6 in the temperature range between 0.77 T_g and 1.15 T_g: In the glassy and supercooled state of this system there is an inelastic intensity, centered at an energy transfer of approximately 2 meV, which does not appear in the crystal (lowest spectrum in fig. 5). In contrast to calcium nitrate the temperature dependence of $S(\vartheta,\omega)$ is fully described by the Bose factor. The observed increase of inelastic intensity of calcium nitrate near T_g might be interpreted in terms of mode coupling theory as scaling effect of the ß-relaxation. A detailed analysis of the scaling properties of the measured line shapes and intensities is in progress.

As the glass dynamics comprises more than 6 decades in frequency, it is desirable to extend the dynamic range of measurements as far as feasible. High resolution measurements of $Ca(NO_3)_2 \cdot 4\ H_2O$ in the supercooled state in the temperature region between 1.06 T_g and 1.47 T_g reveal quasielastic scattering with line shapes that are strikingly non single-Lorentzian. In order to separate the glass transition dynamics from other processes we measured both polycrystalline and supercooled fluid samples of $Ca(NO_3)_2 \cdot 4\ H_2O$ at the same temperature.

In both cases the obtained scattering law of the polycrystalline sample shows non single-Lorentzian lineshapes, a feature which consequently must not be attributed to glass dynamics. The data can be explained microscopically by a fast diffusion process: the macroscopic diffusion constant of aqueous calcium nitrate is well known in a wide concentration range and is less than $2 \cdot 10^{-7}$ cm^2s^{-1} for the supercooled tetrahydrate at 290 K /1/. This slow relaxational diffusion causes a but small broadening of the central line of about 2 μeV at $Q = 1$ $Å^{-1}$ vs 9 μeV FWHM spectrometer resolution. For this reason we treat both crystalline and supercooled data by a superposition of a delta function, describing elastic scattering and slow diffusion, and a Lorentzian describing fast diffusion. Within the experimental error the line width of the Lorentzians are equal in both cases (fig. 6).

The Q dependence of the fast diffusional process was analyzed with better statistics in a supercooled state at 290 K (fig. 7). As at the present stage we

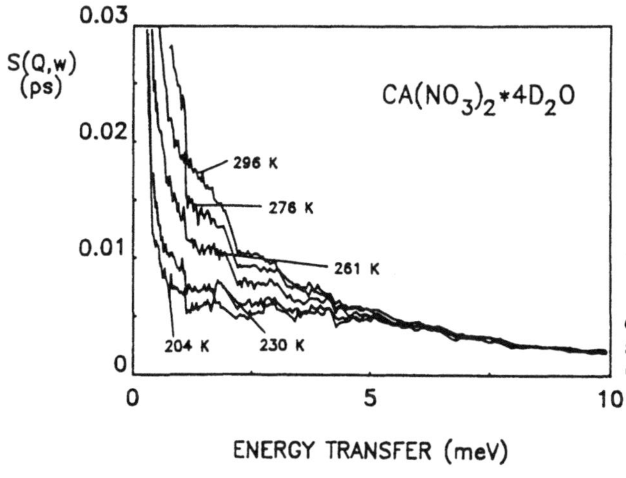

Fig. 4: Dynamic structure factor of $Ca(NO_3)_2 \cdot 4\ D_2O$ at $Q = 1.85$ $Å^{-1}$, normalized as described in the text. Note the agreement of all curves at high energy transfers.

cannot distinguish between purely translational and a superposition of translational and rotational diffusion, the results are described tentatively by the simple diffusion model of Chudley-Elliott /11/ assuming a Gaussian jump length distribution to yield an estimation of the diffusion constant:

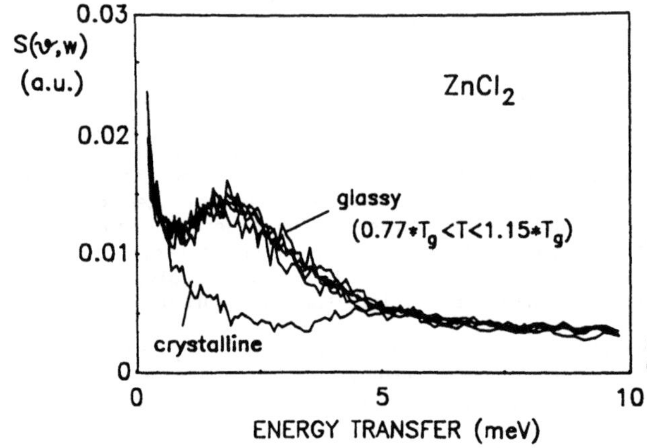

Fig. 5: Dynamic structure factors of glassy and crystalline $ZnCl_2$ at $Q_0 = 2.13$ Å^{-1} near the elastic structure factor maximum. The glassy spectra are normalized as described in the text.

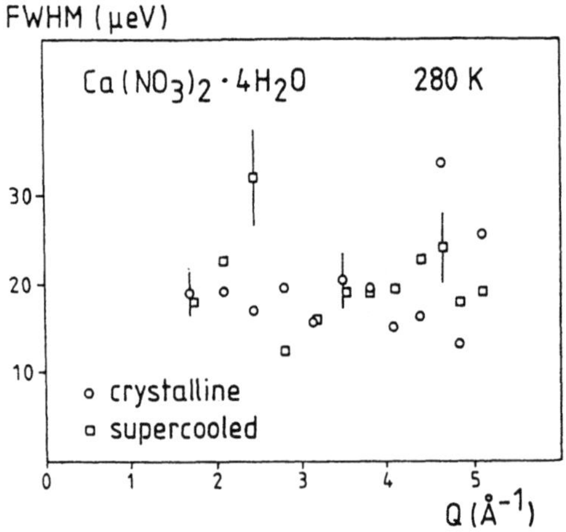

Fig. 6:
FWHM of the fast component of supercooled and crystallized $Ca(NO_3)_2 \cdot 4\ H_2O$ vs Q at T = 280 K.

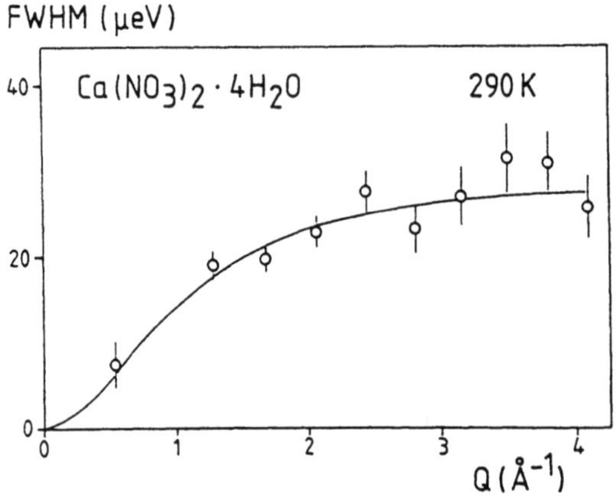

Fig. 7:
FWHM of supercooled $Ca(NO_3)_2 \cdot 4\ H_2O$ vs Q at T = 290 K. The full line is a fit of the model discussed in the text.

$$\Delta\omega = \frac{1}{\tau_0}\left(1 - \frac{1}{1+Q^2(l_0/2)^2}\right) \qquad \Delta\omega = \text{HWHM},\ l_0 \text{ mean jump length}, \quad (2)$$
τ_0 mean time interval between jumps

The fit yields the values $\tau_0 = (4.48 \pm 0.31) \cdot 10^{-11}$ s and $l_0 = (1.94 \pm 0.28)$ Å (fig. 7).

The corresponding diffusion constant, $D = l_0^2/(4\tau_0)$, equals $(2.1 \pm 0.62) \cdot 10^{-6}$ cm^2s^{-1}. The contribution of this fast component to the elastic intensities (fig. 1,2), especially at higher temperatures, masks the square root behaviour near T_g.

As we probe predominantly the motion of the protons (89 % of the incoherent scattering cross section), both the slow and the fast relaxation are due to motions of the hydrogen atoms. Therefore only a part of the hydrogen atoms is bound rigidly, giving rise to the slow diffusive motion signal of structural relaxation. The fast diffusion signal stems from the more freely moving hydrogens. There are several possible explanations for the microscopic nature of this fast process /1,12/. The study of a possible interplay of the fast diffusing component and glass transition dynamics, as recently proposed by theory /7/, will be subject to further work.

We thank the staff of the ILL for the help with the performance of the experiments and gratefully acknowledge valuable discussions with A. Heidemann. This work was supported by the Bundesministerium für Forschung und Technologie.

References:

1. P.M. Gammel, R. Meister: J. Chem. Phys. 64, 4287 (1976)
2. C.A. Angell: J. Elektrochem. Soc. 112, 1224 (1965)
3. C.T. Moynihan: J. Chem. Educ. 44, 531 (1967)
4. R.E. Hester, R.A. Plane: J. Chem. Phys. 40, 411 (1964)
5. U. Bengtzelius, W. Götze, A. Sjölander: J. Phys. C17, 5915 (1984)
6. W. Götze, L. Sjögren: Z. Phys. B65, 415 (1987)
7. J. Bosse, J.S. Thakur: Phys. Rev. Lett. 59, 998 (1987)
8. B. Frick, D. Richter, W. Petry, U. Buchenau: Z. Phys. B70, 73 (1988)
9. F. Fujara, W. Petry: Eur. Phys. Lett. 4, 921 (1987)
10. F. Mezei, W. Knaack: Phys. Scr. T19, 363 (1987)
11. G.T. Chudley, R.J. Elliott: Proc. Phys. Soc. 77, 353 (1961)
12. D.E. Irish, D.L. Nelsen, M.H. Brooker: J. Chem. Phys. 54, 654 (1971)

A Stochastic Model for Phase Transitions

T. Munakata[1] *and J. Bosse*[2]

[1]Faculty of Engineering, Kyoto University, Kyoto 606, Japan
[2]Freie Universität Berlin, Fachbereich Physik, Arnimallee 14, D-1000 Berlin 33, Germany

I. Introduction

Recently, the dynamics near the glass transition gathers a lot of attention and the mode-coupling (M-C) theory is playing an important role in clarifying the ergodic-nonergodic aspects in this transition. /1/ Here we propose a nonlinear stochastic model which shows a structural-like phase transition as the coupling (temperature) becomes large (small). This enables us to study the interesting competition and the interplay between the disordered (both equilibrium and metastable) and the ordered states.

After discussing the phase diagram we study the relaxation function with use of a M-C approximation. Computer experiments reveal some slow relaxation phenomena which are qualitatively in accordance with our theoretical prediction.

II. Model

We consider dynamics of a system consisting of N nonlinear (Duffing) oscillators described by the following Langevin equation:

$$dp_i/dt = -\partial V/\partial x_i - \zeta p_i + f_i(t), \tag{2.1}$$

$$\langle f_i(t)\, f_j(t')\rangle = 2\zeta\, k_B T M \delta(t-t')\delta ij, \tag{2.2}$$

$$V(\{x_i\}) = \sum_i \left[M\omega^2 x_i^2/2 + \alpha x_i^4/4\right] - \beta \sum_{i<j<k} x_i x_j x_k \quad . \tag{2.3}$$

For a theoretical approach it is more convenient to consider the Fokker-Planck equation

$$\partial g^{(N)}\ (\{x_i\},\ \{p_i\};t)/\partial t = D\, g^{(N)} \tag{2.4}$$

$$D \equiv \sum_i \left[-(p_i/M)\ \partial/\partial x_i + (\partial V/\partial x_i)\ \partial/\partial p_i + \zeta(\partial/\partial p_i)p_i + \zeta k_B T\ \partial^2/\partial p_i^2\right]$$

with the stationary distribution

$$g_{st}^{(N)} = C\exp(-H/k_B T) \equiv C\exp\left[-(\Sigma\ p_i^2/2M + V)/k_B\ T\right] \quad . \tag{2.5}$$

Springer Proceedings in Physics, Vol. 37 Dynamics of Disordered Materials
Editors: D. Richter · A.J. Dianoux · W. Petry · J. Teixeira

The linear response to the external force $h_j(t)$ on the j-th oscillator is given by

$$\langle x_i \rangle_t = \langle x_i \rangle + \sum_j \int_{-\infty}^{t} ds \ \langle x_i(t-s)p_j \rangle h_j(s)/(Mk_BT) \quad . \tag{2.6}$$

Here $\langle \ldots . \rangle$ denotes the average over (2.5) and the time evolution $x_i(t) \equiv \exp(\Lambda t)\, x_i$ is produced by the operator Λ, adjoint to D, (2.4). From (2.6) the (adiabatic) response at t = o to the static force $h_j(t) = h_j$ is obtained as

$$\delta\langle x_i \rangle_{t=o} \equiv \langle x_i \rangle_{t=o} - \langle x_i \rangle = \delta\langle x_i \rangle_T - \sum_j \Phi_{ij}(t{=}\infty) h_j \quad , \tag{2.7}$$

where we have introduced the relaxation function by

$$\Phi_{ij}(t) \equiv \langle \delta x_i(t) \delta x_j \rangle/(k_BT) \equiv \Phi(t)\delta_{ij} + \bar{\Phi}(t)[1-\delta_{ij}] \tag{2.8}$$

and the isothermal response by $\delta\langle x_i \rangle_T \equiv \sum_j \langle \delta x_i \delta x_j \rangle \ h_j/(k_BT)$. Here we note that the simple i-j dependence in (2.8) comes from the mean-field nature of our system, (2.3). The relation (2.7) indicates the importance of the asymptotic behavior of $\Phi_{ij}(t)$ which decays to zero for the ergodic system. /2/

III. Theory

We introduce ω^{-1}, $l \equiv (M\omega^2/\alpha)^{1/2}$ and M as unit of time, length and mass, respectively, to obtain

$$H^* \equiv H/(M\omega^2 l^2) = \Sigma(p_i^{*2}/2 + x_i^{*2}/2 + x_i^{*4}/4) - \beta^* \sum_{i<j<j} x_i^* x_j^* x_k^* \quad . \tag{3.1}$$

Thus our system is characterized by two parameters, the coupling constant $\beta^* \equiv \beta/(\omega^2 M\alpha)^{1/2}$ and the temperature $T^* \equiv k_BT/(M\omega^2 l^2)$ in addition to the friction constant ζ. (Hereafter we suppress the asterisk.)

(i) The structural phase transition: The equilibrium distribution $f_{st}^{(N)} \propto \exp(-V/T)$ satisfies the Smoluchowski equation $\sum_i \partial/\partial x_i \ [\partial V/\partial x_i + T\partial/\partial x_i]\ f_{st}^{(N)} = o$. Operating $\int dx_2 \ldots \int dx_N$ and approximating $f_{st}^{(3)}$ by $\simeq f_{st}^{(1)}(x_1)\, f_{st}^{(1)}(x_2)\, f_{st}^{(1)}(x_3)$, we obtain an equation for $f_{st}^{(1)}(x)$ of the form

$$d/dx \ [(x + x^3 - \bar{\beta}\langle x \rangle^2) f_{st}^{(1)}] + T d^2 f_{st}^{(1)}/dx^2 = o \quad , \tag{3.2}$$

where $\bar{\beta} \equiv \beta\ (N-1)\ (N-2)/2$. Equation (3.2) gives $f_{st}^{(1)}(x) = \exp[-(x^2/2 + x^4/4 - \bar{\beta}\langle x \rangle^2 x)/T]$. The $\langle x \rangle$ is to be determined self-consistently from

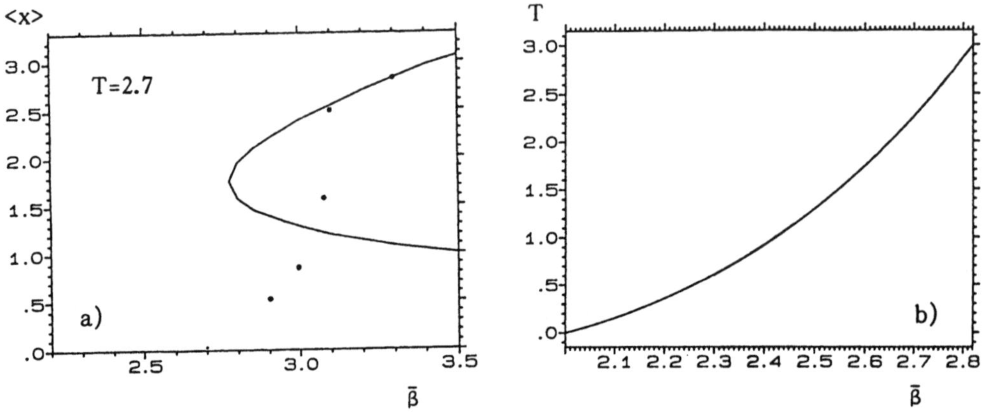

Fig.1: Phase diagram obtained from a mean-field approximation. (a) The bifurcation profile for T = 2.7. (b) The phase-boundary between a liquid and an ordered state. Black circles in (a) show the experimental time average of $\{x_i\}$ for N = 10.

$$\langle x\rangle = \int dx\, f_{st}^{(1)}(x)x / \int dx\, f_{st}^{(1)}(x) \quad . \tag{3.3}$$

For a given T, (3.3) yields two solution $\chi_1 > \chi_2 > o$ in addition to $\chi_3 = o$ for $\bar\beta > \bar\beta_{cr}$ (T) (Fig.1a). Stability analysis shows that χ_1 and $\chi_3 = o$ are stable and χ_2 unstable. In the range $\bar\beta < \bar\beta_{cr}(T)$, to be called a liquid phase, we have only solution $\chi_3 = o$. For $\bar\beta > \bar\beta_{cr}(T)$ we have an ordered state in addition to a metastable state characterized by $\chi_3 = o$. (Fig.1b)

(ii) <u>The ergodic-nonergodic transition:</u> Choosing $A = (\delta x_1, \dots, \delta x_N, p_1 \dots, p_N)$ as a set of dynamical variables we calculate each element in the generalized Langevin eq.

$$\frac{d}{dt} A(t) = i\Omega \cdot A - \int_o^t ds\, \psi(t-s) \cdot A(s) + f(t) \quad . \tag{3.4}$$

Noting that $D_{ij} \equiv \langle \delta x_i \delta x_j \rangle$ and the inverse $(D^{-1})_{ij}$ $(\equiv d\, \delta_{ij} + \bar d\, [1-\delta_{ij}])$ also have the simple structure (2.8), we obtain for $\phi(t) \equiv k_B T\, \Phi(t)/\langle \delta x_1^2\rangle$ and $\bar\phi(t) \equiv K_B T\, \bar\Phi(t)/\langle \delta x_1^2\rangle$

$$\ddot\phi = -\zeta\dot\phi - T\,[d\phi + (N-1)\bar d\bar\phi] - (1/T)\int_o^t ds[\psi(t-s)\dot\phi(s) + (N-1)\bar\psi(t-s)\dot{\bar\phi}(s)],$$

$$\ddot{\bar\phi} = -\zeta\dot{\bar\phi} - T[\bar d\phi + \{d + (N-2)\bar d\}\bar\phi] - (1/T)\int_o^t ds[\psi(t-s)\dot{\bar\phi}(s) + \bar\psi(t-s)\{$$

$$\dot\phi(s) + (N-2)\dot{\bar\phi}(s)\}] \quad , \tag{3.5}$$

where $\langle f_i(t)f_j\rangle \equiv \psi(t)\delta_{ij} + \bar{\psi}(t)\,[1-\delta_{ij}]$ (i,j = N+1,...,2N). Since our system (2.1) ~ (2.3) consists of oscillators we employ the lowest-order perturbation theory (in the nonlinearity) for the calculation of $\psi(t)$ and $\bar{\psi}(t)$ together with the usual replacement of the bare propagator by the full one. Here we only touch the asymptotic values $f \equiv \phi(\infty)$ and $\bar{f} \equiv \bar{\phi}(\infty)$. It turns out that $f = \bar{f}$ solves the equation with f satisfying

$$Bf/(f_o-f) = A\, f^2+3\langle\delta x_1^2\rangle\, f[2f^2-3] \quad , \tag{3.6}$$

where $f_o \equiv [1+r(N-1)]/N$ with $r \equiv \langle\delta x_1\delta x_2\rangle/\langle\delta x_1^2\rangle$, $B \equiv T^2 d_o/\langle\delta x_1^2\rangle^2$ with $d_o \equiv [d+(N-1)\bar{d}]/N$ and $A \equiv \{[3\langle x\rangle-\bar{\beta}]^2+3\langle x\rangle\bar{\beta}\}$. From (3.6) it is generally expected that when the fluctuation becomes large the system is more apt to be driven into a non-ergodic phase.

IV. Simulations

We solved the coupled Langevin equation (2.1) ~ (2.3) numerically with use of a difference scheme valid to $o(h^2)$, h the time mesh. /3/ From the discussions in Sec.III (ii), large fluctuation in $\{\delta x_i\}$ is expected to give rise to slow relaxation. One possibility to make $\langle\delta x_1^2\rangle$ large is to go near the phase transition region and the other to go to a metastable disordered state, avoiding the ordering which reduces the fluctuation as predicted by the theory in III(i). Thus our experiment is classified into the two categories:

(i) $\bar{\beta}$-variation: Fixing the temperature we change $\bar{\beta}$ and study the equilibrium properties. In Fig.1a we show the time average of $\{x_i\}$ at T = 2.7 for N = 10. The deviation from the mean-field picture in the transition region seems to come from the smallness of our system. In the range $3 \lesssim \bar{\beta} \lesssim 3.08$ we observe a slow relaxation of ϕ and $\bar{\phi}(t)$, Fig.2a. (For larger time $\bar{\phi}(t)$ nearly coincides with $\phi(t)$). However, once the system is pulled in an ordered state ($\bar{\beta}$ = 3.1) ϕ and $\bar{\phi}$ decay fast with high frequency oscillation. Fig.2b depicts the slow relaxation of ϕ which is well fitted by $\phi \simeq c\,[1-t/\tau]$ with τ depending on $\bar{\beta}$. We remark that for the system $\bar{\beta}$ = 3.075 (3.6) yields a non-zero physical (i.e., $\psi(t{=}\infty)>o$) solution $f \simeq 0.8$ in addition to f = o. For other systems in Fig.2a (3.6) yields f = o and the unphysical ($\psi(t{=}\infty)<o$) non-zero solutions. We regard the non-ergodic phase for $\bar{\beta}$ = 3.075 predicted by (3.6) as the precursor to the ergodic-nonergodic transition in view of some uncertainties due to insufficient sampling.

(ii) Quenching: We quench our system by changing T by ΔT in (2.2), with the cooling-rate estimatmed by $|\Delta T|/(1/\zeta)$. By measuring the time average of $\{x_i\}$ and

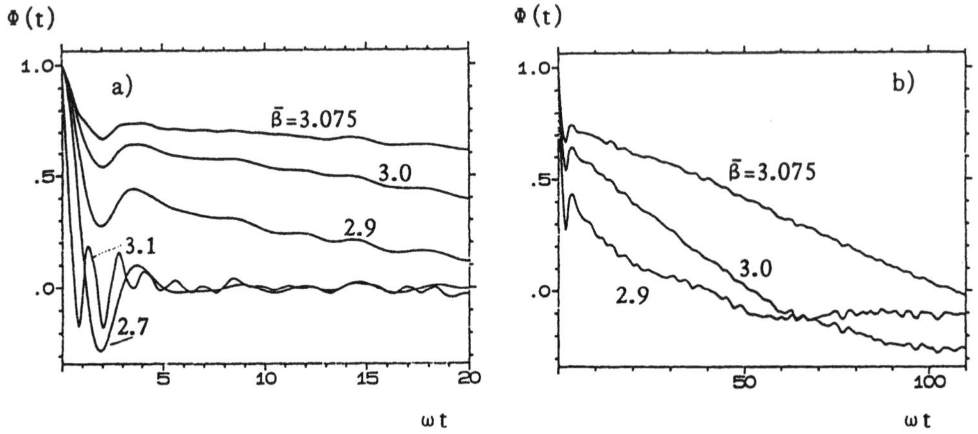

Fig.2: The relaxation functions for various $\bar{\beta}$ are depicted for N = 10 and T = 2.7. $\bar{\beta}$ = 2.7 and 3.1 correspond to a liquid and an ordered phase, respectively. The relaxation function in the transition region is depicted in (b) with a slightly longer time scale.

comparing it with χ_1 in Sec.III (i) we see whether the system is in an ordered or in a metastable state. We calculated the relaxation function of a metastable state produced by quenching from T = 3.5 to T = 0.1. We observe overall slow relaxation in a metastable state. However, as β becomes large the system undergoes crystallization which we cannot avoid at the moment, and the relaxation becomes fast as in the β-variation case (i).

V. Remarks

The mean-field nature of our system enabled us (i) to investigate some interesting phase transition phenomena even for a small system (N = 10) and (ii) to derive a mode-coupling equation (see (3.5), (3.6)) which contains only two components. /4/ Our results, both theoretical and numerical, suggest the importance of the larger-scale and longer-time simulations.

Acknowledgements

This work is supported by Sonderforschungsbereich 337 of Deutsche Forschungsgemeinschaft.

References

1. W. Götze, in: "Amorphous and Liquid Materials", eds. E. Lüscher, G. Fritsch and G. Jacucci, Martinus Nijhoff Publishers, Dordrecht, p. 34 (1987)

2. R. Kubo, J. Phys. Soc. Japan 12, 570 (1957)
3. T. Munakata, in: Noise in Nonlinear Dynamical Systems, Eds.: F. Moss and P. V. E. McClintock (Cambridge University Press, 1988)
4. Two-component (schematic) model is discussed in detail in connection with the glass transition in a simple molten salt by J. Bosse, U. Krieger and J. S. Thakur, contribution to this conference

Kawasaki Dynamics in the Ising Chain

J.M. Nunes da Silva and E.J.S. Lage*

Laboratório de Física da Universidade do Porto, Pç. Gomes Teixeira,
P-4000 Porto, Portugal

*Supported by the Instituto Nacional de Investigação Cientifica of Portugal

Real-space renormalization group techniques are applied to obtain the critical dynamics of the Ising linear chain, subjected to Kawasaki dynamics. The critical dynamic exponent is found to be z=5 and z=2, respectively, for ferromagnetic and antiferromagnetic nearest-neighbour interactions. The case with periodically distributed interactions is also treated confirming results recently obtained by other techniques, namely the non-universality of z.

1. INTRODUCTION

According to the Dynamic Scaling Hypothesis /1/ the relaxation time of the critical mode obeys the relation

$$\tau \sim \xi^{z}\phi(\xi\Delta q)$$

where $\Delta q=q-q_c$ is the shift from the critical wave vector (q_c) and z is the dynamic critical exponent. The critical mode q_c, whose fluctuations diverge as ξ, corresponds to the spin pattern of the ground-state.

In this communication we establish the non-universality of z for the Ising chain with Kawasaki dynamics /2,3/. The use, here, of simple renormalization procedures allows us to extract the value of z. Other methods such as domain wall arguments /4/, approximate linear response theory /5/ and truncated equations of motion /6/ have already been applied to the system under reference, yielding conflicting results.

2. MODEL

We consider a chain of Ising spins $\sigma=\pm1$ and a periodic distribution of nearest neighbour interactions with basis $<J_1, J_2, \ldots, J_b>$. Dynamics is introduced by a stochastic formulation given by the Master Equation (ME)

$$\frac{\partial}{\partial t} P(\{\sigma\},t) = - LP(\{\sigma\},t)$$

Glauber Dynamics /7/ treated by LAGE /8/ corresponds to the liouvillian L:

$$L_G = \sum_i (1-\hat{p}_i)\, W_i^G(\sigma_i)$$

where $\hat{p}_i\, f(\ldots, \sigma_i, \ldots)= f(\ldots, -\sigma_i, \ldots)$ and the normalization flipping rates are given by ($K_i = J_i/K_B T$):

$$W_i^G(\sigma_i) = w_i^G(\alpha_i^G+ \beta_i^G \sigma_{i-1} \sigma_{i+1}) \exp -\sigma_i|K_{i-1} \sigma_{i-1} + K_i\sigma_{i+1}|$$

with

Springer Proceedings in Physics, Vol. 37 Dynamics of Disordered Materials
Editors: D. Richter · A.J. Dianoux · W. Petry · J. Teixeira

$$\begin{matrix}\alpha_i^G \\ \beta_i^G\end{matrix} = \frac{1}{2} \left| \frac{1}{ch(K_{i-1}+K_i)} \pm \frac{1}{ch(K_{i-1}-K_i)} \right|$$

in order to assure normalized rates /7/.

Kawasaki dynamics is given by the liouvillian L /3/:

$$L_K = \sum_i (1-\hat{p}_{i,i+1}) W_i^K(\sigma_i,\sigma_{i+1})$$

where

$$\hat{p}_{i,i+1} f(\ldots, \sigma_i, \sigma_{i+1}, \ldots) = f(\ldots, \sigma_{i+1}, \sigma_i, \ldots)$$

and

$$W_i^K(\sigma_i, \sigma_{i+1}) = w^K \left(\frac{1-\sigma_i\sigma_{i+1}}{2}\right) (\alpha_i^K + \beta_i^K \sigma_{i-1} \ \sigma_{i+2}) \times$$

$$\times \exp - |K_{i-1}\sigma_{i-1}\sigma_i + K_{i+1}\sigma_{i+1}\sigma_{i+2}|$$

with

$$\begin{matrix}\alpha_i^K \\ \beta_i^K\end{matrix} = \frac{1}{2} \left| \frac{1}{ch(K_{i-1}+K_{i+1})} \pm \frac{1}{ch(K_{i-1}-K_{i+1})} \right|$$

The only approximation made in this treatment is the projection of the distribution function into the subspace of functions with the form /9/

$$P(\{\sigma\}, t) \simeq Peq(\{\sigma\}) |1 + \sum_i h_i(t) \sigma_i|$$

This form is expected to give an adequate representation of the shift from equilibrium, valid at long times and with small, time dependent, fields $h_i(t)$. Then, the resulting MEs are linear in the $h_i(t)$ allowing a Fourier decomposition: we write the field at site $l=bn+s$ (where $\underline{n}$ specifies the periodic cell and $\underline{s}$ the position inside the cell) as

$$h_l = H_s e^{iq(bn+s)}$$

3. RENORMALIZATION GROUP PROCEDURE

A partial trace, tr_p, over spins with index nonmultiple of B=b+1, redifines a new ME:

$$\frac{\partial P'}{\partial t} \equiv tr_p \frac{\partial P}{\partial t} = -tr_p(LP)$$

$$= -L'P' \quad , \quad P' = tr_p P$$

In the Glauber case the form of L_G is mantained /8/, but in Kawasaki dynamics /3/ we obtain a mapping into the Glauber one.

3.A <u>Renormalization of the L.H.S.</u> of the ME is common to both cases. It gives the renormalized coupling constants /8/

$$t'_s = t_s \left(\prod_{r=1}^{b} t_r\right) \quad , \quad t_r = \text{th}\, K_r$$

The renormalization of the fields can be obtained as in /8/ or, more easily, by noting that untraced spins must have the same average before and after the partial trace:

$$\sum_{s'=1}^{b} \chi^{ss'}_{Bq,\xi/B} H'_{s'} = \sum_{s'=1}^{b} \chi^{ss'}_{q,\xi} H_{s'}$$

with the, wave-vector dependent, susceptibility matrix

$$\chi^{ss'}_{q,\xi} \equiv e^{-iq(s-s')} \sum_{n'} e^{iqb(n'-n)} \langle \sigma_{bn+s}\, \sigma_{bn'+s'} \rangle$$

After m iterations (with $B^m >> \xi$), the renormalized couplings iterate to zero, the iterated susceptibility reduces to the unit matrix and we easily obtain:

$$H^{(m)}_s = \sum_{s'} \chi^{ss'}_{q,\xi} H_s$$

3.B Renormalization of the R.H.S. of the ME goes as follows. The partial trace in the Kawasaki case determines a new L operator corresponding to Glauber dynamics with flipping rates W^G_s such that

$$w^G_s\, \alpha^G_s\, H'_s \sqrt{1-t'^2_{s-1}}\, \sqrt{1-t'^2_s} = \sum_r A_{sr} H_r$$

where

$$A_{sr} = \delta_{rs}(a_s + a_{s-1}) - a_{s-1}\, e^{-iq}\, \delta_{s,r+1} - a_s\, e^{iq}\, \delta_{s,r-1}$$

and

$$a_r = w^K \alpha^K_i\, e^{-K} / (\text{ch}K_{i-1}\, \text{ch}K_i\, \text{ch}K_{i+1})$$

This expression works like an input in the recursion relation for Glauber flipping rates /3/:

$$w'^G_s\, \alpha''^G_s\, H''_s \sqrt{1-t''^2_{s-1}}\, \sqrt{1-t''^2_s} = w^G_s\, \alpha'^G_s H'_s \sqrt{1-t'^2_{s-1}}\, \sqrt{1-t'^2_s}$$

After m-1 iterations we therefore obtain:

$$w^{(m-1)G}_s H^{(m)}_s = w^G_s\, \alpha'^G_s\, H'_s \sqrt{1-t'^2_{s-1}}\, \sqrt{1-t'^2_s}$$

$$= \sum_r A_{sr} H_r$$

and the iterated ME reads:

$$\frac{\partial P^{(m)}}{\partial t} = -\sum_l h^{(m)}_l \sigma_l\, w^{(m-1)G}_l\, P_{eq}^{(m)}$$

In terms of the original parameters we obtain our basic equation:

$$\bar{H} = -\bar{\bar{\chi}}^{-1}\, \bar{\bar{A}}\, \bar{\bar{H}}$$

4. RESULTS

From the previous equation we conclude the relaxation time of the critical mode near $T_c=0$ can be obtained from:

$$\tau \sim \mathrm{tr}\, \bar{\bar{A}}^{-1} \bar{\bar{\chi}}$$

Interesting cases, confirming the breakdown of universality in critical dynamics at 1D, are for b=2:

i) $J_2 > J_1 > 0$, $q_c=0$ and $z=4+J_2/J_1$ (this result support the non applicability of domain wall arguments /10/ which predict $z=3+2J_2/J_1$)

ii) $J_1 > -J_2 > 0$, $q_c=\pi/2$ and $z=1+J_1/|J_2|$

iii) The other situations give z=2

In particular for $J_1=J_2>0$ we have $z=5$ and the scaling function $\phi(x) \alpha x^2(1+x^2)$. For $J_1=J_2<0$ $z=2$ and $\phi(x)\alpha(1+x^2)$.

The generalization of the results obtained for b=2, 3 and 4 and an analysis of the matrix A suggest that the leading relaxation time is:

$$\tau \sim q^{-2}\xi/a_m$$

when

$$a_m = \mathrm{Min}\left\{ \frac{1-t_s}{\mathrm{ch}(K_{s-1}-K_{s+1})\,\mathrm{ch}(K_{s-1}+K_{s+1})} : s=1,2,\ldots,b\right\}$$

This therefore implies that the dynamical exponent depends on the distribution of the exchange couplings (except for b=2) and constitutes a very strong breakdown of universality in critical dynamics.

5. REFERENCES

1. P.C. Hohenberg, B.I. Halperin: Rev.Mod.Phys., 49, 435 (1977)
2. K. Kawasaki: In Phase Transitions and Critical Phenomena, eds. C. Domb and M. Green (1972) Academic Press
3. J.M. Nunes da Silva, E.J. Lage: J.Phys. C21, 763 (1988)
4. R. Cordery, S. Sarker, J. Tobochnik: Phys.Rev. B24 5402 (1981)
5. W. Zwerger: Phys.Lett. A84, 269 (1981)
6. J.H. Luscombe: Phys.Rev. B36, 501 (1987)
7. R.J. Glauber: J.Math.Phys. 4, 294 (1963)
8. E.J. Lage: J.Phys. C20, 3969 (1987)
9. Y. Achiam, J.M. Kosterlitz: Phys.Rev. Lett, 52, 942 (1978)
10. M. Droz, J.M. Leal da Silva, A. Malaspinas: J.Phys.Lett. A115, 448 (1986)

Signatures of Critical Dynamic Instability Near the Glass Transition: A Neutron Scattering Study

F. Mezei

Hahn-Meitner-Institut, Postfach 390128, D-1000 Berlin 39, Germany

1. Introduction

An onset or change of non-ergodicity is a fundamental feature of all phase transitions. A phase with broken symmetry (some kind of order) settles around one of its equivalent ground states, which is separated from the other ones by a potential barrier of practically infinite height. For example a crystallite never spontaneously changes its orientation or position, or a ferromagnetic domain magnetized parallel to an easy direction does not spontaneously flip over to the opposite direction. Although ordinary phase transitions are per definition related to a break-down of some symmetry in space (e.g. the homogeneity of the system in a gas-liquid transition), one could generalize the notion by maintaining the single criterion of a change in the ergodic behaviour only. This has been first suggested for spin glasses /1/ by a proposal to consider a measure of non-ergodicity as the order parameter of the spin glass phase, which does not differ in its spatial symmetry from the paramagnetic phase. It is of particular importance that the proposed order (i.e. non-ergodicity) parameter, the $t=\infty$ limit of a correlation function, is a quantity which can be directly observed with the help of neutron scattering (in most cases exclusively by this method). This leads to the concept of dynamic phase transition in which the temporal behaviour of the correlations at $t\to\infty$ only distinguishes the various phases.

It has to be kept in mind, however, that the limit of "infinite time" is an unattainable abstraction, and therefore we cannot really test or observe non-ergodicity in practice. We can talk about non-ergodicity if an extrapolation of the variation of the finite time behaviour reveals a divergence of the characteristic relaxation times at a finite temperature. Precisely, this is the signature of a dynamic phase transition as opposed to some natural slowing down of the decay of correlations on approaching absolute zero temperature.

Recent theoretical analysis of generalized non-linear hydrodynamic description of dense liquids /2/ revealed the possible existence of a dynamic singularity corresponding to an instability of the liquid phase against non-ergodicity on cooling towards a critical temperature T_0. Similarly to ordinary critical phenomena this singularity has been found to be governed by power laws. In accordance with the general considerations above the signature of the dynamic phase transition in this case is the singular tendency of the correlations to remain finite for longer and longer times, and not the unobservable strict non-ergodocity itself. Since it has been found that the relevant correlation function in this case is the wavenumber dependent density-density correlation function, neutron scattering is the best suited method for the study of this problem.

In what follows I shall review neutron scattering results in the archetypical fragile (non-network) glassformer $Ca_{0.4}K_{0.6}(NO_3)_{1.4}$ /3-5/ from the particular point of view of identifying a singularity in the tendency towards non-ergodicity. In addition I shall present some new experimental details relevant to a comparison with more specific theoretical predictions /6,7/.

Springer Proceedings in Physics, Vol. 37 **Dynamics of Disordered Materials**
Editors: D. Richter · A.J. Dianoux · W. Petry · J. Teixeira

2. Signatures of non-ergodicity and dynamic instability

Figure 1 shows the temperature dependence of the density-density correlation function $\phi(q,t)$ as observed by neutron scattering at a wavenumber $q \simeq 1.9$ Å^{-1} close to the peak q_0 of the structure factor. In fact, due to the different weighting factors of the various nuclei, in a system containing several atomic species the (intermediate) coherent neutron scattering function $S(q,t)$ is not identical with $\phi(q,t)$ but it is expected to reproduce its behaviour in time. In the figure the data points have been directly obtained by Neutron Spin Echo (NSE) spectroscopy and the continuous lines represent the Fourier transform of Time-of-Flight (TOF) spectra symmetrized following the Schofield prescription /8/ in order to make a comparison with a classical theory feasible. The dashed lines correspond to a simple Kohlrausch law representation of the results for times longer than 10^{-11} s:

$$\phi(q,t) = A(q)\exp(-(t/\tau)^{\beta}) \tag{1}$$

where the convention $\phi(q,t=0)=1$ has been adopted, $A(q)=0.76$ around q_0 and $\beta=0.58$. The effective relexation time τ has been found to follow a power law in a good approximation

$$\tau = \tau_0(q)(T-T_0)^{-\gamma} \tag{2}$$

with T_0 being chosen anywhere between 80 and 100° C. (Actually τ appears to be proportional to the shear susceptibility, thus we can identify the process described by (1) with the primary or α relaxation.) Equations (1) and (2) correspond to a critical instability of the ergodic (liquid) phase against non-ergodicity which would set in at temperature T_0, where the non-ergodicity parameter $f(q)= \phi(q,t\to\infty)$ jumps from 0 to $A(q)$.

The significance of this kind of interpretation of these data is enhanced by evidence for further theoretically predicted signatures of T_0. Thus, the temperature dependence of $f(q)$ and of the minimum of susceptibility $\chi''=\omega S(q,\omega)/kT$ between a first relaxation step around 10^{-13}-10^{-12} s (Fig. 1) and the α relaxation process are expected to display square-root behaviour respectively below and above T_0:

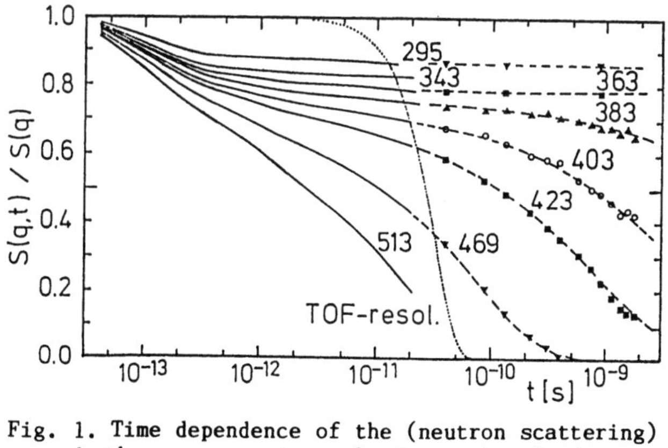

Fig. 1. Time dependence of the (neutron scattering) density-density correlations around the wave vector of the peak of the structure factor $q_0 \simeq 1.9$ Å^{-1} in $Ca_{0.4}K_{0.6}(NO_3)_{1.4}$ at various temperatures indicated in K units /5/. The continuous lines are Fourier transformed symmetrized TOF spectra corrected for the TOF resolution function (dotted line) and the data points are directly measured NSE results. Above 370 K the dashed lines indicate a scaling representation of the second (α) relaxation step using stretched exponential (Kohlrausch) master function as given by (1) and (2).

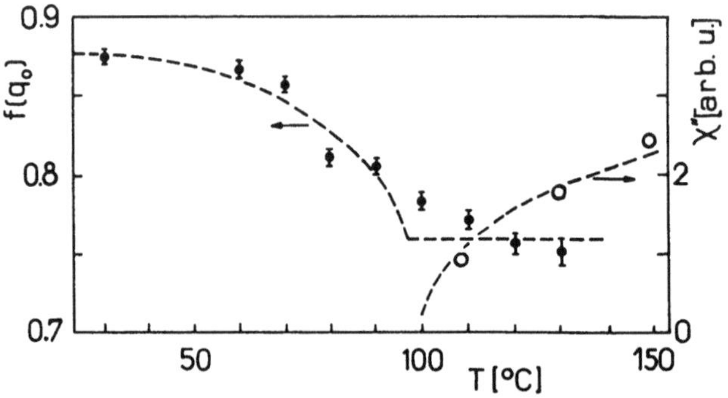

Fig. 2. The temperature variation of the non-ergodicity parameter (left scale) and of the minimum of the susceptibility between the two relaxation steps (right scale). The parameter $f(q_0)$ has been determined from the plateau following the first relaxation step, c.f. Fig. 1. The susceptibility data (open circles) rather indicate a qualitative trend since their determination is not fully independent of the way of matching the TOF and NSE parts of the spectra, as discussed in /9/. For the purpose of qualitative comparison, the dashed lines illustrate the predicted type of behaviour (3) with the square root singularities identifying feasible values of T_0. The horizontal section of the left hand dashed line above T_0 corresponds to the amplitude $A(q_0)$ of the second relaxation step, predicted to be temperature independent, c.f. (1).

$$f(q) = A(q) + h(q)\sqrt{T_0 - T} \ , \qquad T < T_0$$
$$\chi''_{min} \propto \sqrt{T - T_0} \ , \qquad T > T_0 \qquad (3)$$

where h(q) is independent of the temperature. The uncertainties of the experimental data (Fig. 2) do not allow us to confirm these relations quantitatively, but assuming their validity (illustrated by the dashed lines in Fig. 2) we can conclude that, indeed, various observations tend to point to a common $T_0 \simeq 95 \pm 5°$ C. Note that it has been shown elsewhere /9/ that with the choice $T_0 = 93°$ C the predicted fundamental scaling relation between the temperature dependence of the α relaxation and the first relaxation step is also fulfilled.

Thus at this point we can conclude that in our sample the existence of a critical dynamic instability of the liquid phase is substanciated by various features of the evolution of the density correlations, c.f. (1) to (3), as theoretically predicted. The critical temperature T_0 lies significantly above the conventional glass transition temperature $T_g \simeq 60°$ C, showing that the critical instability of the liquid phase does not coincide with the glass transition, and actually the singularity is masked by the onset of theoretically not yet satisfactorily understood proceses near T_0 /10/ (c.f. the lack of any fully developed singularity in the shear viscosity /11/). It is also worth mentioning that no signature of T_0 as found here has been observed in the $q \simeq 0$ dynamics /12/. This is not a contradiction, since theoretically the wavenumber range around q_0 has been found to be the most relevant /6/.

3. Wavenumber dependence of the dynamics

The nature of density correlations can be further clarified by investigating the q dependence of the parameters appearing in (1). By inspection of the TOF spectra

/5/ it becomes clear that there is a basically temperature independent phonon-like scattering contribution which has little relevance to the glass transition. It leads to $f(q)<1$ even at temperatures well below the transition regime. This is indicated by the 25° C data in Fig. 3, where $f(q)$ has been determined by the flat (horizontal) section of the relevant curves in Fig. 1 and of additional similar ones not shown in the figure. This 25° C curve will be considered as the base line in evaluating changes in the glass transition range. (The oscillatory character of the 25° C data is somewhat distorted since in the constant scattering angle TOF and NSE scans the higher energy phonon like processes do not correspond to the elastic nominal q value of the abscissa. On the other hand, the scattering contributions apparently related to the transition involve smaller energy transfers and therefore they correspond well to the q values plotted.) The lower dashed curve in Fig. 3. shows $A(q)$ i.e. the "critical" values of $f(q)$ obtained by extrapolation to T_0 from above. The difference can be identified with the amplitude $h(q)$ of the first relaxation step. In agreement with theory /7/ we find that $h(q)$ is smoothly oscillatory with an absolute minimum at the peak of the static structure factor $S(q)$, schematically illustrated by the bottom line in Fig. 3.

The q dependence of the relaxation time τ has been discussed elsewhere /4/, while the remaining parameter in (1), the Kohlrausch exponent β shows no q dependence beyond our experimental uncertainty of about ± 0.1. This is shown in the special plot in Fig. 4. in which the Kohlrausch function is a straight line with slope β. Again, this observation is in agreement with theoretical predictions, although from this point of view the Kohlrausch law is just a convenient approximation of the line shape. On the other hand, the apparent q and temperature independence of β in the range of our study (c.f. Figs. 1 and 3) is at variance with $q \simeq 0$ light scattering results /12/. As pointed out above, this need not to be a contradiction in view of the lack of a detailed extension of the theory to $q \simeq 0$.

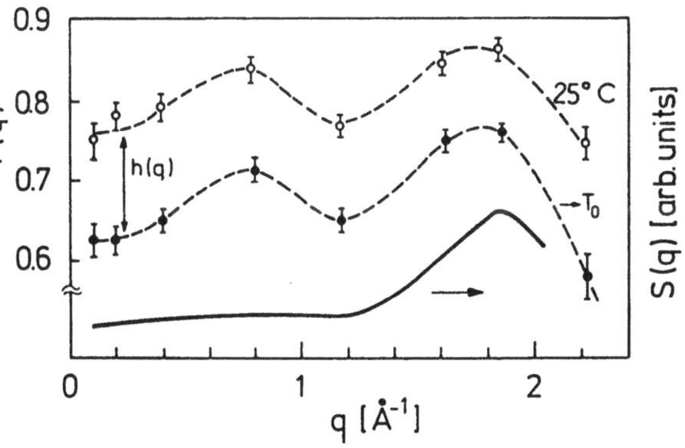

Fig. 3. The wavenumber dependence of the experimentally determined non-ergodicity parameter at room temperature (open circles) and at the temperature T_0 (filled circles). The latter data points have been obtained by extrapolating to T_0 from above and they represent the amplitude of the second relaxation step $A(q)$, c.f. (1). The difference between these two sets of data corresponds to the amplitute $h(q)$ of the first relaxation step at T_0. The continuous line is indicative of the shape of the neutron scattering structure factor with the characteristic peak at $q_0 \simeq 1.9$ Å^{-1}.

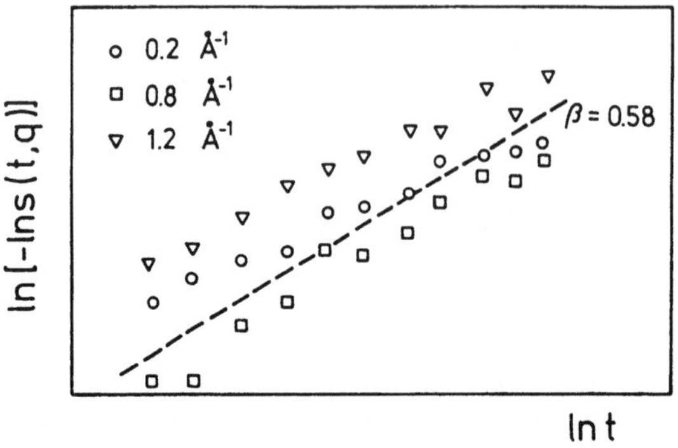

Fig. 4. Approximate wavenumber independence of the shape of the second (α) relaxation step. In this special plot of the normalized spectra $s(q,t)= = \phi(q/t)/A(q)$ the stretched exponential (Kohlrausch) line-shape correponds to a straight line with slope β, c.f. (1). The horizontal and vertical scales are drawn equal, thus the simple exponential decay gives a straight line at 45°. The dashed line indicates the slope of the best fit to the $q=q_0$ data represented by dashed lines in Fig. 1.

4. Conclusion

Neutron scattering investigation of the density-density correlation function in an archetypical fragile glassforming system revealed an impressive list of qualitative agreements with recent theories predicting a critical instability of the liquid phase against a dynamic ergodic to non-ergodic phase transition. Items on this list are the occurance of two relaxation steps at temperatures in the vicinity of the transition, the power law divergence of the characteristic times of these relaxation steps at a critical temperature T_0, a sharp drop of the non-ergodicity parameter at T_0 on approaching from below, a similar drop of the minimum of the susceptibility between the two relaxation steps at the same temperature on approaching from above, finite critical values of the non-ergodicity at T_0 (thus, in contrast to spin glasses, the non-ergodicity order parameter tends to jump at the transition from 0 to finite values, in resemblence to first order phase transitions, c.f. /13,14/), the slight q dependence of the relative amplitude of the first relaxation step with an absolute minimum at the peak of the structure factor and an approximately stretched exponential, wavenumber independent primary relaxation lineshape unchanged over a wide range of temperature.

On this basis it is certainly fair to say that we have convincing experimantal evidence at our disposal for an intrinsic dynamic instability of the liquid phase in our sample with a critical temperature $T_0 \approx 95°$ C, considerably above the conventional glass transition temperature $T_g \simeq 60°$ C. It is therefore also clear that the transition a T_0 is not the glass transition itself. Apparently, the divergence at T_0 is suppressed by additional, non-hydrodynamic processes becoming dominant near T_0 (such as thermally activated jump diffusion /10/). Presumably, these processes are dominating the whole temperature range between T_0 and T_g, and ultimately determine the character of the ordinary glass transition at T_g, which in contrast appears to be a gradual, thermally activated phenomenon. This temperature range is not well understood as yet. Furthermore, one could speculate that the key of this glass transition scenario remains the critical instability at T_0 which leads to a singular, much faster than Arrhenius type slowing down of the density fluctuations in the supercooled liquid. With the intrinsic time scale of relaxations becoming very long near T_0, more complex, "weaker", possibly thermally activated fluctuations become dominant and they delay the transition until T_g is attained, also modifying the apparent character of the transition in the process.

1. S.A. Edwards and P.W. Anderson, J. Phys. F **5**, 965 (1975)
2. T. Geszti, J. Phys. C **16**, 5805 (1983); E. Leutheusser, Phys. Rev. A **29**, 2765 (1984); U. Bengtzelius, W. Götze and A. Sjölander, J. Phys. C **17**, 5915 (1984); S.P. Das, G.F. Mazenko, S. Ramaswamy and F.F. Toner, Phys. Rev. Lett. **54**, 118 (1985)
3. F. Mezei, W. Knaak and B. Farago, Phys. Rev. Lett. **58**, 571 (1987)
4. F. Mezei, W. Knaak and B. Farago, Phys. Scripta, **T19**, 363 (1987)
5. W. Knaak, F. Mezei and B. Farago, Europhys. Lett. **7**, 529 (1988)
6. W. Götze, Z. Phys. B **56**, 139 (1984); ibid **60**, 195 (1985)
7. W. Götze, in this volume
8. P. Schofield, Phys. Rev. Lett. **4**, 239 (1960)
9. F. Mezei, Proceedings in Physics Series (Springer Verlag, Heidelberg) in press
10. W. Götze and L. Sjögren, J. Phys. C **21**, 3407 (1988)
11. G.M. Glover and A. Matheson, Trans. Farad. Soc. **67**, 1960 (1971)
12. C.A. Angell and L.M. Torell, J. Chem Phys. **78**, 937 (1983); L.M. Torell, in this volume
13. F. Mezei, J. Magn. Magn. Mat. **31-34**, 1327 (1983)
14. W. Götze and R. Haussmann, Z. Phys B **72**, 403 (1988)

Part III

Low Frequency Excitations

Low Frequency Excitations in Glasses

U. Buchenau

Institut für Festkörperforschung, KFA Jülich, Postfach 1913,
D-5170 Jülich, Fed. Rep. of Germany

The low temperature glass anomalies seem to be due to vibrational, relaxational and (at very low temperature) tunneling modes which coexist with the sound waves below 1 THz. The eigenvector of the vibrational and relaxational modes can be studied by coherent inelastic neutron scattering. One finds tetrahedra rocking modes in vitreous silica and bond bending modes in amorphous germanium, modes which in the crystal lie at about 2 THz. The softening of these modes in disordered surroundings in explained in terms of a "squeezed-atom" instability which occurs if an atom is compressed between two opposite nearest neighbours. Possible implications of this concept for glasses and liquids are discussed.

1. INTRODUCTION

The thermal and transport properties of crystals at low temperatures are reasonably well understood. The basis of that understanding is the description of the low frequency excitations in terms of sound waves. In a perfect crystal, this description is complete because there are no additional low frequency excitations (by low frequencies we denote frequencies below 1 THz which are low compared to the majority of the vibrational frequencies of the solid).

The situation is different in crystals with defects and even more so in amorphous solids. There one has not only sound waves, but a whole variety of other low frequency excitations, both harmonic and strongly anharmonic ones, which give rise to anomalous low temperature properties /1/. The physical origin of these low frequency excitations is not yet clear, except in cases of crystals with a low concentration of well-defined defects /2/. In the last few years, however, there has also been some progress towards a better understanding of the additional low frequency modes in amorphous solids, essentially stimulated by neutron scattering measurements /3-6/. As we will see, the neutron results seem to support the hypothesis of a common origin of the low frequency excitations in glasses /7,8/.

The present article begins in Section II with a short summary of earlier results on low frequency excitations in amorphous solids. Section III describes some of the neutron work. The final section IV discusses two theoretical concepts for the explanation of the additional low frequency modes and a possible mechanism of mode softening by strains.

2. CLASSIFICATION OF LOW FREQUENCY EXCITATIONS

From the accumulated experimental evidence /1,9/ on glassy low temperature anomalies, one has to distinguish three main groups of excitations which coexist with the sound waves at low frequencies, namely

- two-level or tunneling states
- structural relaxations
- additional harmonic vibrations

which we discuss in this order in the present section. The first two groups are

Springer Proceedings in Physics, Vol. 37 Dynamics of Disordered Materials
Editors: D. Richter · A.J. Dianoux · W. Petry · J. Teixeira

obviously strongly anharmonic excitations which suggest the existence of double minimum potentials in some appropriate configurational coordinate. In this aspect they differ markedly from the third group. Nevertheless we will see experimental evidence for a common origin of the second and the third group in Sec. III.

2.1 Two level states

Let us discuss first the two-level or tunneling states. Though the number of modes in this first group of low frequency excitations in glasses is fairly small (1 ppm of the total number of vibrational states) they are by far the most conspicuous. The reason for this is their low frequency. The seem to have a constant density of states below 100 GHz and give consequently rise to a linear term in the specific heat /1/ which begins to surpass the T^3 Debye contribution of the sound waves around 1 K. They have been very heavily studied in the last fifteen years. Since there are excellent reviews on this topic /1,9,10/ we will only summarize the results. A consistent description of nearly all experiments has been possible in terms of the tunneling model /11,12/. The tunneling model proposes that some atoms or some groups of atoms can occupy one of two (or more) potential minima. At low temperatures the quantum-mechanical tunneling of the atoms from one minimum to another gives rise to a very small energy splitting (of the order of 1 K). Assuming a slowly varying distribution functions of the energy barriers between the minima and the energy differences between adjacent minima (the asymmetry of the double minimum potential) one can explain a wide variety of experimental results. However, though the validity of the tunneling models is well established, it has not been possible so far to go beyond and to analyze the microscopic motion in detail. There are several models, but no generally accepted one /10/. The question wether the density of two level states depends on the glass temperature has given rise to a controversial discussion /13,14/.

2.2 Relaxations

The existence of double minimum potentials with relatively low potential barriers within the glass structre implies not only tunneling states at low temperatures, but also structural relaxation at a somewhat higher temperature. Indeed, structural relaxation has been inferred from accoustic attenuation measurements between 10 and 200 K in many glasses /9/. The best studied case is vitreous silica, where mechanical and dielectric relaxation have been measured in a wide frequency range /15/. One finds a broad relaxation maximum as a function of temperature which shifts from 30 K in the kHz range to about 100 K at 10 GHz. An Arrhenius plot of the frequency dependence yields an average activation energy around 500 K. Since the maximum is fairly broad, one has to assume a broad distribution of barrier heights around this average value.

It is tempting to assume a common origin for both relaxation and tunneling states. This is indeed a very plausible hypothesis /7,8/. One should note, however, that the tunneling of a single oxygen atom over a small distance of 0.05 nm passing a barrier of 500 K occurs much too slowly to explain the observed tunnel splittings around 1 K. Therefore the tunneling states in such a picture would have to bedue to the low barrier part of the distribution /16/.

2.3 HARMONIC LOW FREQUENCY VIBRATIONS

As far as the harmonic vibrations in glasses are concerned, vitreous silica is again by far the most heavily studied example. We begin by discussing the sound wave dispersion. Sound waves are well defined harmonic vibrations in vitreous silica up to about 200 GHz. A linear relation between frequency and wavevector has been measured even farther /17/ up to 500 GHz. On the other hand, the heat transport at low temperatures seems to be restricted to phonons below 200 GHz /18, 19/. Above this frequency, the mean free path decreases dramatically /20/. This decrease cannot be explained by the scattering from the static disorder in the glass, which at these long wavelengths (100 Å) should be relatively weak /18,19/. Thus one has to look for another explanation for the plateau in the thermal conductivity found at about 10 K. It should be mentioned that this plateau in the thermal conductivity has been generally observed in all known amorphous solids.

One possible explanation for the plateau is the resonant interaction of the sound waves with additional localized vibrational states which by chance happen to have a low frequency. This explanation is supported by the specific heat above and around 5 K, which is markedly higher than the pure Debye term calculated from the known sound velocities. The evidence is not fully conclusive; at 20 K also crystalline quartz surpasses the pure Debye heat capacity because of its low-lying TA modes at 2 THz /21/. Cristobalite, another crystalline form of silicon dioxide has even a heat capacity which is fairly similar for $T > 5$ K to that of vitreous silica, probably because in that substance the TA modes are even lower /22/. Nevertheless, though specific heat and thermal conductivity around 10 K do not give an unambiguous picture, they are certainly compatible with a pronounced low-frequency tail in the frequency distribution of short wavelength modes which extends down into the sound wave regime.

3. NEUTRON SCATTERING FROM LOW-FREQUENCY EXCITATIONS IN GLASSES

For a quantitative analysis of the vibrational density of states of disordered solids, neutron scattering is by far the most powerful spectroscopic technique. Each harmonic vibration, sound wave or otherwise, is seen with an intensity which oscillates around a common $Q^2 \cdot \exp(-2W)$ behaviour /23,24,25/ (Q is the momentum transfer of the scattering process and exp (-2W) is the Debye-Waller factor). For a sound wave, the oscillations occur in phase with the oscillations in S(Q) (the Fourier transform of the pair correlation function), because in a sound wave the atomic pair correlations remain essentially conserved during the motion. In some cases like metallic glasses /3/ amorphous polyethylene /6/ or vitreous selenium /26/, the dynamic structur factor of the additional low frequency vibrations does also oscillate in phase with the oscillations in S(Q). Following Suck et al /3/, one can understand this behaviour in terms of an atomic motion which closely resembles the motion of the transverse modes of the corresponding crystals. In other cases like vitreous silica /4/ or amorphous germanium /5/, which will be discussed below, one does find oscillations in the dynamic structure factor which are not in phase with the oscillations in S(Q). However, we will see that the explanation in terms of atomic motions leads again to a picture which is fundamentally consistent with the earlier conclusions /3/.

If the inelastic neutron scattering intensity does indeed reproduce the peaks in the Q-dependence of S(Q), there is a convenient recipe /25/ to evaluate the vibrational density of states g(ν) on an absolute scale using the equation

$$I^{(1)}(Q,\nu)=(k_f/k_i)(hQ^2/4\pi M_{av})\cdot f_B\cdot g(\nu)(\Delta\nu/\nu)\cdot(1/2)\int_{-1}^{+1} dz I_0(Q+q_\nu\cdot z) \quad (1)$$

Here $I^{(1)}(Q,\nu)$ is the inelastic scattering intensity at momentum transfer Q and energy transfer $h\nu$ (restricted to one-quantum processes) k_f and k_i are final and initial neutron wavevector, M_{av} is the average atomic mass, f_B is the Bose factor and $\Delta\nu$ is the frequency width covered by the measurement. $I_0(Q)$ is the integrated elastic intensity measured over the same solid angle as $I^{(1)}(Q,\nu)$ and q_ν is the average wavevector of the vibrational modes. The usefulness of eq. (1) is not restricted to glasses; it is generally applicable to disordered solids and has been used to study vibrations in porous solids. For $q_\nu = 0$, it transforms into the incoherent approximation.

The main disadvantage of the neutron scattering technique as compared to specific heat measurements is its limited sensitivity at high resolution. Though one has the backscattering and spin-echo technique capable to resolve energy differences corresponding to less than 0.1 K, these techniques are completely unable to detect a fraction of 1 ppm of the vibrational density of states. Therefore no successful neutron scattering investigation of the two level states has been possible. On the other hand, neutrons have two main advantages. The first is the possibility to get information about the motional eigenvector as indicated above and as will be seen again in the following examples. In this respect, neutrons are unique. The second advantage, which neutrons share with other spectroscopic techniques like infrared absorption and Raman scattering, is the possibility to check the harmonicity of the observed excitations. Thus one can distinguish between harmonic vibrations and structural relaxation, as will be seen for vitreous silica below.

The first detailed study of the dynamic structure factor in amorphous solids was done by Suck et al /3/. Fig. 1 shows their results for a metallic glass consisting of 70% Mg and 30% Zn at a frequency of 1.3 THz. The inelastic intensities are compared to the theoretical expectations of eq. (1) using the S(Q,0) determined in the same experiment. This kind of comparison, though not treating the multiphonon scattering correctly, does already show the good general validity of eq. (1) and supports the interpretation of transverse modes given by the authors /3/.

The second example is vitreous silica, the most heavily studied glass /1,9/. In vitreous silica, one finds pronounced oscillations around the Q^2e^{-2W} behaviour in the whole frequency range from 300 GHz to 2 THz. Since the curves at different frequencies can be scaled to a common master curve, there seems to be a predominant eigenvector in this whole frequency region. This eigenvector is not a sound wave eigenvector because the peak positions are different from those in S(Q) (Fig. 2).

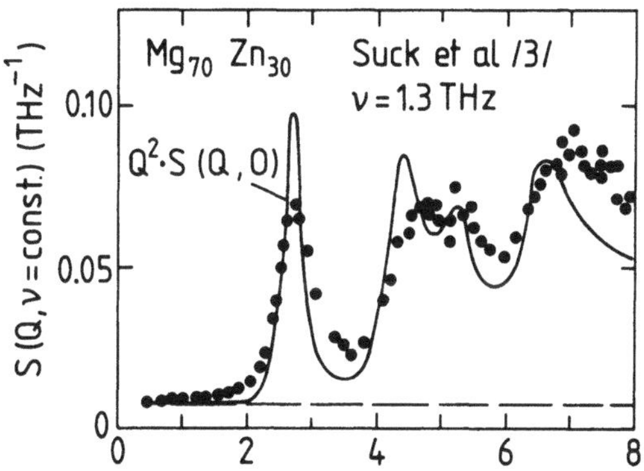

Fig. 1 Dynamical structure factor of a metallic glass compared to $Q^2 \cdot S(Q,0)$ (properly scaled).

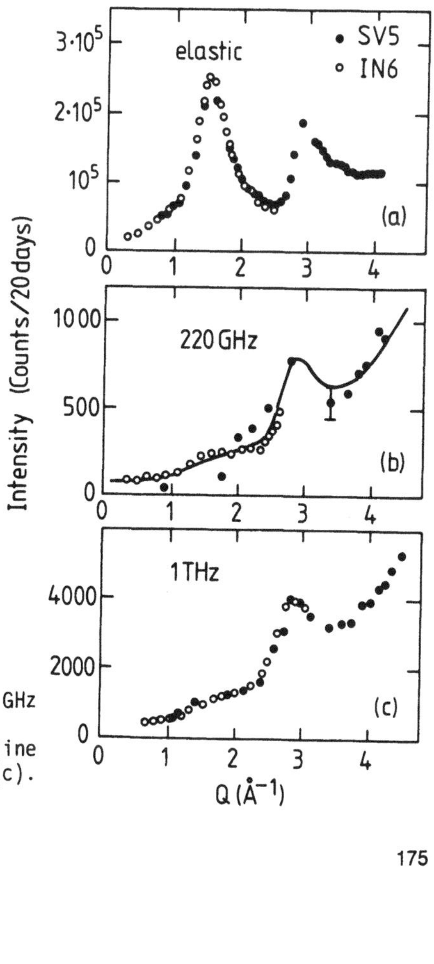

Fig. 2 Momentum-transfer dependence of neutron-scattering intensities in vitreous silica at 300 K: (a) elastic, (b) inelastic at 220 GHz (from 150 to 290 GHz), and (c) at 1 THz (from 0.7 to 1.3 THz). The line in (b) is a fit to the points in (c).

A reasonable fit can be obtained from a model of coupled rotation of five bonded SiO_4 tetrahedra (see insert of Fig. 4). The picture is suggested by the study of the low frequency modes in crystalline quartz /27,28/ or cristobalite /22/, where below 12 THz the modes are either sound waves or involve the rotation of relatively undistorted tetrahedra. In particular, the soft mode at the α-β-transformation at 846 K /29/ is of this type /30/.

The investigation of the temperature dependence (Fig. 3) of the inelastic intensities reveals a nonharmonic contribution below 500 GHz which can be quantitatively explained in terms of classical structural relaxation in an asymmetric well (Fig. 4). The fit uses the barrier distribution determined from the ultrasonic attenuation data /15/ discussed in the preceding section. It even allows an estimate of the jump width of the atoms in the structural relaxation.

An upper boundary of 0.08 nm for the jump width is given by the fact that the dynamic structure factor remains unchanged as compared to the dynamic structure factor of the vibrations (see Fig. 2). Numerical simulations of correlated rotational SiO_4 tetrahedra jumps of the kind shown in the insert of Fig. 4 predict marked changes in the dynamical structure factor for higher jump widths, which are not observed. On the other hand, the observed intensities can only be explained within a reasonably low density of relaxators (~0.01 per atom) for big enough jump widths (~0.05 nm for the oxygen atoms of the central tetrahedron in Fig. 4).

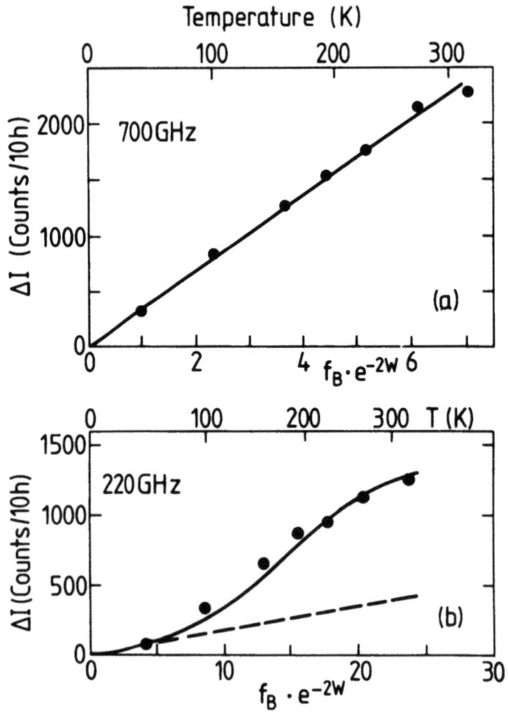

Fig. 3

Temperature dependence of inelastic neutron scattering intensities in vitreous silica at (a) 700 GHz, and (b) 220 GHz (see text). The dashed line in (b) denotes the harmonic contribution.

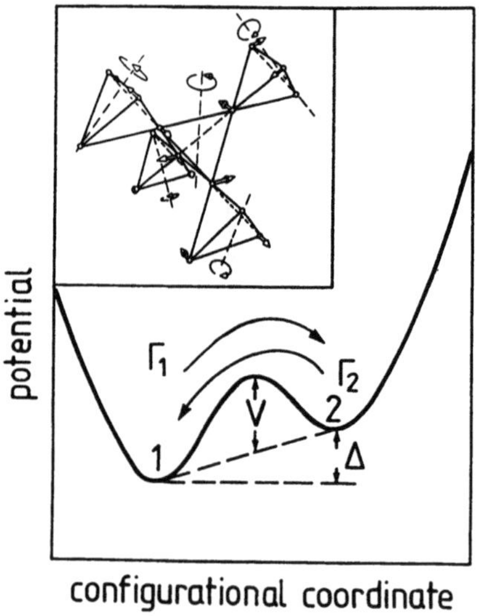

Fig. 4

Asymmetric double-minimum potential. Inset: Model of coupled rotational motion of five tetrahedra as a possible configuration coordinate in vitreous silica.

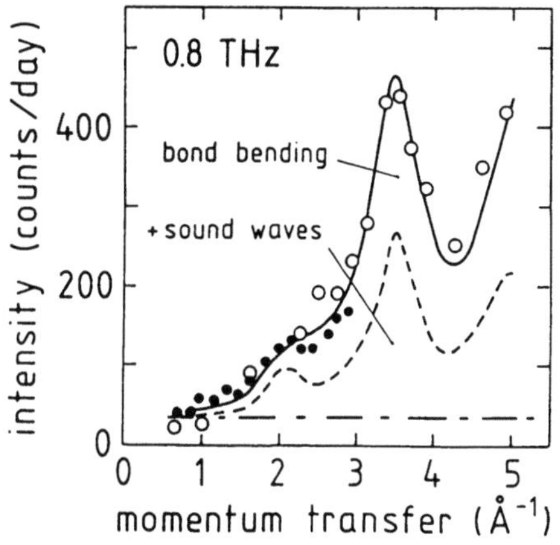

Fig. 5

Inelastic neutron scattering intensities at room temperature in amorphous germanium /5/ as a function of momentum transfer measured on the time of flight spectrometers SV22 (o) and IN6 (•) calibrated to the SV22 scale (calibration factor 1/98) between 0.5 and 1.1 THz. Dash-dotted line: multiple scattering background; dashed line: calculated sound wave contribution; continuous line: sound waves plus a contribution with the measured structure factor of the bond bending modes.

The most important conclusion from the neutron results in vitreous silica is the coexistence of sound waves, additional soft vibrations and structural relaxation below 1 THz. The additional soft vibrations and the structural relaxation have the same dynamic structure factor and must be assumed to have a similar motional eigenvector in an atomic scale, namely a coupled rotation of more or less undistorted SiO_4 tetrahedra.

The last example to be discussed is amorphous germanium /5/, an amorphous solid with comparatively weak low temperature anomalies /31-33/. This is indeed reflected in the neutron results, which at the lowest measured frequency of 300 GHz show only sound waves with no clearly discernible additional vibrational density of states. At a higher frequency of 800 GHz, however, there is an additional density of states as shown in Fig. 5.

From their dynamical structure factor, these modes could be identified as bond bending modes similar to the transverse zone boundary modes in crystalline germanium around 2 THz /34/. Fig. 6 compares the vibrational density of states of amorphous germanium calculated on the basis of the motional model in Fig. 5 to the crystalline density of states /35/. In order to focus on the low frequencies, the density of states divided by frequency squared is plotted, which yields a constant value in the sound wave region of the crystal. Apart from the broadening of sharp features, the main difference between crystalline and amorphous solid is the pronounced tail of the high wave vector TA modes to lower frequencies. The absolute number of these low-frequency tail states is small and would be just barely visible in a plot of the density of states itself, but it is comparable to the number of sound waves at the same frequency (the dashed line in Fig. 6).

The corresponding plot of measured heat capacities /31, 36/ divided by temperature cubed against temperature is shown in Fig. 7. The lines are calculated from the neutron results in Fig. 6. The agreement is fairly good even in the amorphous case where the neutron results were obtained at room temperature and admittedly contain more assumptions and insecurities as in the crystalline case. Thus the picture extracted from the neutron data is confirmed by the measured heat capacity data /31/. There is a small deviation at temperatures below 10 K, where the neutron results predit a slightly higher specific heat. In view of this deviation, one should investigate the temperature dependence of the scattering intensities at the lowest frequencies in more detail in order to see whether one has a small relaxational contribution as in vitreous silica. Though the additional vibrations below 1 THz are much less numerous in a-Ge than in vitreous silica, the main conclusion of a low frequency tail of the low-lying high wavevector modes is the same. Both substances are exceptional in the sense that the character of

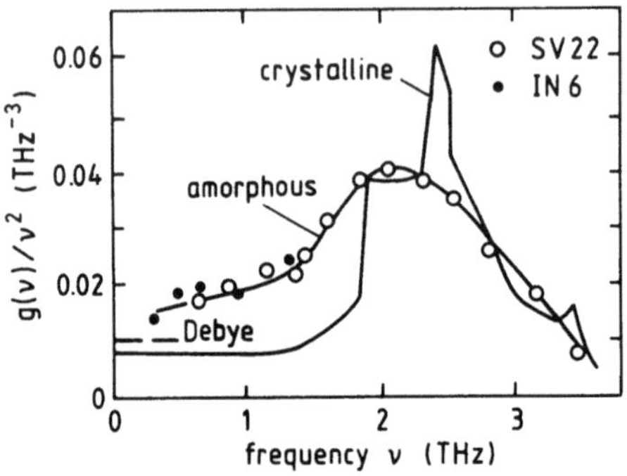

Fig. 6

Vibrational density of states g(ν) over frequency squared as obtained from neutron scattering measurements in crystalline germanium /35/ and amorphous germanium /5/. The points are fits at the corresponding frequencies, the line is an independent global fit. o SV22, • IN6.

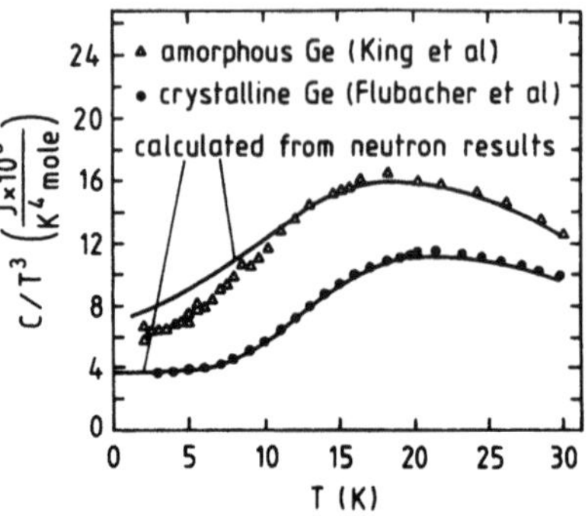

Fig. 7

Specific heat divided by T cubed against temperature T in crystalline (•) /36/ and amorphous germanium (Δ) /31/. The lines are calculated from the neutron results in Fig. 6.

the vibrational eigenvectors changes in the crystalline substance as one proceeds along the sound wave dispersion toward higher frequencies. The difference in eigenvectors leads to a difference in dynamic structure factors which allows to distinguish sound waves and higher wave vector modes, in the present case bond bending modes. This is different in the metallic glass case /3/; but though we cannot distinguish as easily additional modes and sound waves in the metallic glass, the underlying physical picture of a low frequency tail of the first maximum in the vibrational density of states is again the same, independent on the motional eigenvector of the modes in this maximum, be it transverse zone boundary modes, tetrahedra rocking or bond bending.

4. THEORETICAL CONCEPTS

If one wants to explain a low frequency mode in a solid, there are two principally different ways. The first follows a hydrodynamics approach, which yields vanishing frequencies in the long wavelength limit if the mode requires the transport of a conserved physical quantity like momentum or energy. The sound waves belong to this class of so-called Goldstone modes. A recent attempt to explain the additional vibrations in glasses along that line of thought is the so-called fracton concept /37, 38/ which is also discussed at this conference from the more general point of view of the possible vibrational modes of a tenuous structure /39/. Its application to the additional vibrations in glasses requires a marked enhancement of the vibrational density of states at the phonon-fracton crossover. However, a very high resolution neutron investigation of the crossover region in silica aerogels failed to detect such an enhancement /40/. Moreover, the approach fails to explain the dynamic structure factors observed in vitreous silica and amorphous germanium. At the crossover, the average wavelength of the fractons must be decidedly larger than the atomic distances. Else the whole fracton concept becomes meaningless because no further scaling of distances is possible. Therefore the fracton concept predicts of necessity a dynamic sound wave structure factor for the additional vibrations, which is not observed in vitreous silica /4/ and in amorphous germanium /5/.

The same objection must be raised against other explanations based on long-wavelength modes like softening of transverse phonons /41/ or weak localization of phonons /42/.

The second possible way to explain a low frequency mode in the solid is the concept of a resonant mode /2/, which becomes more and more localized as its frequency decreases. Such resonant modes have been studied in crystals with substitutional defects; they result either from a heavier mass (mass defect) or from softer force constants. An example for a mass defect (Ge in Al) is presented at this conference /43/. For glasses, the resonant mode concept cannot be based on mass defects, but has to rely on the local variation of force constants. For low enough restoring force, the local motion decouples from the motion of the neighbouring atoms and becomes a resonant mode of a small group of atoms. In principle, the localization is never complete because there is always the interaction with the sound waves even at the lowest frequencies. That is why these modes are called resonant rather than localized. For still stronger softening, the total restoring force of the mode becomes negative and a local structural instability is created. This marks the transition from the harmonic resonant modes to tunneling and structural relaxation.

The concept is strongly supported by the neutron results described in Section 3. It explains qualitatively a low frequency tail of the first maximum in the vibrational density of states as found in the neutron measurements. It further explains the occurence of harmonic low frequency vibrations and structural relaxation with the same dynamic structure factor as observed in vitreous silica /4/.

The central question to be answered is: how do these modes come down in frequency by a factor of two or more? In terms of restoring forces, this requires a softening of the corresponding force constants by more than a factor of four. From there, it is only a small step to a negative restoring force as required to explain structural relaxations and tunneling states. A possible answer to this question is mode softening by strains. The concept of strains in a glass has been discussed by Egami et al /44/. In the strained glass, neighbouring atoms will be often pressed against each other. The resulting repulsive forces give rise to a negative force constant for a relative motion of the atoms perpendicular to their line of connection. Fig. 8 shows as an example the case of three atoms in a line pressed against each other by the surroundings. If the negative restoring force at the central atom overcompensates the positive restoring forces from the other surrounding atoms, a structural instability is created. This "squeezed-atom-instability" has been estimated to have a reasonably low energy of formation in vitreous silica /8/ and in simple monoatomic liquids /45/.

Qualitatively, the concept of mode softening by strains is also able to explain the formation of amorphous clusters at a critical strain in irradiated crystals /46, 47/ and the occurrence of glassy anomalies in mixed crystals like mixed cyanides /48/, pressure-quenched Al-Si /49,50/ or others /51/. Again as in glasses one would have to expect motional eigenvectors of resonant modes which change from system to system, but which within a given system are common to resonant, tunneling and relaxational modes. Such a picture has been indeed proposed for the mixed cyanides /52/. If such a close parallel between crystalline and glassy systems does indeed exist, one gets the great advantage of being able to study glassy phenomena with scattering methods in single crystals. First attempts in this direction have been reported /53, 54/.

Finally the connections between the low frequency excitations in glasses and the dynamics around the glass transition /55, 56/ should be discussed. The neutron scattering experiment in several glasses /57, 58, 51, 26/ shows an anomalous decrease of the elastic intensity with temperature and a simultaneous increase of the inelastic intensities in the frequency region below 1 THz. Some features of these changes are well explained /59/ within the predictions of the mode coupling approximation /60, 61/. From the experiments at lower temperatures described above, most of these modes are reasonably harmonic modes in the glass. At and above the glass transition, these modes seem to experience a strong additional softening which cannot

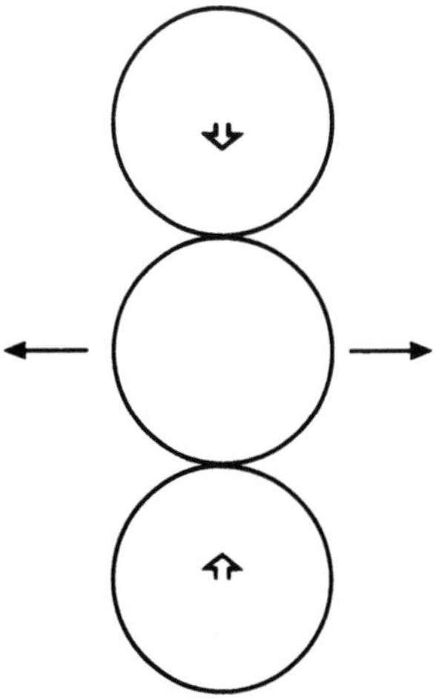

Fig. 8

Structural instability of a central atom squeezed between two opposite nearest neighbours /45/.

be explained solely by the stronger thermal expansion of the liquid /26/. For the special case of polymers, the question is discussed in more detail by Frick in this volume /55/.

1. Amorphous Solids-Low Temperature Properties, (ed. by W.A. Phillips), Topics in Current Physics, Vol. 24 (Springer-Verlag, Berlin 1981)
2. P.H. Dederichs and R. Zeller: In Point Defects in Metals II, Springer Tracts in Modern Physics, Vol. 87 (Springer, Berlin 1980) p. 1 ff
   V. Narayanamurti, R.O. Pohl: Rev.Mod.Phys. 42, 201 (1970)
3. J.-B. Suck, A. Rudin: In Glassy Metals II, ed by H. Beck and H.J. Güntherodt (Springer, Berlin 1983) p. 217
   J.-B. Suck, H. Rudin, H.J. Güntherodt and H. Beck: Phys. Rev. Lett. 50, 49 (1983)
4. U. Buchenau, H.M. Zhou, J. Nücker, K.S. Gilroy and W.A. Phillips: In Phys. Rev. Lett., 60 (1988) 1318 and further references therein
5. U. Buchenau, M. Prager, W.A. Kamitakahara, H.R. Shanks and N. Nücker: Europhys. Lett. 6, (1988) 695
6. U. Buchenau, M. Monkenbusch, M. Stamm, C.F. Majkrzak and N. Nücker: In Polymer Motion in Dense Systems, ed. by D. Richter and T. Springer, Springer Proc. in Physics Vol. 29 (Springer Berlin 1988) p. 138
7. F.N. Ignatiev, V.G. Karpov and M.I. Klinger: J. Non-Cryst. Solids 55, 307 (1983)
8. U. Buchenau: Solid State Commun. 56, 889 (1985)
9. S. Hunklinger and W. Arnold: In Physical Acoustics Vol. XII ed. W.P. Mason and R.N. Thurston (Academic Press, New York 1976)
10. W.A. Phillips: Rep. Prog. Phys. 50, 1657 (1987)
11. W.A. Phillips: J. Low Temp. Phys. 7, 351 (1972)
12. P.W. Anderson, B.I. Halperin and C.M. Varma: Phil. Mag. 25, 1 (1972)
13. U. Reichert, M. Schmidt, S. Hunklinger: Solid State Commun. 57, 315 (1986)
14. J.F. Berret, M. Meißner: Z. Phys. B-Condensed Matter 70, 65 (1988)
15. S. Hunklinger: Proc. Ultrasonic Symp. (IEEE) 1974, 443 (1974)
16. J. Jäckle (private communication)
17. M Rothenfusser, W. Dietsche and H. Kinder: Phys. Rev. B27, 5196 (1983)
18. M.P. Zaitlin and A.C. Anderson: Phys. Rev. B12, 4475 (1975)
19. J. Jäckle: In The Physics of Non-Crystalline Solids, ed. by G.H. Frischat (Transtech, Aedermanndorf 1977) p. 568
20. W. Dietsche and H. Kinder: Phys. Rev. Lett. 43, 1413 (1979)

21. R.O. Pohl: In Ref. /1/, p. 27
22. N. Ahmad: Thesis, Cambridge (1987)
23. J.M. Carpenter and C.A. Pelizzari: Phys. Rev. B12, 2391 (1975)
24. K. Froböse and J. Jäckle: J. Phys. F7, 2331 (1977)
25. U. Buchenau: Z. Phys. B-Condensed Matter 58, 181 (1985)
26. W.A. Phillips, N. Nücker, U. Buchenau: unpublished
27. N.M. Elcombe: Proc. Phys. Soc. London 91, 947 (1967)
28. H.K. Barron, C.C. Huang and A. Pasternak: J. Phys. C9, 3925 (1976)
29. J.D. Axe and G. Shirane: Phys. Rev. B1, 342 (1970)
30. H. Grimm and B. Dorner: J. Phys. Chem. Solids 36, 407 (1975)
31. C.N. King, W.A. Phillips and J.P. De Neufville: Phys. Rev. Lett. 32 (1974) 538
32. R. van den Berg and H. van Löhneysen: Phys. Rev. Lett. 55 (1985) 2463
33. J.E. Graebner and L.C. Allen: Phys. Rev. B29 (1984)5626
34. W. Weber: Phys. Rev. B15 (1977) 4789
35. G. Nelin and G. Nilsson: Phys. Rev B 5 (1972) 3151
36. P. Flubacher, A.J. Leadbetter and J.A. Morrison: Philos Mag., 4 (1959) 273
37. R. Orbach: Science 231 (1986) 814
38. A.J. Dianoux, J.N. Page and H.M. Rosenberg: Phys. Rev. Lett. 58, 886 (1987)
39. O. Entin-Wohlman: (this volume)
40. H. Conrad, J. Fricke, G. Reichenauer, R. Schätzler: this volume
41. D.P. Jones, J. Jäckle and W.A. Phillips: In Phonon Scattering in Condensed Matter, ed. by H.J. Maris (Springer, Berlin 1981) p. 65
42. E. Akkermans and R. Maynard: Phys. Rev. B32, 7850 (1985)
43. A.I. Kolesnikov, M. Prager, U. Buchenau, O.I. Barkalov, I.T. Belash, E.G. Ponyatovskii: this volume
44. T. Egami, K. Maeda and V. Vitek: Phil. Mag. A41, 883 (1980)
45. U. Buchenau (unpublished)
46. C. Laermans and A. Vanelstraete: Phys. Rev. B35, 6399 (1987) and further references therin
47. A. Seidel, S. Massing, B. Strehlau and G. Linker: Phys. Rev. B38, 2273 (1988)
48. J.J. de Yoreo, W. Knaak, M. Meissner and R.O. Pohl: Phys. Rev. B34, 8828 (1986)
49. J. Chevrier, J.B. Suck, J.J. Capponi and M. Perroux: Phys. Rev. Lett. 61, 554 (1988)
50. J. Chevrier, J.B. Suck, J.C. Lasjaunias, F. Zougmere and J.J. Capponi: E-MRS Meeting (Strasbourg, June 1988)
51. D.A. Ackerman, D. May, R.C. Potter and A.C. Anderson: Phys. Rev. B23, 3886 (1981)
52. E.R. Grannan, M. Randeria and J.P. Sethna: Phys. Rev. Lett. 60, 1402 (1988)
53. H. Grimm: this volume
54. P. Wochner, E. Burkel, J. Peisl, C.M.E. Zeyen, W. Petry: this volume
55. B. Frick: this volume
56. W. Götze: this volume
57. B. Frick, D. Richter, W. Petry, U. Buchenau: Z.Phys. B-Condensed Matter 70, 73 (1988)
58. F. Fujara, W. Petry: Europhys. Lett. 4, 921 (1987)
59. W. Knaak: this volume
60. E. Leutheusser: Phys. Rev. A29, 2765 (1984)
61. U. Bengtzelius, W. Götze and A. Sjölander: J.Phys. C17, 5915 (1984)

Structural Relaxation in a Metallic Glass

J.-B. Suck

Kernforschungszentrum Karlsruhe,
Institut für Nukleare Festkörperphysik, Postfach 3640,
D-7500 Karlsruhe, Fed. Rep. of Germany

The influence of structural relaxation on the atomic dynamics of the metallic glass $Ni_{33.3}Zr_{66.7}$ was investigated at room temperature using cold neutron inelastic scattering (IN6, ILL). Changes of the atomic dynamics of the same kind as observed after crystallization but nearly an order of magnitude weaker were found. A narrowing of the generalized vibrational density of states on structural relaxation is observed for the first time.

1. Introduction

Glasses are far from thermal equilibrium. They therefore try to adjust their frozen-in structure towards a structure corresponding to their actual temperature. This relaxation rate is normally very slow compared to experimental times and one therefore has to add energy to the system to accelerate this structural relaxation (SR), e.g. by heating the glass well below its glass temperature T_G or its crystallization temperature T_X. Remeasuring the physical properties of the relaxed glass one finds nearly all of them changed in comparison with the properties of the as-quenched glass [1]. Quite a few of these properties, measured with integral methods, are relate to interactions with atomic vibrations [1]. The aim of the present investigation is therefore to study the influence of SR on the atomic dynamics of a metallic glass and, in a second step, to correlate quantitatively the changes observed with the results obtained by integral methods. Here first results are presented from an investigation of the dependence of the generalized vibrational density of states (GVDOS) on annealing of the metallic glass $Ni_{33.3}Zr_{66.7}$ well below the crystallization temperature. In contrast to $Mg_{70}Zn_{30}$, which has been investigated before [2] this glass is very well suited for such a study because T_X >> room temperature ($T_X \approx 660$ K) and several other physical properties have been investigated for NiZr glasses before and after SR.

2. Sample Treatment, Experiment and Data Evaluation

The melt spun sample material was divided into three samples which were measured at room temperature before and after heat treatment on the cold neutron time focussing time-of-flight spectrometer IN6 at the HFR of the ILL in Grenoble, using an incident energy of 4.8 meV and scattering angles between 10.7 and 114.3 degrees, thus covering momentum transfers of the elastically scattered neutrons from 3 to 25.5 nm^{-1}. As the energy gain spectra were used, the momentum transfers for higher energy transfers extended out to considerably higher values. Thus a deficit in sampling volume in reciprocal space is only expected for energy transfers below 10 meV. This changes the absolute intensity of the GVDOS but not the differences which are discussed here. The three samples were remeasured under the same experimental conditions after their heat treatment: glass 1: 450 K ($\sim$0.68 T_X) for 155 min., glass 2: 580 K ($\sim$0.85 T_X) for 150 min. and sample 3 after crystallization at 750 K. As demonstrated in Fig. 1 for the glass sample 2 before and after heat treatment and sample 3 after crystallization no trace of a crystalline peak can be observed in the smooth static structure factor of the metallic glass, which was annealed strongest.

Springer Proceedings in Physics, Vol. 37 Dynamics of Disordered Materials
Editors: D. Richter · A.J. Dianoux · W. Petry · J. Teixeira

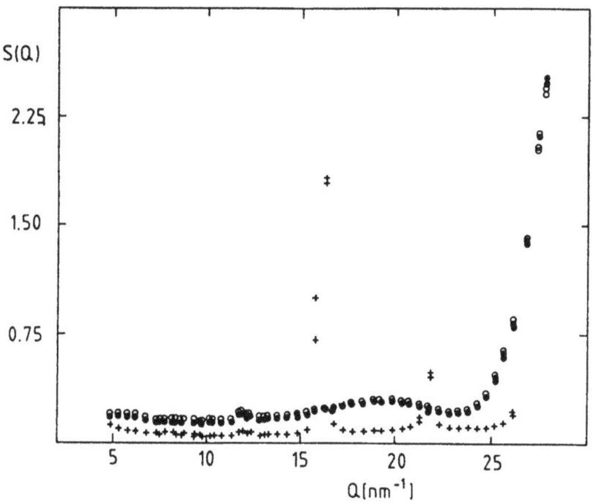

Fig. 1: Static structure factors as determined from the integrated t-o-f spectra of $Ni_{33.3}Zr_{66.7}$ glass (sample 2) before (o) and after (•) heat treatment and of the crystallized sample 3 (+)

As irreversible relaxation processes related to changes in the topological short range order (TSRO) were investigated here, a new freshly quenched sample was used for each relaxation step.

$Ni_{33.3}Zr_{66.7}$ is a binary, predominantly coherently scattering glass. Therefore the GVDOS was determined from the sum of all t-o-f spectra [3]. The GVDOS is the weighted sum of the partial density of states of each element in the sample [4]. The normalized weighting factors in this case are 0.675 for Ni and 0.325 for Zr. The GVDOS were normalized to 1 with a cut-off frequency of 32 meV. The resolution of the spectrometer at focussing energy ($\hbar\omega$=0) was 125 µeV. The inelastic spectra could be separated from the foot of the elastic peak slightly below 0.8 meV. The gap between 0 and 0.8 meV was bridged by fitting a Debye spectrum to the data at 0.8 meV. All corrections were applied as described elsewhere [5], except those for multiple scattering and the complicated resolution function of the time focussing spectrometer. Multiphonon contributions were corrected iteratively using the harmonic and incoherent approximation in the phonon expansion of the inelastic scattering cross-section up to the 4th phonon term until self consistency was obtained.

3. Results

Since the relaxational situation of the as-quenched samples is unknown the GVDOS of all samples before the heat treatment was determined. Figure 2 shows the resulting GVDOS of glass 2 and 3 divided by the GVDOS of glass 1.

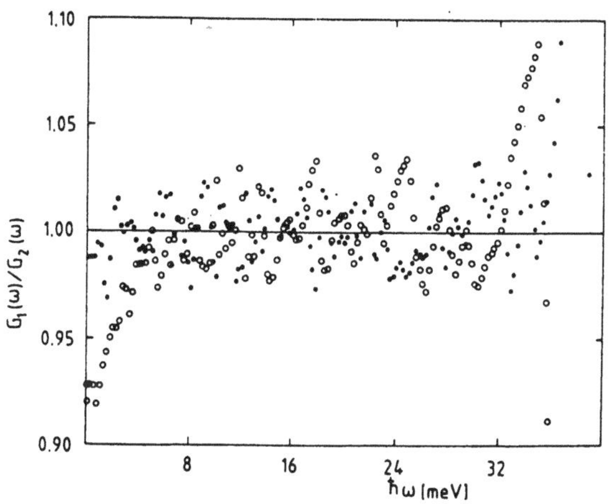

Fig. 2: Ratio of the GVDOS of two of the as-quenched metallic glass $Ni_{33.3}Zr_{66.7}$ samples divided by the GVDOS of the third. $G_2(\omega)/G_1(\omega)$ (o) and $G_3(\omega)/G_1(\omega)$ (•) show that only sample 1 and 3 were in the same relaxational state before heat treatment.

While G(ω) of glass 1 and 3 are equal within experimental errors (the ratio fluctuates by ±1.5% in the energy region $\hbar\omega < 32$ meV of the GVDOS), glass 2 has already less low energy modes (LEM) in the as-quenched state than G(ω) of the two other glasses. In the following therefore only the GVDOS of the *same* sample before and after the heat treatment are compared. It is however assumed that the generalized phonon density of state (GPDOS) of $NiZr_2$ would have been the same irrespective which of the 3 glassy samples would have been crystallized.

As the GVDOS of the metallic glasses and the GPDOS of the crystallized sample 3 are only slightly more structured but otherwise very similar to those determined for $Ni_{24}Zr_{76}$ [6] and these G(ω) are still convoluted with and therefore strongly influenced at energies above 15→20 meV by the resolution function of the time focussing spectrometer, the *differences* of the GVDOS of glass 2 before and after annealing and the GPDOS of the crystallized sample 3 are shown in Fig. 3 as one example. The positive parts of this difference between 0 and 19 meV and 29 and 35 meV are due to additional excitations with these energies in the GVDOS of the glass. Conversely, the phonons grouped in the peak of the GPDOS of crystalline $NiZr_2$ at 26 meV appear as a negative peak in Fig. 3.

Finally the difference of the GVDOS of glass 2 before and after annealing at 580 K are shown in Fig. 4 as one example. The positive parts at the extrema of the difference spectrum are due to the vibrational modes which are lost during the annealing process, while the negative parts are modes which are added in the GVDOS.

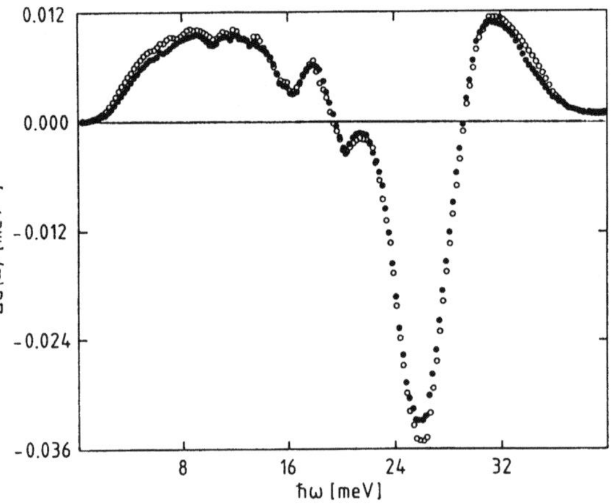

Fig. 3: Differences of the GVDOS of the metallic glass (sample 2) $Ni_{33.3}Zr_{66.7}$ before (o) and after annealing (•) (155 min at 580 K) and of the GPDOS of the crystallized sample 3 ($NiZr_2$). Positive part: additional modes in the GVDOS of the glass of low (LEM) and high energies.

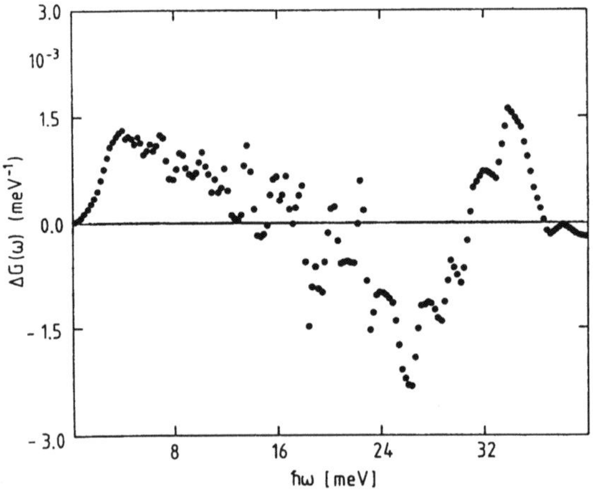

Fig. 4: Difference spectrum $\Delta G(\omega) = G(\omega)_{as\text{-}qu.} - G(\omega)_{ann.}$ of the GVDOS of the as-quenched metallic glass (sample 2) $Ni_{33.3}Zr_{66.7}$ and of the same glass after heat treatment at 580 K (=0.85 T_x) for 150 min.

4. Discussion

Figure 1 demonstrates that the effects of SR on the atomic dynamics of the metallic glass are intrinsic and not due to partial crystallization of the samples. The increase of intensity near the maximum (not covered here) of the first peak in S(Q) after annealing is due to a sharpening of this first peak as a consequence of the structural relaxation and has been observed before [7]. Connected with this is a sharpening of the first peak of the pair distribution function g(r) which was interpreted as a reduction of extreme (short or long) distances of atoms in the nearest neighbour shell around a given atom.

Figure 4 most clearly shows the influence of SR on the atomic dynamics of a metallic glass: the decrease of intensity due to low and high energy modes in the glass and an increase of the most pronounced modes in the GPDOS of the crystallized sample near 26 meV. Very similar results but less pronounced, especially in the negative part of the difference spectrum, were obtained after heat treatment at 450 K (glass 1). In general, SR in a metallic glass *changes the GVDOS in a similar manner as does crystallization, however the effects are much smaller* (here about a factor of 8) than observed after crystallization. There are, however, differences in the shape of the difference spectra: the positive parts of $\Delta G(\omega)$ in Fig. 4 have their extrema at the edges of the distribution, while the maxima appear near the middle of the positive parts of $\Delta G(\omega)$ in Fig. 3. This effect is also seen very clearly in Fig. 3: The strongest changes on structural relaxation are observed at lowest and highest energy transfers. *The GVDOS narrows during structural relaxation. The modes with lowest and highest energy are annealed out first and most effectively*. Such a behaviour can be interpreted in a simple free volume model: the strongest density *fluctuations* are annealed out first, leaving the mean density nearly unchanged ($\Delta\rho$ is typically of the order of 10^{-3} after SR). A similar interpretation of the narrowing of the frequency distribution in terms of annihilation of atomic level stresses, here positive and negative hydrostatic pressure, can be given also [2]. But most likely both interpretations apply only to part of the effects observed, since these mechanisms should mainly affect the longitudinal modes in the glass. SR, however, changes most drastically the LEM, and these are believed to be dominated by transverse type of modes [4]. In fact an increase of the shear modulus has been observed after structural relaxation of the metallic glass $Ni_{24}Zr_{76}$ [8] which supports the idea that an appreciable part of the modes affected by structural relaxation are of predominantly transverse type.

I would like to thank Professor H.-J. Güntherodt (University of Basel) for providing me with the sample material and P. Reimann for the melt spinning of the metallic glass.

References

1. e.g. T. Egami: In Amorphous Metals and Semiconductors, ed. by P. Haasen and R.I. Jaffee (Pergamon Press, Oxford 1986) p. 222
2. J.-B. Suck, H. Rudin, H.-J. Güntherodt, H. Beck: J. Non-cryst. Sol. 61+62, 295 (1984)
3. V.S. Oskotskii: Sov. Phys. - Solid State 9, 289 (1974)
4. J.-B. Suck, H. Rudin: In Glassy Metals II, ed. by H. Beck and H.-J. Güntherodt, Topics in Applied Physics, Vol. 53 (Springer 1983) p.217
5. J.-B. Suck, H. Rudin, H.-J. Güntherodt, H. Beck: J. Phys. C: Solid State Phys. 14, 2305 (1981)
6. J.-B. Suck, H. Rudin, H.-J. Güntherodt: In Rapidly Quenched Metals, ed. by S. Steeb and H. Warlimont (Elsevier Science Publishers 1985) p. 471
7. Y. Waseda, T. Egami: J. Mat. Sci. 14, 1249 (1979)
8. E. Balanzat: J. Physique Lett. 45, L639 (1984)

Low-Energy Excitations and Vibrational Modes in Amorphous Zr-Cu Alloys

F. Zougmoré, J.C. Lasjaunias, and O. Béthoux

Centre de Recherches sur les Très Basses Températures,
C.N.R.S., B.P. 166 X, F-38042 Grenoble Cedex, France

The low-temperature specific heat of a series of Zr-Cu amorphous (sputtered) alloys has been investigated. In comparison to samples prepared by the melt-spinning technique, densities of states of both low-energy excitations (TLS) and vibrational modes (in their regime $g(\omega) \propto \omega^2$) are largely enhanced.

1. Introduction

Amorphous materials are characterized by typical properties of the low-energy/frequency vibrational excitations at the origin of two features in the low-temperature specific heat : a large excess from the Debye prediction in the liquid-helium range, characteristic of amorphous insulators, and an anomaly below ~ 1 K ascribed to tunneling or two-level states (TLS), which exist in all amorphous (metallic and insulating) materials [1,2]. Up to now, the direct relation between the two kinds of excitations is a matter of debate. This has been recently suggested in the special case of vitreous silica from low-frequency inelastic neutron scattering [3], a technique which directly pointed out the existence of low-frequency vibrational modes specific of the amorphous state, for energy above a few meV [3,4]. We have investigated here both the "lattice" (vibrational) and TLS contributions to the low-temperature specific heat of a series of sputtered Zr-Cu alloys, and compared to those of corresponding less-disordered liquid-quenched (L.Q.) alloys. We have also studied the role of structural relaxation in the *amorphous state* on these two contributions.

2. Characterization

The amorphous $Zr_{100-x}Cu_x$ ($19 \leq x \leq 64$ at.%) samples have been prepared by the sputtering technique (a "vapor-quenching" technique) with a deposition temperature of 77 K. This method leads to a more disordered structural state than the liquid-quenching (L.Q.) one (in general melt-spinning), as verified by numerous fine structural investigations (see in [5] and references cited therein). However, all characterization data such as electrical resistivity, superconducting transition T_c, glass-transition T_G and recrystallization T_χ temperatures, and density (see Fig. 1) are in excellent agreement with the corresponding L.Q. alloys [5,6]. For $x \geq 30$ at.%, T_G increases monotonously with x.

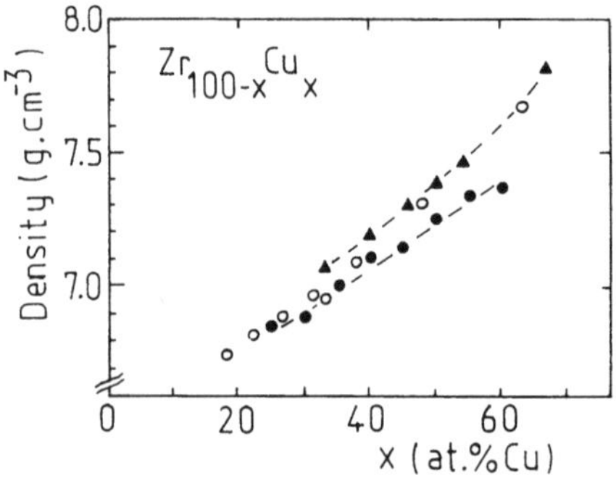

Fig. 1 : Density of Zr-Cu alloys. Sputtered samples : (o). Melt-spun samples : (●) ref. 8, (▲) ref. 9. The agreement between the three series of alloys is better than 2 %.

Springer Proceedings in Physics, Vol. 37 **Dynamics of Disordered Materials**
Editors: D. Richter · A.J. Dianoux · W. Petry · J. Teixeira

About 1 g of sample was used for the specific heat (C_p) measurements, by means of a transient heat-pulse technique [6,7] in the T-range from 0.1 to 7 K. Samples were firstly measured in their "as-prepared" state, after a stay of a few hours at 300 K, and then after a thermal treatment at temperatures T_a (200-250°C) well below T_χ or T_G [5,6] in order to ensure a relaxation of the amorphous structure.

3. Results and analysis

The superconducting transition ($1 \lesssim T_c \lesssim 4$ K for $48 \leq$ [Zr] ≤ 81 at.%) divides the T-range in two zones :

(i) Above T_c, C_p is well fitted by $\gamma T+\beta T^3$, sum of electronic and lattice βT^3 terms (see Fig. 2). Similar good fits are obtained for L.Q. Zr-Cu [11,12], and more generally for metallic glasses. First, we notice that such a T^3 Debye-like variation is at variance with insulating glasses. That means that the vibrational spectrum can be analysed as $g(\omega)\propto\omega^2$ in the corresponding frequency range, although it is not excluded that it can involve modes in excess to propagative phonon modes [10]. Anyway, we will use the Debye-like temperature θ_D deduced from β [$\beta\propto 1/\theta_D{}^3$] for quantitative comparisons with the L.Q. Zr-Cu alloys analysed in the same way [11-13]. θ_D values are reported in Fig. 3, together with the corresponding melt-spun and recrystallized phases, from [12].

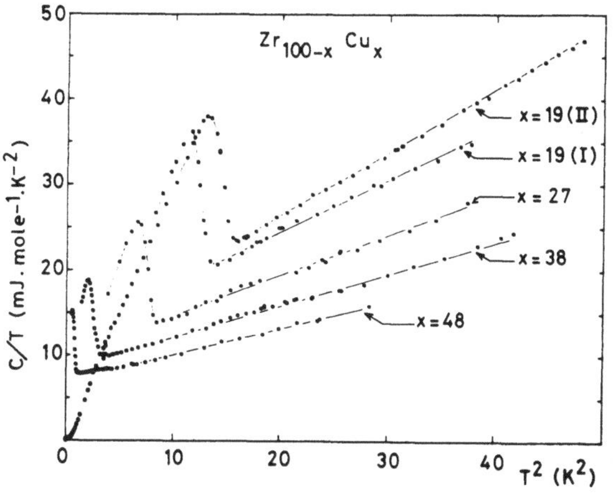

Fig. 2 : Specific heat of the sputtered samples between 1 and 7 K, as C/T versus T^2. (All data, except for x = 19 (sample II), correspond to annealed samples).

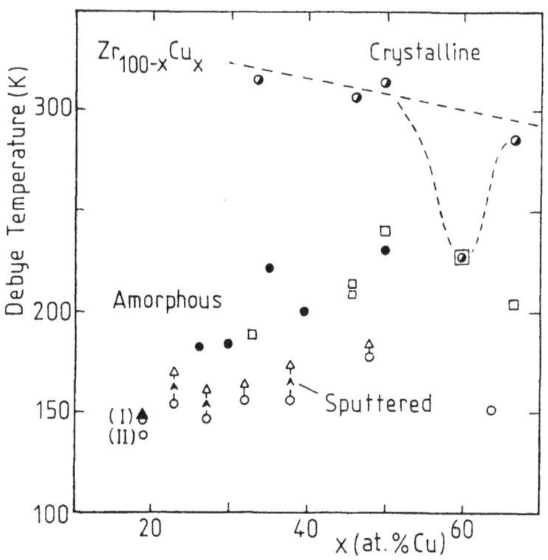

Fig. 3 : Debye temperature (from the βT^3 contribution). Crystalline samples : (◐) ref. 12 - Amorphous, melt-spun : (●) ref. 11 , (□) ref. 12 - Amorphous, sputtered : (o) as-prepared, (Δ) annealed, (▲) highly relaxed : $Zr_{81}Cu_{19}$(I).
Note the drop of θ_D for cryst. $Zr_{40}Cu_{60}$ ascribed to the complexity of the Zr_7Cu_{10} phase (ref. 12).

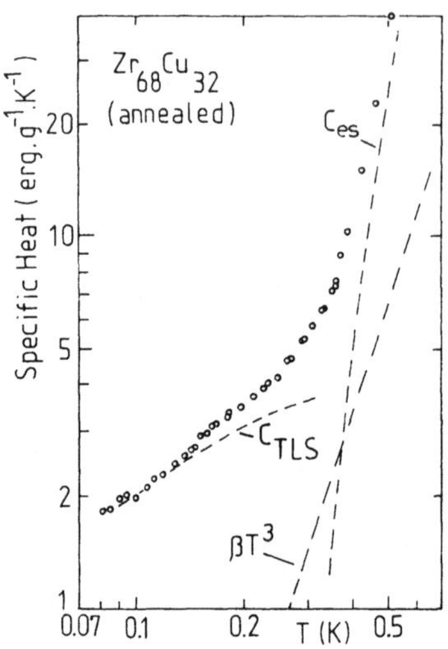

Fig. 4

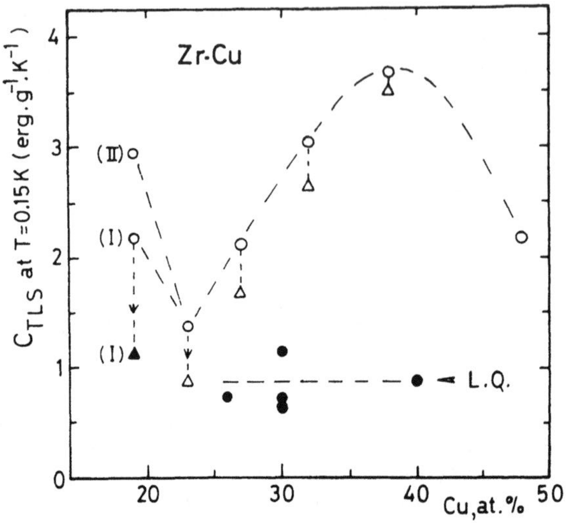

Fig. 5

Fig. 4 : Low-T specific heat, below 0.5 K, of a sputtered alloy. The different contributions are indicated by the dashed lines : lattice (βT^3), superconducting electronic (C_{es}) and T.L.S.

Fig. 5 : Amplitude of the TLS specific heat at T = 0.15 K for sputtered samples successively in the as-prepared (o) and annealed (Δ) state, or highly relaxed (▲ : annealing + ageing at 300 K for $Zr_{81}Cu_{19}$-I). Liquid-quenched (melt-spun) alloys (●) from ref. [14].

(ii) Well below T_c, when the electronic condensation process is achieved (exponential decay of C_{es}), i.e. below 0.3-0.5 K, two contributions remain : that of lattice and the TLS [see a typical decomposition of total C_p in Fig. 4]. At temperature below about 0.2 K, the lattice term also becomes negligible compared to the TLS contribution, which varies smoothly with T ($\sim T^{\alpha}$ for $T \lesssim 0.25$ K, with $\alpha < 1$ [5]), so that the amplitude of this contribution is defined with a good precision around T = 0.1 K. The results (at T = 0.15 K) are reported in Fig. 5, together with the only other reported data for L.Q. alloys from [14]. Note that the origin in TLS excitations for this low-T residual contribution is clearly established, in particular by its good quantitative correlation with the T^2 regime of the thermal conductivity [15], also a characteristic of these tunneling states.

4. Discussion

We first consider the θ_D values (Fig. 3). The general property of softening of the transverse acoustic phonons when going from crystalline to amorphous state is well demonstrated here by a large decrease of θ_D [16] (at the exception of the $Zr_{40}Cu_{60}$ composition where the L.Q. and crystalline values are similar). Moreover, there is a further decrease by about 20 % when going from L.Q. to sputtered amorphous alloys ; this is obviously related to the increasing structural disorder, in particular on the short-range atomic scale, as proved by the broadening of the first X-ray diffraction halo. The overall variation of θ_D with composition is preserved in these two series of amorphous alloys, with a maximum for $Zr_{50}Cu_{50}$. At variance with the L.Q. alloys where no systematic variation of θ_D is detected [13], a *systematic* hardening of the structure is obtained by thermal treatment in the sputtered, corresponding to an increase of θ_D by about 10-15 K (for $19 \leq x \leq 48$ at.%).

We consider now the TLS excitations, which correspond to configurational tunneling excitations intrinsic to the structural disorder ; their density of states is directly related to the specific heat signal. The first evidence from Fig. 5 is its larger amplitude for sputtered alloys in comparison to L.Q. which can reach a factor of 4 for x = 40 at.%. This is also a direct consequence of a more disordered structure. We observe a strong composition dependence, contrary to L.Q., with a local minimum at x = 23 at.%, which could be the counterpart of the local maximum of θ_D (annealed state).

Among several theoretical attempts for a structural origin of TLS, that one based on a free-volume model for the glass-transition [17] predicts an inverse relationship between the (constant) density P of TLS measured at low temperature, and T_G : $P \propto N_t T_G^{-1}$, with N_t the total density of TLS, which is assumed to depend weakly on materials. Untill now, this relation has only been verified for a few insulating glasses [18]. Our present results don't agree with this prediction since the variation of P with composition for x ≥ 30 at.% is non monotonous contrary to that of T_G [5], which, in addition, is almost similar to the value of L.Q. alloys. At this step, we can outline that these two kinds (vapor-quenched and liquid-quenched) of alloys correspond to "effective" quenching-rates for the atomic arrangements which differ by several orders of magnitude, and therefore are probably characterized by different amounts of frozen-in free volume, larger in V.Q. alloys [19]. In a more general approach than in ref. [17], the density P of TLS reflects these differences in the as-prepared states ($P \propto v_f$), and is in agreement with the subsequent reduction of free-volume upon structural relaxation for the sputtered samples.

Indeed, the second point is, like for θ_D and anew at variance with L.Q. alloys (where no detectable (< 20 %) change was detected [14]), the sensitivity to structural relaxation. The relative variation of C_{TLS} on annealing, which decreases for increasing x, we have related [5] to the decrease of the ratio kT_a/kT_G or kT_a/kT_χ (T_a : annealing temperature), in other words, to the decrease of the efficiency of the thermal treatment for the structural relaxation. Hence the amplitude of C_{TLS} is a very sensitive probe to the degree of structural relaxation (the behaviour of θ_D is different, which shows variations almost independent of the composition). At high Zr content, i.e. for larger T_a/T_χ, the TLS density of the relaxed state comes in rough agreement with the extrapolation of the constant L.Q. value : for the TLS "defects", both structures have probably the same degree of disorder.

The main conclusion of this study is the *simultaneous occurrence* of higher densities of states for both low-frequency vibrational and low-energy TLS excitations when increasing the degree of structural disorder, demonstrated here for the metallic Zr-Cu system. This correlation is also verified, but with less magnitude in the effects, by a structural relaxation obtained by thermal annealing of the highly disordered (sputtered) state. A precise structural origin, and eventually a structural connection as proposed in the covalent glass SiO_2 [3], for these two kinds of excitations is far from being understood in these metallic systems. In a very general statement for amorphous materials (insulating glasses, amorphous metallic alloys, ...) the occurrence of localized TLS is related to the softening of transverse acoustic phonons in comparison to the equilibrium crystalline phases. Moreover, this property has recently been extended to the case of metastable crystalline solid solutions Al-Si (6 at.%) where the "disorder" character appears to be due to the highly unstable electronic state of Si atoms (metallic against covalent) [20].

References

1. Amorphous Solids - Low Temperature Properties, ed. by W.A. Phillips (Springer-Verlag, Berlin, 1981)
2. Glassy Metals I and II, ed. by H.J. Güntherodt and H. Beck (Springer-Verlag, Berlin, 1981 and 1983)
3. U. Buchenau, N. Nücker and A.J. Dianoux, Phys. Rev. Lett. 53, 2316 (1984)
4. J.B. Suck and H. Rudin, in Glassy Metals-II, ibid. chap. 7
5. J.C. Lasjaunias, F. Zougmoré and O. Béthoux, Solid State Commun. 60, 35 (1986)

6. F. Zougmoré, J.C. Lasjaunias and O. Béthoux, to be published
7. A. Ravex, J.C. Lasjaunias and O. Béthoux, J. Phys. F : Met. Phys. 14, 329 (1984)
8. Z. Altounian and J.O. Strom-Olsen, Phys. Rev. B 27, 4149 (1983)
9. Y. Calvayrac, J.P. Chevalier, M. Harmelin, A. Quivy and J. Bigot, Phil. Mag. B 48, 328 (1983)
10. One needs systematic transverse sound velocity measurements to check this hypothesis. In vitreous silica, time-dependent specific heat experiments were analysed *below 1 K* including a T^3 contribution in excess to the Debye value (M.T. Loponen, R.C. Dynes, V. Narayanamurti and J.P. Garno, Phys. Rev. Lett. 45, 457 (1980)
11. K. Samwer and H.v. Löhneysen, Phys. Rev. B 26, 107 (1982)
12. P. Garoche and J. Bigot, Phys. Rev. B 28, 6886 (1983)
13. J. Tebbe and K. Samwer, Z. Phys. B 63, 163 (1986)
14. S. Grondey, H.v. Löhneysen, H.J. Schink and K. Samwer, Z. Phys. B 51, 287 (1983)
15. J.C. Lasjaunias, A. Ravex and O. Béthoux : in Phonon Scattering in Condensed Matter, Springer Ser. Solid-State Sci., Vol. 51 (Springer-Verlag, Berlin 1984)
16. θ_D is mainly determined by the transverse sound velocity ($\theta_D \propto \rho^{1/3} v_t$, ρ the mass density, and $v_t = \sqrt{\mu/\rho}$, μ the shear modulus)
17. M.H. Cohen and G.S. Grest, Solid State Commun. 39, 143 (1981).
18. U. Reichert, M. Schmidt and S. Hunklinger, Solid State Commun. 57, 315 (1986) and papers cited therein.
19. A. Van Den Beukel and S. Radelaar, Acta Metall. 31, 419 (1983). Note the insensitivity of T_G to the initial free-volume amount.
20. J. Chevrier, J.B. Suck, J.C. Lasjaunias, F. Zougmoré and J.J. Capponi, Proc. of Europ. M.R.S. Conf. (Spring 1988, Strasbourg), to be published in J. Less Common Metals.

Low Frequency Excitations in Al-Ge Alloys Quenched Under High Pressure

A.I. Kolesnikov[1;*], *M. Prager*[1], *H.R. Schober*[1], *U. Buchenau*[1], *O.I. Barkalov*[2], *I.T. Belash*[2], *and E.G. Ponyatovskii*[2]

[1]Institut für Festkörperforschung, KFA Jülich, D-5170 Jülich, Fed.Rep.of Germany

[2]Institute of Solid State Physics Ac. Sci. USSR, SU-142432 Chernogolovka, USSR

*Alexander von Humboldt Research Fellow, from the Institute of Solid State Physics Ac. Sci. USSR, SU-142432 Chernogolovka, USSR

The vibrational density of states of nonequilibrium Al-Ge alloys quenched under high pressure was studied by inelastic neutron scattering. The increase in the low frequency part of the spectrum as compared to pure Al is markedly stronger then that found recently in the Al-Si system. Partial annealing leads to a drastic reduction of the additional low frequency vibrational density of states even though most Ge atoms are still dissolved in the Al lattice.

1. INTRODUCTION

The origin of the low temperature anomalies in glasses /1/ is still an open question. One of the attempts to attack this question involves the study of mixed crystals which in some cases /2, 3/ exhibit glasslike low temperature anomalies. A very recent example of such a crystalline system with glassy anomalies is Al-Si, a nonequilibrium substitutional alloy, which was obtained after quenching from high to room temperature under high pressure /4, 5/. The system is appealing because of its simple fcc structure. It shows a glasslike linear term in the specific heat /5/ below 0.5 K (in the superconducting regime). Another glasslike feature /6-8/ is the softening of the lowest zone boundary mode /4/. One of the possible explanations of this feature is a softening and localization of high-wavevector modes by local strains /9/. In the present paper we report on experimental evidence in the system Al-Ge which is similar to the Al-Si system but has stronger effects because of the heavier mass of Ge.

2. EXPERIMENT

Two samples with 4 and 13 at. % of Ge in Al (solubility under normal conditions 2.6 % /10/) were quenched at pressures of 40 and 60 kbar, respectively, from 300°C to liquid nitrogen temperature. The pressure was released, but the samples kept cold in the metastable (M) state until the first measurement. The measurement was repeated after ten days of annealing at room temperature (intermediate (I)-state). Finally the samples were annealed to equilibrium ((E)-state) at 200°C for two hours and measured a third time.

The effect of these heat treatments can be judged from Figs. 1 and 2. Fig. 1 shows the decrease of the superconducting transition temperature with annealing /11, 12/. In the intermediate state, the transition temperature is reduced by nearly a factor of two. However, most of the Ge atoms are still dissolved in the lattice. Fig. 2 shows a comparison of the Debye-Scherrer reflections of the inter-

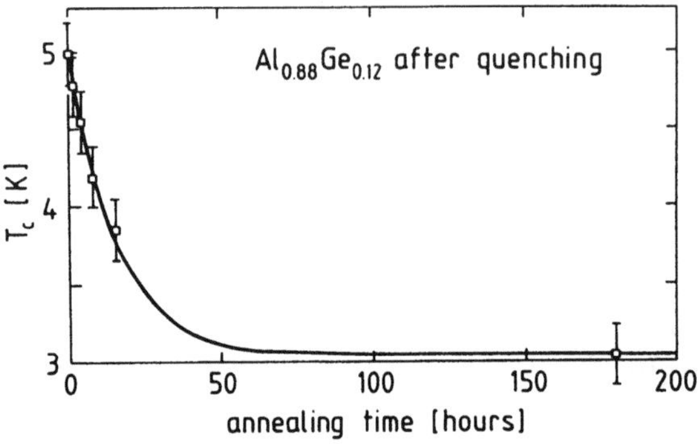

Fig. 1 Superconducting transition temperature vs. annealing time at room temperature for high-pressure quenched $Al_{0.88}\,Ge_{0.12}$ /12/.

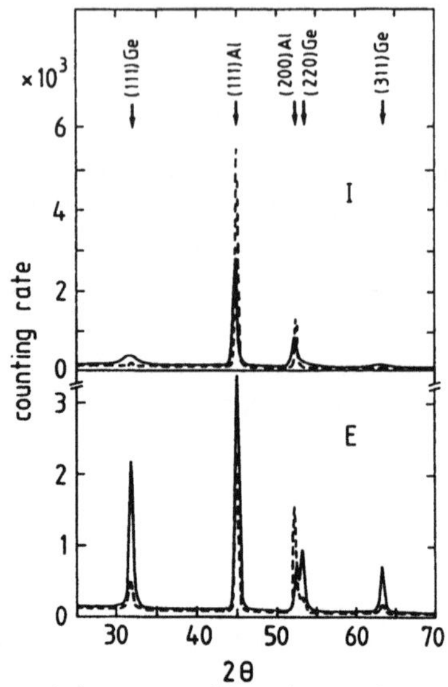

Fig. 2 Debye-Scherrer reflections of $Al_{0.87}\,Ge_{0.13}$ (—) and $Al_{0.96}Ge_{0.04}$(---) in the intermediate (upper part) and the equilibrium state (lower part).

mediate and the equilibrium state. In the intermediate state, there is a broad feature at the location of the Ge (111) reflection. Its integrated intensity is only one third of the final reflection in equilibrium.

The neutron measurements were done on the time-of-flight spectrometer SV22 at a thermal beam of the reactor DIDO at Jülich. The incident energy was 63 meV, thus ensuring the wide momentum transfer range necessary for the application of the incoherent approximation (both Al and Ge are coherent scatteres). All measurements were done at a sample temperature of 120 K. In addition to the alloys, a pure aluminium sample was measured. Each sample consisted of several pills of 12 mm diameter and 2.5 mm thickness.

3. RESULTS AND DISCUSSION

Fig. 3 a) and b) shows the measured intensities corrected for background, detector efficiency and spectrometer resolution summed over all detectors. The insert shows the region between 5 and 15 meV in energy loss of the neutron. The annealing behaviour of the additional intensity is clearly seen.

A generalized vibrational density of states $G(\omega)$ was calculated from the corrected data. Multiple and multiphonon corrections were estimated to be negligible. The density of states was normalized to one in the frequency range below 40 meV. The results are shown in Fig. 4 in the low frequency region including the transverse zone boundary modes around 20 meV. A comparison to pure Al is shown in Fig. 5. As seen from Fig. 4, statistically meaningful changes by the heat treatment appear only below 15 meV. In contrast to the findings in the Al-Si system, the additional excitations do not look like a low frequency tail of the transverse zone boundary modes, but rather like a broad peak at 10 meV. This is qualitatively explained in terms of the mass ratio of 2.69 between Ge and Al which tends to give rise to resonant heavy defect modes around 10 meV, similar

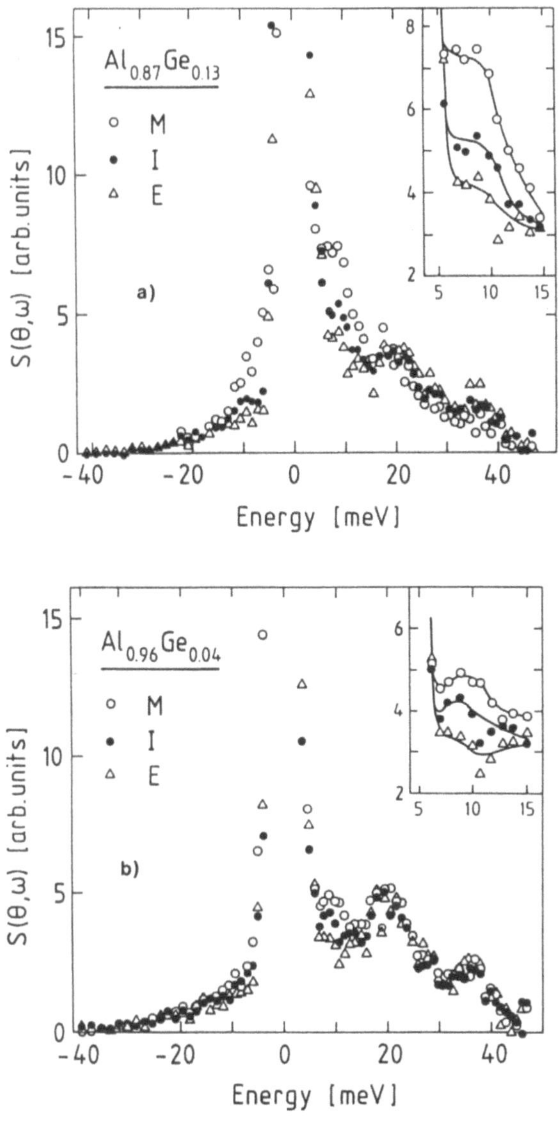

Corrected neutron scattering intensities $S(Q,\omega)$, from Al-Ge measured at 120K on the time of flight spectrometer SV22 at Jülich. M metastable, I intermediate and E equilibrium state as described in the text.
(a) $Al_{0.87}Ge_{0.13}$
(b) $Al_{0.96}Ge_{0.04}$

to those observed in the system Al-Ag /13/. Fig. 6 shows a theoretical estimate of the expected neutron intensity on the basis of a pure mass defect picture for the metastable alloy with a concentration of 13 % germanium. The calculation is based on a simple incoherent approximation scheme at constant momentum transfer and resolution. The semiquantitative agreement with the left side of Fig. 5 shows that one cannot argue for glassy features from that measurement alone.

This explanation, however, fails to account for the strong decrease of the 10 meV modes at partial annealing, where the majority of the Ge atoms stays dissolved in the lattice. Therefore this decrease as well as the strong decrease of the superconducting transition temperature in Fig. 1 could be possibly due

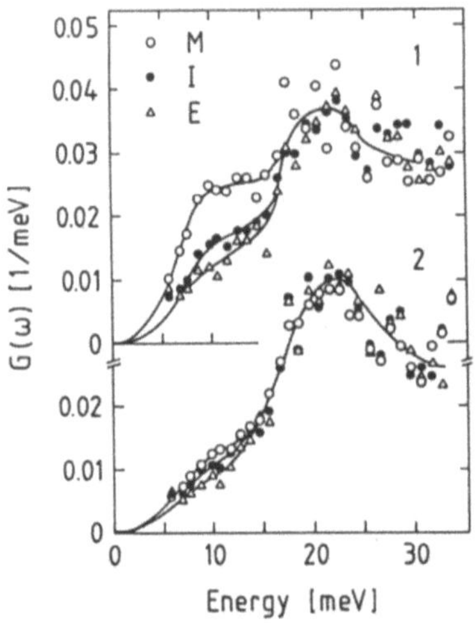

Fig. 4

Generalized vibrational density of states G (ω) vs. frequency. M metastable, I partially annealed intermediate and E equilibrium state.
Upper part $Al_{0.87}Ge_{0.13}$
lower part $Al_{0.96}Ge_{0.04}$

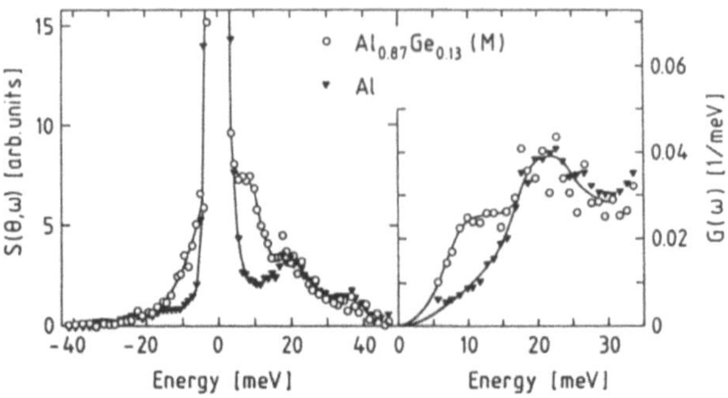

Fig. 5

Comparison of metastable $Al_{0.87}Ge_{0.13}$ and pure aluminium.
Left side: corrected neutron intensities vs. energy transfer, right side: generalized vibrational density of states.

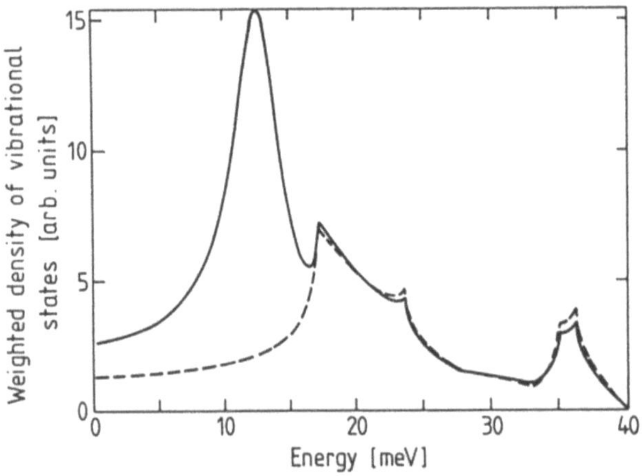

Fig. 6

Calculated density of states for $Al_{0.87}Ge_{0.13}$ weighted with the scattering lengths and the frequency and temperature factors for inelastic neutron scattering at 120 K.

to another mechanism. A possible mechanism is the relaxation of strains which must be expected to be present in the metastable state. Mode softening /9/ and lattice instabilities /14/ by strains have been considered earlier as essential mechanisms in the formation of the amorphous state. Further investigations in this system should clarify this point.

ACKNOWLEDGEMENTS

The authors thank Dr.W. Reichert for x-ray diffraction measurements. One of us (A.I.K) would like to thank the Alexander von Humboldt Foundation for the research grant.

REFERENCES

1. W.A. Phillips: In Amorphous Solids-Low Temperature Properties, (Springer, Berlin 1981)
2. D.A. Ackerman, D. May, R.C. Potter, A.C. Anderson: Phys. Rev. B23, 3886 (1981)
3. J.J. De Yoreo, W. Knaak, M. Meissner, R.O. Pohl: Phys. Rev. B34, 8828 (1986)
4. J. Chevrier, J.B. Suck, J.J. Capponi, M. Perroux: Phys. Rev. Lett. 61, 554 (1988)
5. J. Chevrier, J.B. Suck, J.C. Lasjaunias, F. Zougmore, J.J. Capponi: E-MRS-Meeting 1988
6. J.B. Suck and H. Rudin: In Glassy Metals II, ed. by H. Beck and H.J. Güntherodt, Topics in Appl. Phys., Vol. 53, p. 217 (Springer Verlag, Berlin 1983)
7. U. Buchenau, M. Prager, N. Nücker, A.J. Dianoux, N. Ahmad and W.A. Phillips: Phys. Rev. B34, 5665 (1986)
8. U. Buchenau, M. Prager, W.A. Kamitakahara, H.R. Shanks and N. Nücker: Europhysics Letters 6, 695 (1988)
9. U. Buchenau: Solid State Commun. 56, 889 (1985)
10. M. Hansen, K. Anderko: Constitution of Binary Alloys, 2nd ed. (Mc Graw-Hill, New York 1958)
11. V.F. Degtyareva, G.V. Chipenko, E.G. Ponyatovskii, V.I. Rashchupkin: Sov. Phys. Solid State 26, 733 (1984)
12. V.F. Degtyareva, G.V. Chipenko, I.T. Belash, O.I. Barkalov, E.G. Ponyatovskii: Phys. Stat. Sol. (a) 89, K 127 (1985) and private communication
13. A. Zinken, U. Buchenau, H.J. Fenzl and H.R. Schober, Solid State Commun. 22, 693 (1977)
14. A. Seidel, S. Massing, B. Strehlau and G. Linker, Phys. Rev. B38, 2273 (1988)

Opposite Extreme Structural Relaxation Behaviour in Glassformers; a Brillouin Scattering Study of B_2O_3 and $Ca_{0.4}K_{0.6}(NO_3)_{1.4}$

M. Grimsditch[1] *and L.M. Torell*[2]

[1]Argonne National Laboratory, 9700 South Cass Avenue, Argonne, IL 60439, USA

[2]Department of Technology, Uppsala University, Box 534, S-751 21, Uppsala, Sweden

Brillouin scattering in B_2O_3 and $Ca_{0.4}K_{0.6}(NO_3)_{1.4}$ shows that the two liquids may be taken as opposite extremes in a representation of the general behaviour of structural relaxation in glass forming liquids. In B_2O_3, which is a typical network glassformer, transverse Brillouin modes are observed above the glass transition over the large temperature range $\Delta T \approx 800$ K, which is the widest range reported so far for shear waves in liquids. From the ability of B_2O_3 to support shear stresses it is concluded that the intermediate range order persists in the liquid state over the wide temperature range $\Delta T \approx 1.4\ T_g$. In case of $Ca_{0.4}K_{0.6}(NO_3)_{1.4}$, which represents a non-network ionic glassformer, transverse modes were only observed in the limited range $\Delta T \approx 0.4\ T_g$ in which their velocity drastically decreased. The observation reflects the rapid collapse of the structure with increase in temperature typical for liquids which are weakly bonded. It is also supported by the finding that in the non-network $Ca_{0.4}K_{0.6}(NO_3)_{1.4}$ system, longitudinal structural relaxation processes become fast enough ($\sim 10^{-11}$ s) to be observed by Brillouin scattering already at temperatures $\sim 1.5\ T_g$, while in the network glassformer B_2O_3 structural relaxation is still not observed at a temperature of $2.4\ T_g$. The average shear relaxation times were obtained in the wide time range (10^{-9} s - 1 s) from the shifts of the transverse modes. In an Arrhenius plot of the relaxation times, using scaled (T_g/T) temperatures, the data of the two liquids fall close to the opposite curves for relaxation behaviour. The structural differences of the two systems are discussed in terms of some characteristics of the relaxation time such as its non-Arrhenius temperature dependence and its non-exponential time dependence. The rapid loss of exponentiality in $Ca_{0.4}K_{0.6}(NO_3)_{1.4}$ at times longer than those of the Brillouin window may partly be explained by a decoupling of the NO_3-reorientational relaxation from the average shear/structural relaxation which, at even longer times, is followed by a decoupling of the shear and structure relaxations close to the glass transition.

1. INTRODUCTION

It has been proposed that the various routes by which a high-temperature fluid liquid supercools to form a glass are characteristic of the structural properties of the glassformer/1/. Experimental results from a wide variety of glassforming systems show that a general pattern for the structural relaxation time τ versus T seems to be followed independent of the detailed internal molecular structure of the liquids/1/. The shape of the τ vs. T plots seem however to be related to the ability of the systems to form medium range order so that network and non-network glassformers represent the opposite extreme cases as discussed in what follows.

Network systems generally show an almost Arrhenius behaviour of the structural relaxation times. They are characterized by a viscosity which only slowly increases as the temperature decreases through

Springer Proceedings in Physics, Vol. 37 **Dynamics of Disordered Materials**
Editors: D. Richter · A.J. Dianoux · W. Petry · J. Teixeira

the supercooled range and by small changes of the heat capacity C_p and expansion coefficient α as the structure becomes arrested at the glass transition. Examples of network systems are for instance the familiar group of SiO_2-glasses formed by inter-connected silica tetrahedra/1,2/.

The current attention on glassforming materials by both theoreticians and experimental physicists focuses mainly on the behaviour of the non-network systems, expecially the extreme cases of weakly bonded liquids like molecular or simple ionic liquids, since they are more suited for comparisons with recent theoretical models. Examples in this category are for instance tri-a-naphtylbenzene and $Ca_{0.4}K_{0.6}(NO_3)_{1.4}$ which are currently being used for comparisons between the dynamic description of the glass transition proposed by mode-coupling theories/3,4/ and recent neutron scattering results/5-7/. Contrary to the network glassformers, the non-network systems are characterized by large changes of C_p and α at the glass transition and by a rapidly changing viscosity. In the τ vs. T plot the dramatic behaviour of a non-network glassformer is demonstrated by a non-Arrhenius curved-like shape which seems to increase in curvature with decreasing number of configurational restrictions of the network/1/.

Extreme cases of non-network forming systems have previously been investigated in light scattering studies of structural relaxation processes by one of us (L.T.)/9-11/. Using the Brillouin scattering technique structural relaxation times of the simple ionic liquids $Ca_{0.4}K_{0.6}(NO_3)_{1.4}$ /8/, Bi-Cl/9/ and $Ca(NO_3)_2$ - $8H_2O$/10/ were obtained over a wide time scale from 10^{-12} s - 1 s. This is the broad time range for structural changes of the material from the short time/high temperature behaviour of the very fluid liquid to the long times characteristic of processes close to the glass transition. Scaled behaviour of the average relaxation time τ was found for the various ionic systems independent of detailed molecular structure when using T_g as a scaling parameter/9,10/. The resulting relaxation time curves showed the largest (with the exception of o-terphenyl/11/) curvature reported so far, which makes the ionic materials convenient model systems for representing non-network glassformers. Moreover, it was found that the "activation energy" for the process, as given by (δ log τ/∂ 1/T), changed most dramatically in a temperature range where crossover from exponential to non-exponential decay of the time dependence of the relaxation process was observed/10/. The observation supports the suggestion of a common origin of the non-Arrhenius relaxation time behaviour and the existence of a distribution of relaxation times/12/.

In the present Brillouin scattering study we present results for two liquids, which represent the opposite extremes of structural relaxation behaviour; on one hand from B_2O_3 which is well-known to form a network of interconnected units of three-fold coordinated boron atoms/13/ and, on the other, for the non-network $Ca_{0.4}K_{0.6}(NO_3)_{1.4}$ system which has previously been studied in this laboratory/8/ and for which we will add som new data. The aim of the work is to investigate the structural stability and the shear relaxation time from observations of the transverse modes. In B_2O_3 the slowly changing viscosity/14/ implies that shear restoring stresses can be supported over a wide temperature range contrary to the case in $Ca_{0.4}K_{0.6}(NO_3)_{1.4}$ where the viscosity/15/ drops rapidly just above T_g. In the latter the viscosity changes thirteen decades from the value of ~ 10^{13} P typical close to T_g to a value of ~ 1 P at about 70 K above T_g, whereas in B_2O_3 the corresponding fluidity is only reached after a temperature increase of more than 1000 K above the glass transition temperature. In Brillouin scattering from a liquid the propagation of shear waves in the low temperature/high viscosity range is demonstrated by transverse modes in the spectra. The two systems of the present study were then choosen with the intention to follow shear waves, on one hand, over the widest possible temperature range (i.e. in B_2O_3) and, on the other , in a system in which they dissappear with temperature in the most dramatic way (i.e. in $Ca_{0.4}K_{0.6}(NO_3)_{1.4}$). The overall aim is to compare the temperature dependence of the average shear relaxation times of the two systems with the different structural properties in mind. In particular the suggested relation between non-Arrhenius temperature dependence and non-exponential time decay will be investigated to see if there is a general pattern.

2. EXPERIMENTAL RESULTS

2.1. Brillouin Scattering Spectra

Brillouin spectra of B_2O_3 were recorded over the wide temperature range of about ^ :J -1300 K, while the limited range 300 - 660 K was sufficient for observations in case of $Ca_{0.4}K_{0.6}(NO_3)_{1.4}$. Some typical spectra recorded in the 90° scattering geometry for the two systems are displayed in Figs 1 and 2 respectively. The (HH + HV)-spectra in Fig. 1 show both the longitudinal (L) and transverse (T) Brillouin doublets, while in the HV spectra in Fig. 2 only the transverse modes are present.

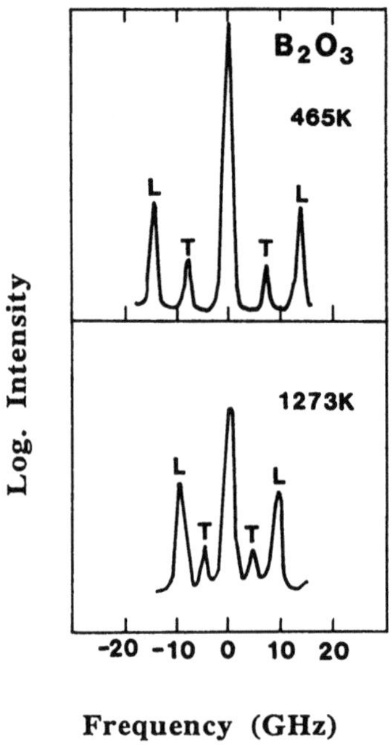

Fig. 1. Brillouin spectra (HH + HV) of B_2O_3 at 465 K and 1273 K respectively. The longitudinal (L)and transverse (T) modes are indicated.

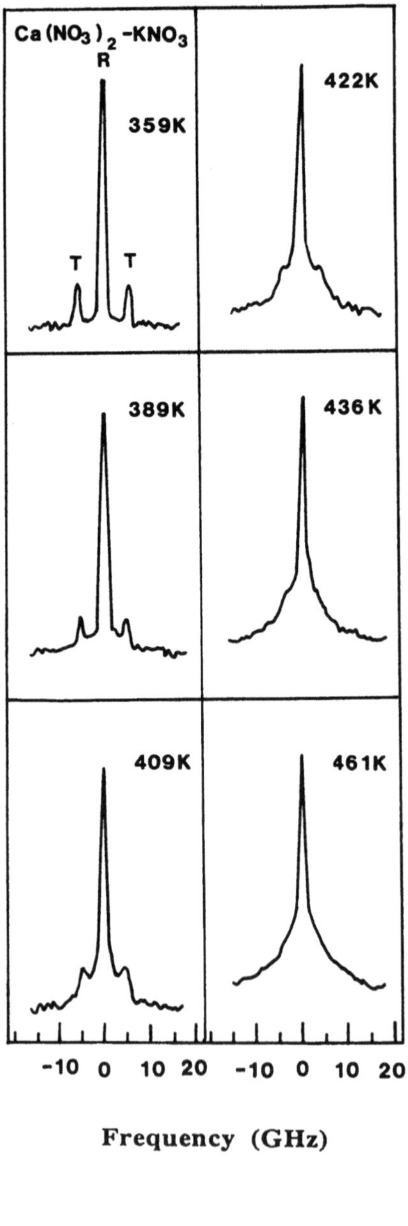

Fig. 2. Brillouin spectra (HV) of $Ca_{0.4}K_{0.6}(NO_3)_{1.4}$ at various temperatures.The transverse (T) and Rayleigh (R) modes are indicated.

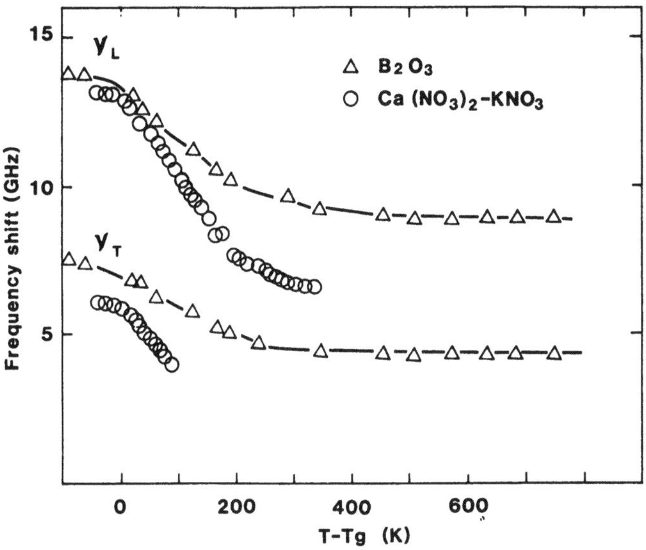

Fig. 3. Temperature dependence of the longitudinal (ν_L) and transverse (ν_T) Brillouin frequency shifts in B_2O_3(Δ) and in $Ca_{0.4}K_{0.6}(NO_3)_{1.4}$ (o).

In Fig. 3 the corresponding longitudinal and transverse frequency shifts are shown for the two materials. The data are plotted versus the scaled temperature $T\text{-}T_g$ for clarity and kinks in the temperature dependences are clearly seen at T_g as expected. Above T_g the frequency shift decreases more rapidly with temperature and in a more dramatic fashion in $Ca_{0.4}K_{0.6}(NO_3)_{1.4}$ than in B_2O_3, as will be discussed below.

The outstanding differences beween the two glassforming materials are immeditely recognized from the temperature evolution of the spectra in Figs. 1 and 2. From the presence of transverse modes it can be seen that shear stresses are supported in liquid B_2O_3 up to a temperature of at least ~ 1300 K, which is far above not only the glass transition temperature (T_g = 526 K) but also far above the temperature of melting (T_m = 723 K). Moreover, in Fig. 3 it can be seen that, apart from in a small temperature region just above T_g, the frequency shift of the transverse mode is little affected by temperature even at the highest temperatures of observation. Similarly the intensity of the transverse component is hardly affected over the same temperature range, see Fig. 1, which implies that shear waves in B_2O_3 persist to still higher temperatures. To our knowledge this is the largest temperature range reported so far for observations of shear waves in the liquid state of a material. In contrast, shear waves in case of $Ca_{0.4}K_{0.6}(NO_3)_{1.4}$ are only clearly observed in the limited temperature range of ~100 K above the glass transition (T_g = 333 K), i.e. in the supercooled region (T_m = 438 K), see Fig. 2. In this temperature range the frequency shift of the transverse mode decreases rapidly as demonstrated in Fig. 3. Apart from the behaviour of the transverse modes, the spectra of $Ca_{0.4}K_{0.6}(NO_3)_{1.4}$ show dramatic effects in both the longitudinal mode and the central line, which in case of B_2O_3 are more or less unaffected by temperature.

It was found that the line width of the longitudinal component of the $Ca_{0.4}K_{0.6}(NO_3)_{1.4}$ spectra increased and later decreased with temperature. Maximum damping was observed at a temperature of ~ 500 K as seen in Fig. 4, where the measured full widths at half maximum are plotted. The effect was

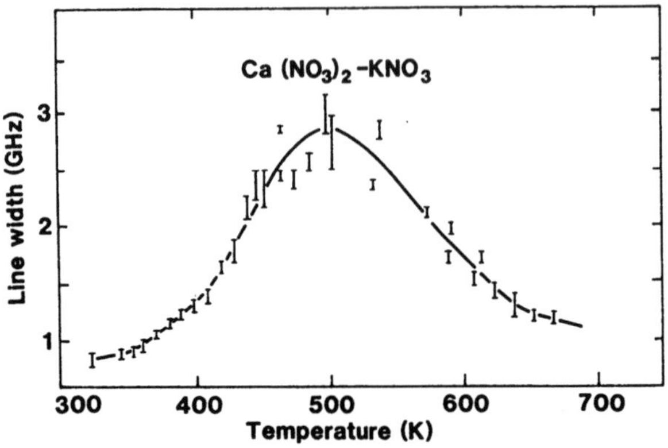

Fig. 4. Temperature dependence of the Brillouin line width of longitudinal modes in a $Ca_{0.4}K_{0.6}(NO_3)_{1.4}$ at hypersonic frequencies.

obtained in the same temperature range in which changes in the longitudinal frequency shifts were found, see Fig. 3. Both phenomena reflect the presence of structural relaxation processes with characteristic times in the time window of the Brillouin scattering technique, i.e. ~ 10^{-11} s. In contrast, no such effects were observed in B_2O_3, in which the line width of the longtudinal mode was unaffected, even at the highest temperature of observation.

The broadening of the central line of the spectra of $Ca_{0.4}K_{0.6}(NO_3)_{1.4}$ is more difficult to evaluate quantitatively. The effect is an indication of molecular reorientations, and it is attributed to reorientational motions of the nitrate ion. It is clearly observable in the spectra over the temperature range presented in Fig. 2, where the central line broadens till it envelops the transverse modes. The true spectral shape of the reorientational mode is therefore not only affected by the instrumental profile but is strongly influenced by the transverse modes. A crude estimation of the reorientational contribution to the broadening of the central component can be obtained by simply taking into account the instrumental line width and the widths of the transverse modes. Results of such calculations in the representative temperature range give estimates of the increasing line width from a value of ~ 1.5 GHz at 390 K to ~ 5 GHz at 460 K. Thus in this temperature range the reorientational relaxation processes are in the frequency/time range of Brillouin scattering.

2.2. Relaxation Data of B_2O_3

In this section we use the experimental results, which differ widely in character for the two systems, to calculate their relaxation properties needed later for a general discussion of typical network and non-network glassforming systems.

From the observed frequency shifts of the longitudinal (ν_L) and transverse (ν_T) modes of B_2O_3, see Fig. 3, values of the longitudinal (v_L) and shear (v_T) sound velocities were determined using the Brillouin equation

$$v_{L,T} = \nu_{L,T}\,\lambda_o/(2n\,\sin(\theta/2)), \qquad (1)$$

where n is the refractive index, λ_o the wavelength of the incident light, and θ the scattering angle. In the

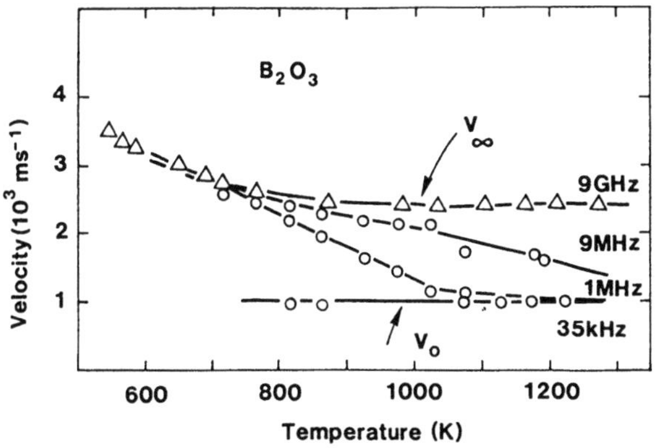

Fig. 5. Temperature dependence of sound velocity in B_2O_3; Brillouin results Δ, ultrasonic results o from Ref. 17. Limiting high and low frequency velocities are denoted v_∞ and v_0 respectively.

calculations, values for the refractive index were taken from Ref. 16. The derived data are shown versus temperature in Fig. 5 along with previously reported ultrasonic data/17/. The latter work yielded sound velocities v_0 for the "limiting low frequency region" in which the configurations of the liquid structure can change and follow the passage of the propagating sound wave. Such relaxed behaviour was observed in the ultrasonic study for a probe frequency of 35 kHz while relaxation effects were present for larger frequencies, as can be seen from the velocity dispersion curves in the MHz range in Fig. 5. In the present hypersonic (GHz) frequency range the spectra showed no sign of either velocity dispersion or damping of the longitudinal modes which implies that in this case the probe frequency is far above the frequency range of relaxation. Thus structural relaxation in B_2O_3 in the GHz range has to be sought at much higher temperatures than the range investigated in this work. The measured hypersonic velocities can therefore be taken as identical to v_∞, i.e. to values of the "limiting high frequency region". Here the liquid structure changes too slowly to follow the high frequency perturbation and hence exhibits the elastic properties of a glassy solid.

Since v_∞ is obtained directly from the longitudinal frequency shifts, which are accurate to ~1 %, precise data for the high frequency longitudinal modulus M_∞ are provided using the relation

$$M_\infty = \rho v_\infty^2 . \tag{2}$$

Results for M_∞ are shown in Fig. 6. In the calculations density data were taken from Ref. 14. Values have previously been reported from ultrasonic velocity and absorption data/17/. In the latter study, since the measurements were carried out in the relaxation range, M_∞ was obtained by a fitting procedure in which single relaxation theory was used in the high temperature $T > 1000$ K range, whereas a distribution of relaxation times was needed to account for the behaviour at lower temperatures. Results from such analyzing methods are also shown in Fig. 6. It can be seen that the hypersonic and ultrasonic data are in agreement at temperatures $T < 800$ K while large deviations of ~ 40 % are found in the high temperature range. The strength of the present result is that the longitudinal modulus data are calculated directly from observed results which removes the uncertainty associated with a fitting procedure involving many parameters. The correspondence found at low temperatures is explained by the fact that in this range the ultrasonic data are almost frequency independent and therefore close to the true v_∞, see Fig. 5, while at higher temperatures the measured ultrasonic velocities are far from v_∞ and the fitting procedure becomes more delicate.

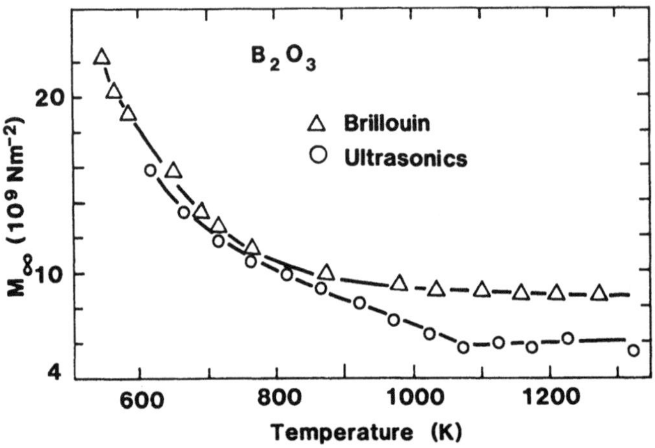

Fig. 6. Temperature dependence of high frequency longitudinal modulus in B_2O_3 from ultrasonic (o, Ref. 17) and present high frequency (Δ) measurements.

Since relaxation processes are absent in the temperature range studied the Brillouin frequencies may be regarded as "limiting high frequencies", and values of the average shear relaxation time $<\tau>$ are then immediatly obtained using the relations

$$<\tau_s> = \eta/G_\infty \tag{3}$$

and

$$G_\infty = \rho\, v_T^2 . \tag{4}$$

Present results for G_∞ are shown in Fig. 7 together with ultrasonic data/18/ and good agreement is found. The broad temperature range over which shear modes were observed in the present study implies that the available G_∞ range is now extended to much higher temperatures. Using viscosity data of ref 14 the corresponding $<\tau_s>$ values were obtained over the wide range from the fluid liquid region (~ 10^{-9} s) to the long relaxation times (~ 10^3 s) typical of processes close to T_g, see Fig. 8. Also included in the figure are ultrasonic results of structural relaxation times τ_L based on observations of the longitudinal modes/17/. Since convincing evidence of single relaxation time behaviour is provided in the ultrasonic study of the high temperature range T > 1000 K/17/, we expect $\tau_s = \tau_L$ in this range. The deviation between ultrasonic and hypersonic results at high temperatures in Fig. 8 is attributed to the miscalculated M_∞-values in the delicate fitting procedure of the ultrasonic analysis as discussed above. More interesting is to note the close similarity of $<\tau_s>$ and $<\tau_L>$ at lower temperatures where a distribution of relaxation times was noted in the ultrasonic report. The present findings therefore support the assumption of the same time distribution for both processes and that their average relaxation times are equal as previously proposed/17/. This may be subjected to an even more severe test by comparing photon correlation/19/ data with the present results. In the digital correlation study the average relaxation time of the central Mountain line $<\tau_M>$ was observed. It is related to the average longitudinal stress relaxation time through the equation

$$<\tau_M> = (M_\infty/M_0) <\tau_L> . \tag{5}$$

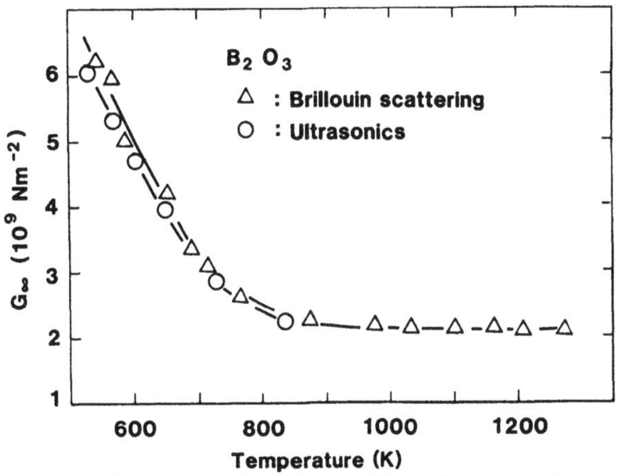

Fig. 7. Temperature dependence of high frequency shear modulus in B_2O_3 from ultrasonic (o, Ref. 18) and present high frequency (Δ) measurements.

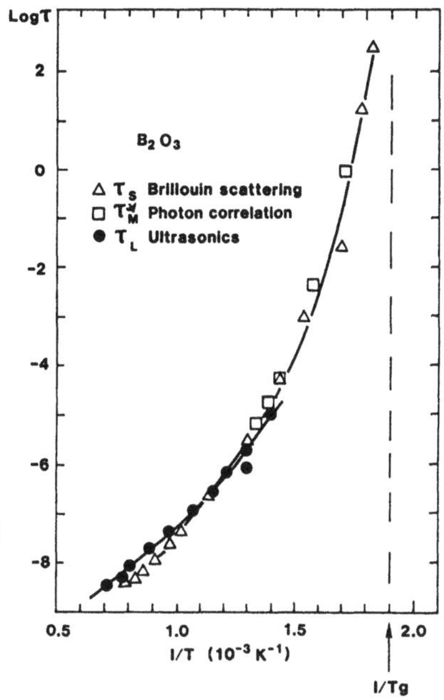

Fig. 8. Temperature dependence of the relaxation time in B_2O_3. Present results (Δ) are for shear relaxation times τ_s. Longitudinal relaxaion times τ_L(●) are from Ref. 17. $\tau_M{}^*$ (□) are longitudinal relaxation times calculated from the relaxation times of the central Mountain line τ_M (Ref. 18) using $\tau_M{}^* = (M_0/M_\infty)\tau_M$.

Using present findings for M_∞ and ultrasonic values for M_0, τ_L was calculated from the data of the photon correlation study and results (denoted τ_M*) are shown in Fig. 8. The close resemblance of the photon correlation and Brillouin scattering relaxation data again show that the average longitudinal and shear relaxation times are at least approximately equal. This is especially notable in view of the low temperatures involved in the comparison since in this range the time distribution function is considerably broader than in the ultrasonic range discussed above. Thus decoupling of τ_s and τ_L, observed close to T_g in some weakly bonded glassforming systems (see discussion below), is not found in the case of B_2O_3.

2.3. Relaxation Data of $Ca_{0.4}K_{0.6}(NO_3)_{1.4}$

The spectra of the nitrate glassformer show much more complex behaviour than those of the B_2O_3 system. For the nitrate glass, several relaxation processes are present which dramatically change the spectrum with temperature. They are observable in the rapidly decreasing frequency shifts of the transverse modes, the damping of the longitudinal modes and the evolution of a central component as temperature increases. In the following the various relaxation processes and their characteristic times will be discussed.

First we will discuss shear relaxation. The rapid shift of the transverse mode towards the central line with increasing temperature shown in Figs. 2 and 3 mirrors a shear stress relaxation time τ_s which changes by 12 orders of magnitude in a limited T-range of 90 K from the very long times (10^3 s) close to T_g to the short times of ~ 10^{-9} s for structural rearrangements at $T \approx 420$ K. The relaxation times, obtained from the present transverse frequency shifts using (3) and (4) and the viscosity data of Ref.15, are presented in Fig. 9.

Next we turn to the longitudinal relaxation. The broadening of the longitudinal mode shown in Fig. 4 is a measure of the sound absorption and is related to the relaxation of compressional stresses. Since an absorption maximum is observed, the characteristic relaxation time can immediatly be obtained from the peak condition

$$\omega \langle\tau_L\rangle \approx 1 \qquad (6)$$

which relates the average longitudinal relaxation time to the inverse of the Brillouin probe frequency, $1/\nu_B = 2\pi/\omega$, at the temperature of maximum absorption. In the present case the absorption peak is observed at $T \approx 500$ K (see Fig. 4) for a Brillouin frequency of $\nu_B \approx 8.46$ Hz which corresponds to a relaxation time $\tau_L \approx 1.9 \times 10^{-11}$s. This result is also incorporated in Fig. 9 and it can be seen that the shear relaxation times obtained in the viscous range just above T_g extrapolate to the longitudinal relaxation time obtained ~ 170 K above T_g.

Finally the reorientational relaxation will be examined. The new central mode, which emerges in addition to the elastic component in the HV-spectra at intermediate viscosities, signals molecular reorientational relaxation processes, in this case due to NO_3^--reorientations. Fig. 2 demonstrates that the reorientational motions mix with the propagating shear waves in the short time range typical of Brillouin scattering, which results in a composite spectrum. Analysis of the detailed HV-spectra thus requires a curve fitting procedure which at present is in progress and will be presented separately/20/. Here we

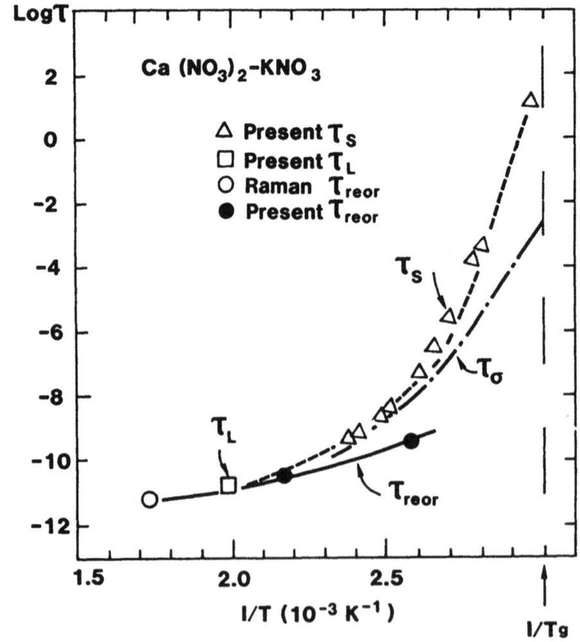

Fig. 9. Temperature dependence of the relaxation times in $Ca_{0.4}K_{0.6}(NO_3)_{1.4}$. Present results are for the shear (τ_s ; Δ), the longitudinal (τ_L ; □), and the NO_3^- reorientational (τ_{reor}; ●) relaxation times. The Raman value (o) of τ_{reor} is from Ref. 21 and the conductivity relaxation time τ_σ (dot-dashed line) is from Ref. 23, the dashed line represents τ_s and τ_L values from Ref. 8.

only estimate the characteristic time τ_{reor} for NO_3^--reorientations. It is obtained from the half-width at half-maximum of the reorientational component Γ_{reor} using the relation

$$\tau_{reor} = 1/2\pi\,\Gamma_{reor} \tag{7}$$

and is a measure of collective reorientations. Results, representing the temperature range in which the phenomenon has been detected with the present Brillouin technique, are presented in Fig. 9. Also shown in Fig. 9 is the reorientational time obtained at shorter times/higher temperatures using Raman scattering/21/. Note that present results extrapolate to the Raman value despite the fact that in Brillouin scattering collective behaviour is observed while the Raman technique measures the individual molecular reorientations. The agreement of the results implies similar average relaxation times. It is even more interesting to note that the characteristic times for reorientational motions, i.e. even the average collective times τ_{reor}, become decoupled from the relaxation times for longitudinal and shear stresses at lower temperatures. This is concluded from the present observations of the three relaxation processes shown in Fig. 9 and also by comparing present τ_{reor} with previous results for τ_L and τ_s obtained in this laboratory/8/. In the latter study it is reported that $\tau_L \approx \tau_s$ in the short time of Brillouin scattering and that both processes follow approximately the same temperature dependence as demonstrated by the dashed curve in Fig. 9. Now we observe in the case of the reorientational times at high temperatures that $\tau_{reor} \approx \tau_L \approx \tau_s$ whereas decreasing the temperature implies an increasing decoupling of τ_{reor} from the τ_L-τ_s-curve (see Fig. 10). The phenomenon may be the first one in a series of decoupling processes which subsequently occur as the temperature is lowered, see discussion below.

3. DISCUSSION

3.1. Scaled Relaxation Times

To be able to compare the temperature dependence of the relaxation times for the two systems B_2O_3 and $Ca_{0.4}K_{0.6}(NO_3)_{1.4}$ in a common graph, the data are plotted versus scaled temperatures T_g/T in Fig. 10. This also allows for a comparison of the τ vs. T behaviour of every other glass forming system from the common long relaxation time determining the glass transition (~ 10^3 s) at $T_g/T = 1$ to the shortest possible times at $T_g/T \rightarrow 0$ determined by the typical liquid quasilattice vibrational times (~ $10^{-13.5}$ s). The manner in which various types of systems reach the approximately similar high temperature limit from the common low temperature point varies considerably and seems to depend on structural properties of the system investigated/1/.

Expressed in terms of the Vogel-Tammann-Fulcher (VTF) law the various relaxation curves can be identified by a single parameter D according to

$$\tau = 10^{-13.5} \exp\,[D\,T_o/(T - T_o)] \tag{8}$$

where D may take any value in the wide range 4-100 depending on the system under investigation. For systems represented by extremely curved VTF-relations, i.e. small D values, T_o in (8) can be taken as the ideal glass transition temprature. As D increases, the deviation from the glass transition temperature increases and $T_o \approx 0.3\,T_g$ for the extreme case of $D \approx 100$.

We find that the present data of B_2O_3 correspond to a D value of ~ 35 and $T_o \approx 0.5\,T_g$. This places B_2O_3 on the upper part of the diagram in Fig. 10, which is limited by the straight (dashed) line. This is the region of classical network oxide glasses for which only relatively weak departures from Arrhenius behaviour are observed. The latter systems are characterized by a tetrahedral network which seems to

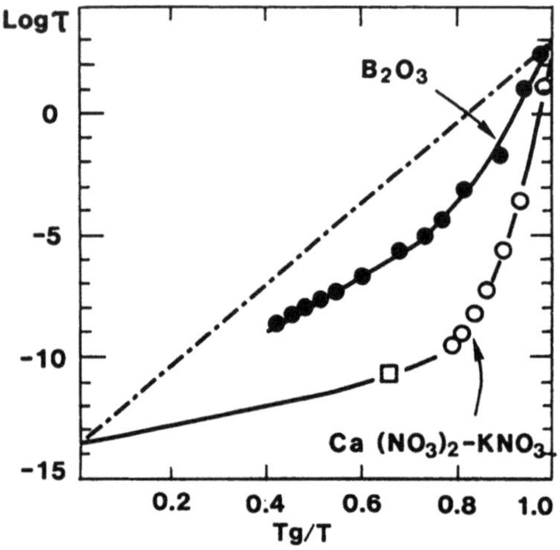

Fig. 10. Arrhenius representation of the logarithm of the mean relaxation time vs. scaled temperatures T_g/T. Shear relaxation times of B_2O_3 are denoted ● and for $Ca_{0.4}K_{0.6}(NO_3)_{1.4}$, o; the longitudinal relaxation time value for $Ca_{0.4}K_{0.6}(NO_3)_{1.4}$ is shown as □.

keep the intermediate range order of the liquid rather uneffected by temperature, and accordingly, such systems are denoted "strong". Similarly we find for the three-coordinated B_2O_3 system spectra which are more or less uneffected by temperature over the wide range of $T - T_g \approx 800$ K and in which the presence of shear modes of unchanged intensity are especially notable.

At the opposite extreme in Fig. 10, we find that the present results for $Ca_{0.4}K_{0.6}(NO_3)_{1.4}$ correspond to a D value of ~ 4 and $T_0 \approx 0.9\ T_g$. This represents the lower limit of non-Arrhenius relaxation time behaviour observed in liquids with the only exception, as far as we know, being o-terphenyl/11/. Liquids of extreme non-Arrhenius behaviour, the so-called "fragile" liquids, are associated with lack of directional character in the molecular bondings, weak interacting forces, a high configurational degeneracy and as a consequence rapid degradation of the intermediate range order as temperature increases. Such properties are all characteristic of the present ionic glassforming systems. Apart from the present results of a rapidly changing τ, the temperature evolution of the spectra themselves immediatly reveals the rapid structural changes and the collapse of the intermediate range order of the nitrate glassformer.

The more complex behaviour of the structural relaxation process in $Ca_{0.4}K_{0.6}(NO_3)_{1.4}$ than in B_2O_3 is also evident in the present observation of the decoupling (see Fig. 9) of the reorientational modes of motions of NO_3^- from those characterizing the longitudinal and shear relaxation. There are also indications that the latter two processes, which follow the same τ vs. T curve in the low and intermediate viscosity range show slightly divergent average relaxation times in the temperature range close to T_g, i.e. for $\tau > 100$ s/22/. In this high viscosity/slow process range there is evidence, based on the observation of a generally shorter relaxation time for the viscosity than for enthalpy in other fragile systems, of a decoupling of the shear modes from the configurational modes/22/. A direct implication would be that the longitudinal relaxation time differs from the shear relaxation time since bulk viscosity is likely to be closely related to the process leading to comfigurational equilibrium. This supports the indicated decoupling of τ_s and τ_L in $Ca_{0.4}K_{0.6}(NO_3)_{1.4}$ at low temperatures, even though the data by themselves are too inaccurate to confirm such a behaviour. It is interesting to note that there is also another type of decoupling which is observed in the intermediate viscosity range in $Ca_{0.4}K_{0.6}(NO_3)_{1.4}$ namely the decoupling of the electrically conducting modes from the viscosity modes. The phenomenon can be seen from the results of the conductivity relaxation times/23/ which also have been incorporated in Fig. 9.

From the behaviour of various relaxation processes discussed above we conclude that there is no indication that the different processes are strongly coupled in $Ca_{0.4}K_{0.6}(NO_3)_{1.4}$ except in the high temperature/short time range. The observation may be general for fragile systems and it may explain the broader distribution of relaxation times compared to that observed in "strong" liquids such as B_2O_3, as will be discussed below.

3.2. Non-Arrhenius Behaviour and Non-exponential Time Decay

In most glassforming liquids a distribution of structural relaxation times is observed. It is suggested that since the structural breaking and re-arrangement processes associated with the relaxation are co-operative in the liquid, they do not follow a simple rate process. Thus, the rate by which the degree of order decreases in time cannot be expressed by an exponential decay. Experimental results show that in most systems it is well described by the Kohlrausch-Williams-Watts relaxation function, i.e.

$$\phi(t) = \exp\left[-(t/\tau)^{\beta}\right] \tag{9}$$

in which β is a measure of the apparent distribution of relaxation times, being 1 in case of a single relaxation time and decreasing as the distribution increases. In the following, by discussing the results of Fig. 10 in terms of reported β-values for B_2O_3 and $Ca_{0.4}K_{0.6}(NO_3)_{1.4}$, we will consider how this non-exponentiality in time (9) is related to the non-Arrhenius in temperature (8).

A general observation of most liquids is that the relaxation width increases with decreasing temperature from a single relaxation time ($\beta = 1$) in the high temperature/low viscosity range to a broad distribution of relaxation times ($\beta < 1$) in the intermediate viscosity range. In some cases (e.g., glycerol and propylene glycol) β saturates and becomes independent of temperature near T_g/24,25/ in other cases (e.g., o-terphenyl) it continues to decrease/26/, however, in most systems the temperature dependence is unknown. In fact in many liquids constant β values have been assumed in the analyses of the data. On the other hand, from the recent study of o-terphenyl it is suggested that a strong temperature dependence of β may be typical of "fragile" glass formers in general/26/. One may therefore expect that the temperature dependence is a function of the structural properties of the liquid. To analyze such a dependence we plot reported β values versus T in Fig. 11a for the present two glassformers which in structure represent opposite extremes.

From Fig. 11a we note that in the case of the "strong" glass former B_2O_3 β changes gradually over a wide temperature range of $\Delta T \approx 2T_g$ from exponential decay at high temperatures/17/ to a distribution which saturates at a value of $\beta \approx 0.6$ close to T_g/19/. In $Ca_{0.4}K_{0.6}(NO_3)_{1.4}$, much more rapid changes are observed/8,27/; the final value ($\beta \approx 0.4$) is considerably lower and it is reached in a narrower temperature range of $\Delta T \approx 0.5\ T_g$. Thus stronger temperature dependence seemes indeed to be related to a more fragile system in support of the suggestion in Ref. 26

Next, we compare the width to the average time of the relaxation. In Fig. 11b we therefore plot the reported β values for B_2O_3/17,19/ and $Ca_{0.4}K_{0.6}(NO_3)_{1.4}$ /8,27/versus our results for log τ. On the left of the figure are the shortest relaxation times typical for liquids, the dashed line on the right is the long time limit as determined by the glass transition. It can be seen from the figure that single relaxation time behaviour, which is found in the short time range of most liquids, is extended to much longer times in the network system B_2O_3 and that it only gradually changes into a distribution of relaxation times. In contrast, the transition from $\beta = 1$ to $\beta < 1$ in the non-network nitrate system occurs more abruptly, at considerably shorter times, and results in a broader distribution function. However, it is interesting to

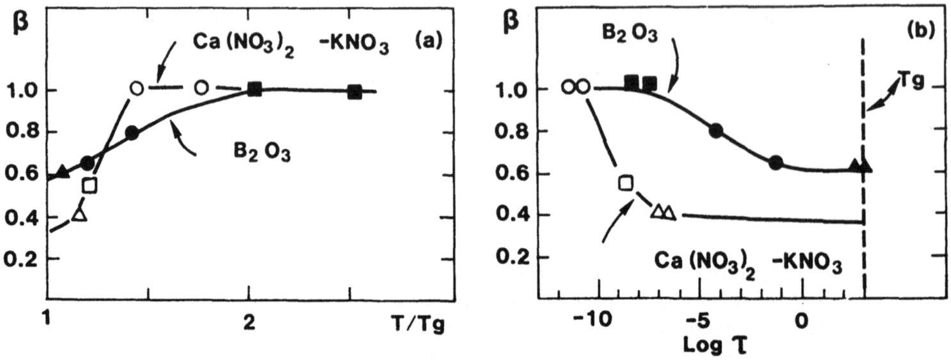

Fig. 11. The Kohlrausch-Williams-Watts fitting parameter β vs. (a) 1/T and (b) log τ. Ultrasonic results (■ and □) are from Refs. 17 and 29, photon correlations (●) and pressure jump relaxations (▲) data are from Ref. 18; hypersonic results (o) are from Ref. 8; enthalphy relaxations results (△) from Ref. 27.

note that in both systems, the crossover from exponential to non-exponential decay in Fig. 11 occurs in the relaxation time range where there is a bend in the structural relaxation time curves of Fig. 10, which separates the high temperature/low activation energy range from the low temperature/high activation energy range. The τ versus T curves in Fig. 10 of the borate and nitrate systems, represented by the VTF-relations (8) with D ~ 35 and ~ 4 respectively, are, in the high temperature range, equally well represented by an Arrhenius curve, whereas at intermediate and low temperatures they are non-Arrhenius-like. For B_2O_3, ultrasonic/17/ and viscosity/14/ results show that Arrhenius behaviour is relevant for T > 1000 K, which corresponds to the relaxation time range limited by log τ < -7.5 in Fig. 10, i.e. by the value at which β starts to decrease, see Fig. 11. Similarly it is found for $Ca_{0.4}K_{0.6}(NO_3)_{1.4}$ that while crossover from the high temperature Arrhenius' regime is observed at much higher temperatures/shorter times, in this case at log $\tau \approx$ -10.5 (see Fig. 10), it still corresponds to the crossover in time from exponential to non-exponential decay shown in Fig. 11. These findings thus indicate that the non-exponential time decay and the non-Arrhenius temperature depence seem to have a common origin, as has been suggested.

In conclusion, the observations from the network and non-network glassforming systems B_2O_3 and $Ca_{0.4}K_{0.6}(NO_3)_{1.4}$ respectively, indicate that

* the temperature dependence of both τ and β increases as "fragility" increases
* the relaxation broadens as the "fragility" of the system increases
* single relaxation time behaviour persists to longer times/lower temperatures as the "strength" of the system increases
* crossover in time decay is associated with crossover in temperature dependence (i.e. ativation energy).

Recent reports from polymers show that for macromolecules like poly (propylene glycol) β is independent of T and that single relaxation time behaviour cannot be reached in the T range of an intact polymer chain/28/. Thus the findings above are not representative for polymeric glassforming systems. Of non-polymeric glasses, apart from the present oxide and ionic glass former, results from simple ionic liquids in the "intermediate" to "fragile" range, i.e. $ZnCl_2$, BiCl and $Ca(NO_3)_2$ - $8H_2O$, are in support of these findings/8-10/ as are also results from the molecular "fragile" system o-terphenyl/11,26/. To see if the present observations are typical for non-polymeric glassforming systems in general, we have to await experimental evidence from a broader variety of glassformers and especially other classes of viscous liquids.

The above findings are based on observations using relatively low probe frequencies, i.e. the macroscopic (low q) responses of the glass forming systems are probed. It is then interesting to note that recent neutron spin echo studies on $Ca_{0.4}K_{0.6}(NO_3)_{1.4}$ /7/, while demonstrating almost the same $<\tau>$ as for macroscopic kinetics, the microscopic behaviour seems to scale with temperature (i.e. constant β). The phenomenon is at present the subject of an extended spin echo study covering a broad time/temperature range /30/.

This work was supported in part by the U.S. Department of energy (Office of Basic Energy Sciences, Divisions of Materials and Chemical Sciences) under contract N° W-31-109-ENG-38 and in part by the Swedish Natural Science Research Council. We wish to thank R. Bhadra for recording some of the Brillouin spectra.

References

1. C.A. Angell: Proceedings of the Workshop on Relaxation Effects in Disordered Systems, Blacksburg, VA, 1983, edited by K. Ngai and T.K. Lee (Mc Gregor and Werner Inc) 1984, p. 3
2. S. Brawer: Relaxation in Viscous Liquids, Am. Ceram. Soc. (1985) p. 4-28
3. E. Leutheusser: Phys. Rev. A29, 2765 (1984)
4. U. Bengtzelius, W. Goetze, A. Sjölander: J. Phys. Chem. 17, 5915 (1984); U. Bengtzelius, A. Sjölander: Ann. N.Y. Acad.Sci. 484, 229 (1986)
5. F. Fujara, W. Petry: Europhys. Lett. 4, 921 (1987)
6. B. Frick, D. Richter, W. Petry, U. Buchenau; Z. Phys. B. 70, 73 (1988)
7. F. Mezei, W. Knaak, B. Farago: Physica Scripta T19, 363 (1987)
8. L.M. Torell: J. Chem. Phys. 76, 3467 (1982); C.A. Angell and L.M. Torell: J. Chem. Phys. 78, 937 (1983)
9. L.M. Torell, D.C. Ziegler, C.A. Angell: J. Chem. Phys. 81, 5053 (1984)
10. M. Elmroth, L. Börjesson, L.M. Torell (in manuscript)
11. W.T. Laughlin, D.R. Uhlmann; J. Phys. Chem. 76, 2317 (1972)
12. G.Mc Duffie, T.A. Litovitz: J. Chem. Phys. 37, 1699 (1962)
13. A.H. Silver, P.J. Bray: J. Phys. Chem. 29, 984 (1958); P.V. Johnsson, A.C. Wright, R.N. Sinclair: J. Non. Cryst. Solids 42, 107 (1980)
14. P.B. Macedo, W. Capps, T.A. Litovitz: J. Chem. Phys. 44, 3357 (1966)
15. H. Tweer, N. Laberge, P.B. Macedo: J. Am. Ceram. Soc. 54, 121 (1971)
16. J. Lörosch, M. Couzi, J. Pelous, R. Vacher, A. Levasseur: J. Non. Cryst.Solids, 69, 1 (1984)
17. P. Macedo, T.A. Litovitz: Phys. Chem. Glasses 6, 69 (1965)
18. W. Capps, P.B. Macedo, B. O'Meara, T.A. Litovitz: J. Chem. Phys. 45, 3431 (1966)
19. J.A. Bucaro, H.D. Dardy, R.D. Corsaro: J. Appl. Phys. 46, 741 (1975)
20. M. Grimsditch, (to be published)
21. J.H.R. Clarke, S. Miller: Chem. Phys. Lett. 13, 97 (1972)
22. C.A. Angell, (to be published)
23. F.S. Howell, R.A. Bose, P.B. Macedo, C.T. Moynihan: J. Phys. Chem. 78, 639 (1974)
24. N.O. Birge, S.R. Nagel: Phys. Rev. Lett. 54, 2674 (1985); Rev. Sci. Instrum. 58, 1464 (1987)

25. Y.H. Jeong, S. Bhattacharya, S.R. Nagel: Phys. Rev. A 34, 602 (1986)
26. P.K. Dixon, S.R. Nagel: Phys. Rev. Lett. 61, 341 (1988)
27. C.T. Moynihan, H. Sasabe, J.C. Tucker: In "Molten Salts", Ed. J.P. Pemsler, The Electrochem. Soc. Inc., 188 (1976)
28. L. Börjesson, L. Stevens, L.M. Torell: Polymer, 28, 1803 (1987)
29. R.A. Werler, R. Bose, P.B. Macado: J. Chem. Phys. 53, 1258 (1970)
30. F. Mezei, W. Knaak, C.A. Angell, L.M. Torell, (work in progress)Figure captions

Line Shape Analysis of Brillouin Spectra from Glassforming Liquids

S. Loheider[1], I. Petscherizin[2], M. Soltwisch[1], J. Sukmanowski[1], G. Vögler[1], and D. Quitmann[1]

[1]Institut für Atom- und Festkörperphysik, Fachbereich Physik, Freie Universität Berlin, Arnimallee 14, D-1000 Berlin 33, Germany
[2]Chemical Institute, Leningrad State University, Leningrad, USSR

Glass forming liquids have characteristically two rather different sound velocities, viz. the adiabatic and the high frequency values c_0 and c_∞, respectively: The energy exchange between the collective, large scale sound waves $\omega=c\cdot q$ and the microscopic or mesoscopic heat reservoirs, i.e. degrees of freedom, takes place quickly with some degrees of freedom, but it has a finite rate $1/\tau_s$ with others /1/. This leads to frequency dependent viscosity, to the two sound velocities (c_0 for $\omega\tau_s<1$, c_∞ for $\omega\tau_s>1$) and also to a new contribution to the spectrum /2/,/7/. The time constant τ_s is ascribed to "structural relaxation", and it is of central importance in the liquid to glass transition diverging as $T\to T_{g+}$. (Usually it is extracted from measurements of overdamped motions (e.g. dielectric relaxation), where actually a distribution of τ, $g(\tau)$, is found upon approaching the glass transition temperature T_g from above. For a survey, see e.g. /3/.) Here we ask, to which extent precise measurements of the ω,q and T dependence of Brillouin scattering, together with a detailed analysis of the line shape, may supply correspondingly detailed information about relaxation of density waves in glassforming liquids.

Lineshape analysis requires stable spectrometer, high resolution and good statistics. We are using a Brillouin spectrometer based on a singlemode Ar ion laser (λ=514 nm), and a five pass Fabry–Perot–Interferometer (finesse about 30) which is scanned by three piezoelectric elements. For one spectrum, we sweep the FP of the order of 100 times through about 1.2 orders placing the first elastic peak at 0.10 of the full sweep (a chopper reduces the intensity, while the FP is on the elastic lines). We use normally 1024 channels, the dwelltime per channel is 20 ms, and a full sweep is performed within 20 to 30 seconds. The moving mirror is automatically reset on each back sweep to the same position and the finesse is also optimized ensuring that the profile of the elastic line (at 0.10 on the forward sweep) is very stable and as narrow as possible. In the cases to be discussed here, there was strong elastic scattering from impurities in the samples so that the elastic line – because it was essentially not broadened dynamically – could be used as the apparatus profile. In the fitting procedure, this profile is numerically

Editors: D. Richter · A.J. Dianoux · W. Petry · J. Teixeira

convoluted with the theoretical line shape before comparing with the measured spectra. – Broadening of the momentum transfer $q=4\pi n/\lambda \cdot \sin\theta/2$ (θ=full scattering angle, n=refractive index), due to the finite opening of the scattering aperture, was reduced by a slit diaphragm when necessary, and included in the analysis by summing up three weighted spectra calculated for slightly different angles $\theta \pm \Delta\theta$.

Measurements were performed on $ZnCl_2$ amd $GeSBr_2$. Both are good glassformers of the network type, glass temperatures are $T_g=100^0C$ and $T_g=-40^0C$, respectively. Samples, contained in sealed quartz ampoules, were studied at T=80 ... 600°C, and T=–60 – 200°C, respectively, and at scattering angles θ between 5° and 180°. Details are reported elsewhere /4, 5/. Brillouin scattering from $ZnCl_2$ had been studied earlier /6/, and analyzed in terms of line position and line width.

The Brillouin spectrum is given by the Fourier transform of the density–density correlation function $S(q,\omega)$ and is obtained theoretically by solving the linearized equations for a visco–elastic fluid, see e.g. /7,8/.The slow relaxation rate $1/\tau_S$ affects sound velocity and damping via the generalized viscosity $D_V=(\eta_V+4/3\eta_S)\ (m\rho_0)^{-1}$; η_V, η_S= volume, shear viscosity; $m\rho_0$=mass density. D_V is to be replaced by $D_V(\omega)$:

$$D_V(\omega) = D_V+(c_\infty^2 - c_0^2)\ \tau_S/(1 + i\omega\tau_S). \qquad (1)$$

For most liquids, a 3 line–approximation to $S(q,\omega)$ is used which is a sum of the Rayleigh–line, the Brillouin doublet and an additional line. The full solution on the other hand is a closed analytical expression also, though more complex; see e.g. /8/ eqs. 10.4.39 and 10.5.6. We mention here that there are other formulations of the viscoelastic spectra in the literature, e.g. /9/.

In $ZnCl_2$ the Brillouin lines become extremely non–Lorentzian in the region $\omega\tau_S=1$. As a first step we have tried to check the viscoelastic theory by comparing the spectra with both forms, the 3 line approximation and the full solution. Both were fitted to one set of five spectra taken at T=450°C θ=10...135°. The fit parameters were c_∞, c_0, and τ_S for the whole set of five angles, plus for each spectrum individually S(q) and a frequency independent background. The two sets of best fit parameters are listed in table 1, and are seen to differ somewhat. Rather obvious differences occur in the line shapes, as seen in fig. 1 (left). We summarize the following results of the analysis /4/ which pertain to the line shape:

(a) A good fit is in fact obtained using the viscoelastic theory, for a total of about 100 spectra, covering momentum transfers 0.3 ... 5.7. 10^6/m and temperatures 290 ... 600°C, fitted indvidually or (usually) in sets of 6 ... 10 spectra.

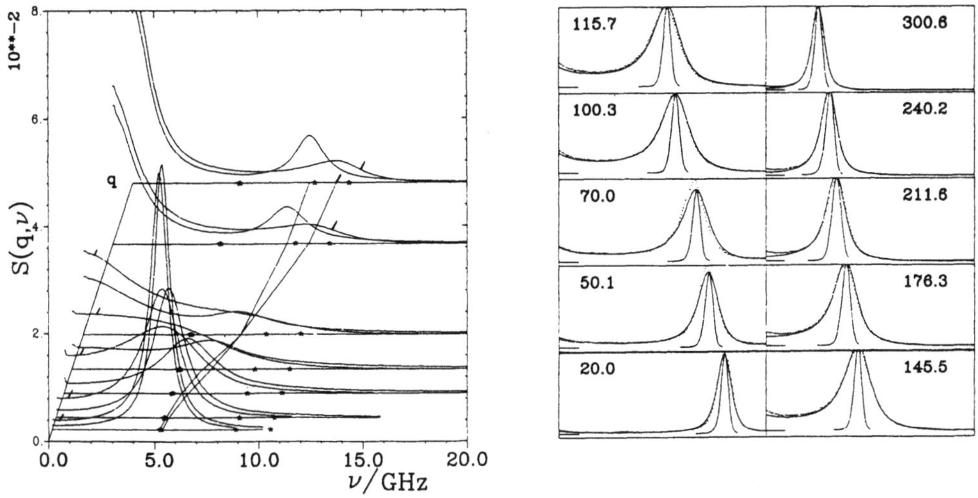

Fig. 1 (Left) Comparison of theoretical line shapes as given by full solution (left part) of viscoelastic theory, with the shape from the three–line approximation (right part), for the optimal parameter sets (from table 1). $ZnCl_2$, T=450°C, q(θ)=0.54–5.73·10^6/m as indicated. Full lines are theroretical curves, convoluted with the experimental apparatus profile. The spectra are the Stokes lines in first order, and the region roughly 0.1–0.5 FSR is displayed; FSR were different for different q. For the largest momentum transfer, however, the Antistokes–line of the next order has moved into the region displayed. Note that in the fits the intensity scale factor is a free parameter.
(Right) Brillouin spectra from $GeSBr_2$, θ=90°±1.5°, q=2.75·$10^6 m^{-1}$, temperatures in °C. Full lines are fits using parameters from table 1, including three relaxation times for eq.(1). The background level is indicated by horizontal lines.

(b) The fit is somewhat better for the full form, than for the said approximation (χ^2 is 1.5 ... 2.5 for the full form, and typically a factor of 2 worse for the 3 line approximation) Fig. 1 (left) shows that the differences in shape are quite remarkable in the region $\omega\tau_s \approx 1$.
(c) This full form corresponds to the one derived from the Mori–formalism, which fullfills the second moment rule and has two relaxation times (the longer one being τ_s, the short τ is contained in D_v).
(d) One single value for τ_s, and an Arrhenius temperature dependence were found sufficient in this case. Note that $(T-T_g)/T_g > 0.5$.
(e) When precise values for the parameters are important, or comparisons are made between different works, the form of solution chosen will become important.

In $GeSBr_2$, the line shape remains rather close to a Lorentzian in the whole regime of $\omega\tau_s \approx 1$. Fig. 1 (right) gives an example. Here we have only used the full solution of the

Table 1 Results from fitting viscoelastic theory to Brillouin spectra: a) $ZnCl_2$. 3 line approximation; b) $ZnCl_2$ full solution five 450°C spectra with $q=0.54...5.73\cdot10^6$/m were used; c) $GeSBr_2$, three relaxation times introduced into full solution. For spectra and fit curves see fig. 1 (right). $D_V=3\cdot10^{-2}$cm s^{-1}, $D_T=1\cdot10^{-3}$ cm^2 s^{-1}, $\gamma=1.05$.

$c_\infty/10^3$ m s^{-1}	$c_0/10^3$ m s^{-1}	$\tau/10^{-15}$s
a) 1.68	0.91	$4.5\cdot10^4$
b) 1.80	0.88	$4.0\cdot10^4$
c) $1.49-0.00345\cdot T/°C$	$0.92-0.0012\cdot T/°C$	$0.8/4.7/18.8\cdot\exp(3900°K/T)$

hydrodynamic equations. Since the two sound velocities in this case merge within the temperature range studied, a smooth crossover of the curve $c_\infty+c'_\infty\cdot T$ to $c_0+c'_0\cdot T$ was introduced (a crossing and $c_\infty<c_0$ would not make sense). Arrhenius form $\tau_s=\tau_0\exp(E_A/k_BT)$ was assumed. Acceptable fits were obtained; however, the apparent activation energy E_A changed with q by almost a factor of 2; the equivalent change of τ_s by an order of magnitude appears improbable at the q values in question /7/. We therefore ascribe this inconsistency to the approximation of using one single τ_s value.
To study the effect of a distribution $g(\tau_s)$, fits were performed using three values of τ_s (with 1/3 weight each) in eq. (1), leaving the ratios between them as additional fit parameters. The new fits had the same quality as before. The preliminary results are presented in table 1. The present fits are not yet satisfactory (fig. 1 (right), $\chi^2=3$); note also that the fit placed the three τ's about evenly on a log τ scale, contrary to expectation.
The examples given demonstrate that in spite of the seemingly simple shape(s) of Brillouin spectra the search for quantitative analysis of the line shape connects it closely to the details of viscoelastic theory and of the relaxation processes.

The help of C. Paulick is gratefully acknowledged. – This work is supported by Deutsche Forschungsgemeinschaft through Sfb 337.

References

1. R. Zwanzig: J. Chem. Phys. 43 (65) 714
2. R. Mountain: Rev. Mod. Phys. 38 (66) 205
3. J. Wong and C.A. Angell: Glass, M. Dekker, New York 1976

4. M. Soltwisch, J. Sukmanowski and D. Quitmann: J. Chem. Phys. 86 (87) 3207
5. S. Loheider, I. Petscherizin, G. Vögler, M. Soltwisch and D. Quitmann: to be published
6. G. Knape: J. Chem. Phys. 80 (84) 4789
7. J.P. Boon and S. Yip: Molecular Hydrodynamics, McGraw Hill, New York 1980
8. B.J. Berne and R. Pecora: Dynamical Light Scattering, Wiley, New York 1976

   In eq. (10.5.6), the expression for $N_2(\omega)$ is $\ldots -\gamma D_T b_1 \tau q^4 \ldots$ and the denominator of $S(q,\omega)$ reads $(D_1^2+D_2^2)$.
9. Y.H. Liu and C.H. Wang: J. Chem. Phys. 70 (79) 681

Acoustic Glass Relaxations and Mesomorphic Transitions in Polymer Liquid Crystals

J.K. Krüger[1], *C. Grammes*[1], *and J.H. Wendorff*[2]

[1]Fachbereich Physik, Universität des Saarlandes, D-6600 Saarbrücken, Fed. Rep. of Germany
[2]Deutsches Kunststoff-Institut, Schloßgartenstr. 6R, D-6100 Darmstadt, Fed. Rep. of Germany

1 Introduction

Classical liquid crystals of low molecular weight (CLC's) show significant optical anisotropies /1a-b/, but only small elastic anisotropies are observed within the monodomain state /2/. This holds true even at high frequencies /e.g. 3/, where the shear stiffness constants become significant. This means that CLC's show only a very weak coupling between the static order parameter /1b/ and the classical elastic strain tensor components ε_i (Voigt notation).

In nematic CLC's the anisotropy of the sound velocity of quasilongitudinal polarized phonons is typically of the order 10^{-2} with a maximum along the optical axis /2/. In general no acoustic attenuation anomalies are observed at the nematic - isotropic transition (T_{ni}) at hypersonic frequencies. Only recently Brillouin spectroscopic investigations of side chain polymer liquid crystals (SPLC's) have been reported for the smectic and nematic state of polydomain samples PA6BCN and PA6BOCH3 /4a/. Preliminary results on SPLC's obtained from ultrasonic investigations have been published recently /4b/. In contrast to the behaviour of CLC's, monodomains of SPLC's show a rather strong elastic anisotropy with a maximum along the optical axis and the inverse shear stiffness constants do not diverge. Moreover, all elastic constants behave continuously at the nematic-isotropic transition. From the behaviour of certain optoacoustic properties and refractive index functions, order parameter fluctuations turned out to play no role at T_{ni}, but the glass transition dominates the dynamic elastic properties at hypersonic frequencies. From these results a significant coupling between the static order parameter and the elastic strain ε_3 was concluded. Tentatively this coupling was attributed to the dynamic glass transition, implying an unusual frequency / temperature dependence of the coupling constant.

However, neither the quasistatic glass transition nor the dynamic glass transition and their influence on the hypersonic properties of SPLC's were investigated yet.

The aim of this work is to present and discuss first results on the static and dynamic hypersonic properties at the glass transition and the nematic - isotropic transition of a SPLC with a polymethacrylate backbone, a biphenyl type mesogenic side-group, a cyanide end-group and a n-

Springer Proceedings in Physics, Vol. 37 **Dynamics of Disordered Materials**
Editors: D. Richter · A.J. Dianoux · W. Petry · J. Teixeira

Table 1 The chemical formula, the transition temperatures and elastic stiffness coefficients at room temperature of PMA6BCN

Chemical formula:

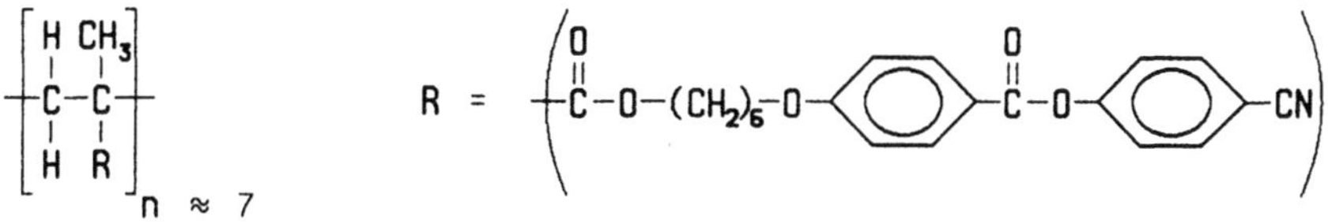

Transition temperatures:

T_g = 273.9 K T_g^{dyn} (f = 4.4 GHz) = 390 K T_{ni} = 330.2 K

Elastic stiffness coefficients: (ρ = 1100 kg/m^3)

$c_{11} = c_{22}$ = 6.72 GPa c_{33} = 13.69 GPa $c_{44} = c_{55}$ = 1.34 GPa

$c_{13} = c_{23}$ = 5.04 GPa

alkane spacer (Table 1). For simplicity we use the short notation "PMA6BCN" in the following to denote the material.

2 Experimental

Thin film-like samples of thickness d ~ 25 µm have been prepared using glass slides as support for the film. To prepare thin homogeneously aligned monodomain samples, the surfaces of the glass slides were coated with a polyimide (Merck, Liquicoat PI) and rubbed along an arbitrary direction within the film plane which defines the director and the x_3-axis of an orthogonal coordinate system, the x_2-axis lying perpendicular to the film plane. The transition temperatures obtained from Brillouin spectroscopy and differential scanning calorimetry (DSC) are given in Table 1.

The Brillouin measurements were made with a high performance Brillouin spectrometer using an argon-ion-laser at the vacuum wavelength λ_o = 514.5 nm. The characteristics of this spectrometer and the measuring technique were discussed elsewhere /5/. The principle of the 90A-scattering geometry is given in ref. /6/.

To avoid heating or burning of the sample, the effective laser power within the scattering volume was kept below 5 mW.

3 Results and Discussion

The 90A-scattering geometry was used to determine sound velocity polar plots for PMA6BCN at room temperature by rotating the phonon wave vector

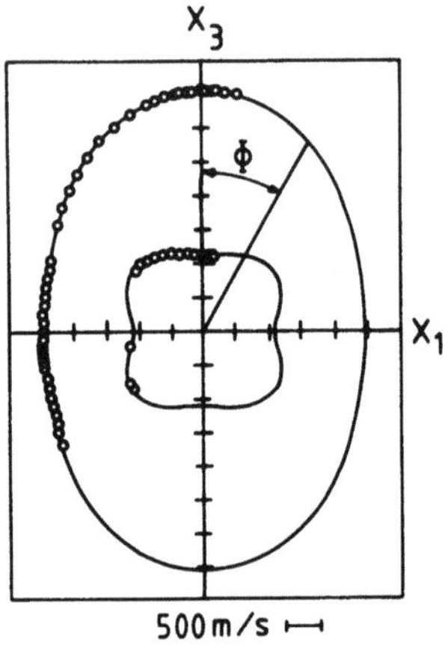

<u>Figure 1</u> Sound velocity polar plot of PMA6BCN for **q** oriented in the (x_1, x_3)-plane measured at room temperature. I: sound velocities of quasi-longitudinal sound waves, II: sound velocities of quasi transverse sound waves. **Φ**: Angle between the wave vector **q** and the x_3-axis

q within the (x_1, x_3)-plane. Figure 1 shows the measured data points for the quasi-longitudinal (I) and quasi-transverse (II) sound waves.

The basic relation for the determination of the elastic stiffness coefficients from the measured data is the Christoffel equation. Assuming fiber symmetry for the monodomain samples, this leads to the following cubic eigenvalue equation:

$$\det[\, \mathbf{l}\,\mathbf{c}\,\mathbf{l}^T - E\, c_{\mathbf{q}}] = 0 \quad \text{with} \quad c_{\mathbf{q}} = \rho\omega^2/q^2 = \rho v \tag{1}$$

and

$$\begin{vmatrix} l_1 & 0 & 0 & 0 & l_3 & l_2 \\ 0 & l_2 & 0 & l_3 & 0 & l_1 \\ 0 & 0 & l_3 & l_2 & l_1 & 0 \end{vmatrix} = \mathbf{l} \qquad \begin{vmatrix} c_{11} & c_{12} & c_{13} & & & \\ c_{12} & c_{11} & c_{13} & & & \\ c_{13} & c_{13} & c_{33} & & & \\ & & & c_{44} & & \\ & & & & c_{44} & \\ & & & & & c_{66} \end{vmatrix} = \mathbf{c} = \mathbf{c}^T \tag{2}$$

and $$\hat{\mathbf{q}} = (l_1, l_2, l_3) = l_1\,\mathbf{x}_1 + l_2\,\mathbf{x}_2 + l_3\,\mathbf{x}_3 \tag{3}$$

$\mathbf{x}_i$ (i=1, 2, 3): unit vectors in the coordinate directions of the sample coordinate system, $\mathbf{q} = q\,\hat{\mathbf{q}}$: acoustic wavevector, l_i (i=1, 2, 3): direction cosines of **q**, c_{ij} (i, j = 1, 2, ...,6): elastic stiffness coefficients, $f = \omega/2\pi$: sound frequency, v: sound velocity, ρ: mass density, $c_{\mathbf{q}}$: eigenvalues which have been measured by Brillouin spectroscopy for different directions $\hat{\mathbf{q}}$ of phonon propagation within the (x_1, x_3)-plane. An appropriate data analysis procedure using (1) - (3) and Marquardt's nonlinear least squares fit yields the elastic stiffness coefficients c_{11}, c_{33}, c_{44} and c_{13}. The results are given in Table 1. The coefficient c_{33} is at room temperature about twice as large as c_{11}, indicating a pronounced anisotropy of the longitudinal elastic stiffness with the maximum along the preferential orientation of the mesogenic side chains. The influence of a preferential orientation of the short backbone molecules is not known, but believed to be negligible or overcompensated by the mesogenic side chains.

We have also investigated the temperature dependence of longitudinal sound modes propagating along the x_3-axis (Fig. 2). The transverse sound velocity v_4 behaves comparably to that found in amorphous polymers /e.g. 7 /. The remarkable size of these values is a consequence of the frequency dependent imaginary part of the complex shear viscosity, which leads to a real elastic shear constant at high frequencies ($\omega\tau >> 1$, τ : relaxation time of the corresponding shear viscosity). A strong influence of the order parameter is reflected by the temperature behaviour of the longitudinal sound velocity v_3, whereas v_4 shows nearly no anomaly at the nematic-isotropic transition temperature T_{ni}. As expected, the quantity v_3 shows a slight discontinuity, whereas v_4 behaves continously at T_{ni} (within the error limits).

Well below the transition temperature T_{ni} a further sound velocity anomaly (kink) appears in the sound velocity curves of v_3 and v_4 at a temperature T_g. This anomaly is attributed to the quasistatic glass transition observed in DSC measurements at the same temperature. The corresponding dynamic glass transition (α-process) appears well above the temperature T_{ni} as a maximum in the temporal sound attenuation coefficient Γ at the temperature T_g^{dyn} = 390 K (Fig. 3). Qualitatively, the same hypersonic anomalies are observed at the quasistatic and dynamic glass transitions of amorphous polymers, e.g. in polymethylmethacrylate (PMMA) /7/. From Fig. 2 and 3 it turns out, that the nematic - isotropic transition appears about 60 K below the dynamic glass transition,

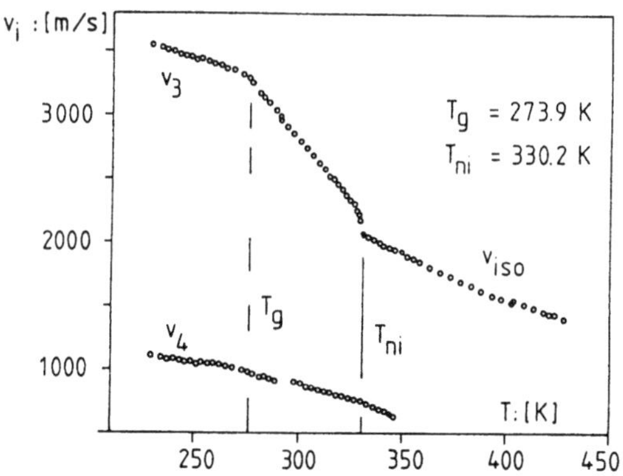

Figure 2 Longitudinal sound velocities v_3 and v_{iso} and transverse sound velocity v_4 of PMA6BCN as a function of temperature T. For further explanations see text

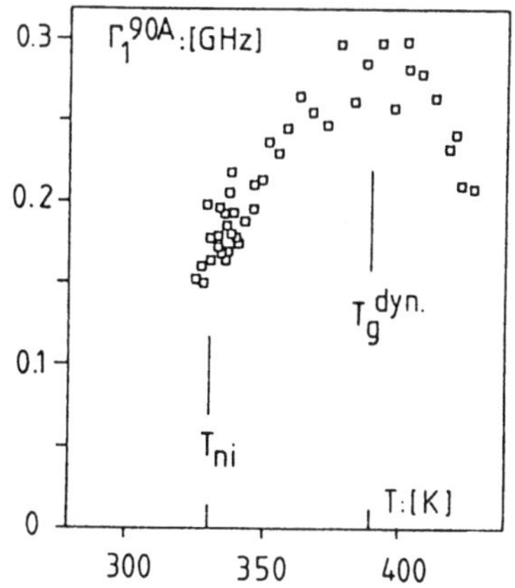

Figure 3 Sound attenuation Γ_1^{90A} of the longitudinal sound wave in the isotropic phase of PMA6BCN as a function of temperature T. For further explanations see text

indicating, that the sound velocities v_3 and v_4 reflect already the high frequency static properties of a dynamically frozen liquid. This is believed to be the origin for the observed elastic anisotropy below T_{ni}, reflecting the coupling between the order parameter and the strain component ε_3. This idea is supported by the fact, that the $v_3(T)$- curve shows a flex point between T_g and T_{ni}, indicating, that the strong decrease of the sound velocity v_3 just above T_g is not a consequence of the order parameter saturation. From preliminary opto-acoustic data /8/ it turns out, that order parameter fluctuations do not contribute significantly to the sound attenuation at high frequencies.

4 References

1. a) H. Kelker, R. Hatz, "Handbook of Liquid Crystals", Verlag Chemie, Weinheim (1980); b) P. G. de Gennes, "The Physics of Liquid Crystals", Clarendon Press, Oxford (1979)
2. K. Miyano, J. B. Ketterson, in Physical Acoustics, Vol. XIV, p.93 ff (1979)
3. A. Wergin, W. Krasser, H. H. Stiller, C.G.B.Frischkorn, Phys. Rev. A20, 1120 (1979)
4. a) J. K. Krüger et al., Phys. Rev. A37, 7 (1988)
   b) L. Benguigui, P. Ron, Mol. Cryst. Liq. Cryst. 153, 241 (1987)
5. J. K. Krüger, R. Kimmich, J. Sandercock, H.-G. Unruh, Polym. Bulletin 5, 615 (1981)
6. J. K. Krüger, A. Marx, L. Peetz, R. Roberts, H. G. Unruh, Colloid & Polym. Sci. 264, 403 (1986)
7. J. K. Krüger, R. Roberts, H.-G. Unruh, K. P. Frühauf, J. Helwig, H. E. Müser, Progress Colloid & Polym. Sci. 71, 77 (1985)
8. C. Grammes, J. K. Krüger, to be published

Very Low Frequency Light Scattering in Amorphous Silica

A. Fontana, L.M. Gratton, and C. Tosello

Dipartimento di Fisica, Università di Trento, I-38050 Povo(TN), Italy

In this work we present the first Raman and Brillouin measurements in amorphous silica performed at the same time without changing the experimental conditions.
By the help of a new double-monochromator it has been possible to perform the measurements in a spectral region never studied at present by light scattering in solids: 0.40-8.0 cm^{-1}.

These new experimental data agree with the "arguments" reported in a previous work, and demonstrate that the Light Scattering Excess in amorphous silica is not due to Brillouin doublets.

Recently, there has been a considerable contribution to the understanding of disordered solids by means of light scattering investigations. For istance, in recent years, the fractal description has been intensively used to treat the structure and the dynamics of several inhomogeneous materials [1]. For providing some informations about the density of states, the light scattering technique seems to be a very promising tool [2], unfortunately the comprehension of the nature of the light scattering at low frequency (0-100 cm^{-1}) is not complete [3].

As a matter of fact a quasielastic scattering, called "Light Scattering Excess", is superimposed to the scattering, induced by the disorder, due to acoustic sound waves (phonons and/or fractons). This scattering can mask and modify the shape of the disordered induced light scattering, making less general and sure the theories employed to explain the structure and the dynamics of amorphous systems.

A lot of causes can originate the L.S.E., without taking into account a possible and often probable presence of some spurious elastic components: firstly the presence of topological defects in the structure allows the jump of the ions in equivalent sites; the jumps modulate the system polarizability inducing a light scattering centered at zero frequency [4]; secondly the two phonon processes could be relevant in our systems; finally a third cause could be that the L.S.E. is the tail of Brillouin scattering lines.

Editors: D. Richter · A.J. Dianoux · W. Petry · J. Teixeira

In a previous work, performed on amorphous silica, WINTERLING [5] measured the light scattering from ω = 4-5 cm^{-1}, finding a L.S.E. in a measured range of about 5-10 cm^{-1}. By the analysis of the behaviour of the extrascattering intensity as a function of the temperature, he excluded that the L.S.E. could be due to two phonon processes; further, by the analysis of Brillouin and the L.S.E. intensities he estimated that the L.S.E. was not due to the tails of Brillouin lines.

In the present work, we report a detailed investigation of the Light scattering in amorphous silica from 0.10 to 35 cm^{-1}, measuring at the same time the Brillouin and the Raman scattering and covering for the first time, in our knowledge, the frequency range necessary to measure the shape of the L.S.E.

The light scattering has been analysed by a new conception double-monochromator [6] with very high resolution (0.12 cm^{-1} FWHM with the three slits opened at 100 μm).

The sample was synthetic amorphous silica (wet sample ~1200 p.p.m.). The VV polarised scattering was collected in usual 90° geometry. The 5145 Å line of an Argon-ion laser has been used.

In Fig. 1 we report the Brillouin scattering obtained with the three slits opened at 100 μm in the -1 to +1 cm^{-1} frequency range.

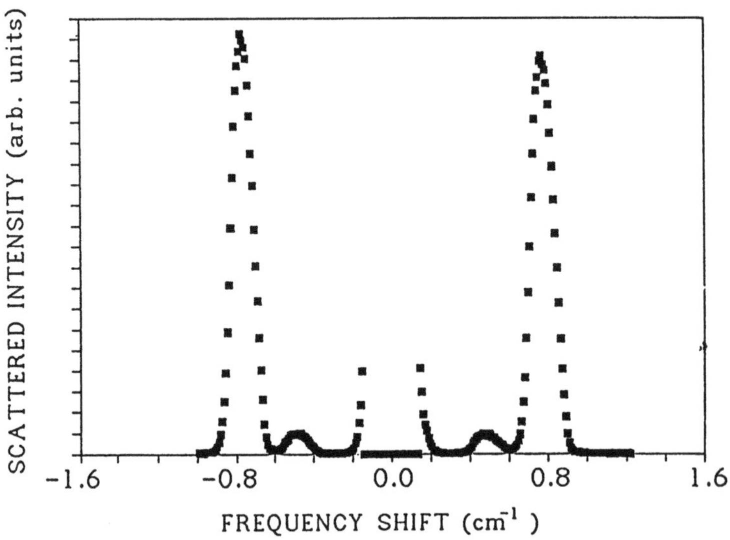

Figure 1 Brillouin scattering of amorphous silica. The instrumental resolution was 0.12 cm^{-1}.

The shape and the Longitudinal-Transverse peak intensity ratio are those usual for the amorphous silica already reported in literature. Obviously the line width is only due to instrumental resolution (0.12 cm^{-1}).

In order to achieve an appreciable Raman intensity it is necessary to increase the slit widths; in fact the Brillouin intensity is 3 or 4 orders of magnitude greater than the Raman intensity due to the very different widths of the two processes (without taking into account other causes).

In Fig. 2a we report the scattering between 0.40 and 7.7 cm^{-1}, with the three slits opened at 500 μm. The resolution was 0.30 cm^{-1}. The resolution was so low that the Longitudinal and Transverse doublet is degenerate. We can note the presence of a low scattering at the end of Brillouin tail with a completely different slope; this fact is evidenced in Fig. 2b where we changed the ordinate scale by three order of magnitude.

In Fig. 3 we report the same datum of Fig. 2 in the frequency range 0-35 cm^{-1}, in order to show the presence of a minimum in the spectrum at about 10 cm^{-1}; after this frequency the scattering again increases assuming the typical shape already reported in literature [5].

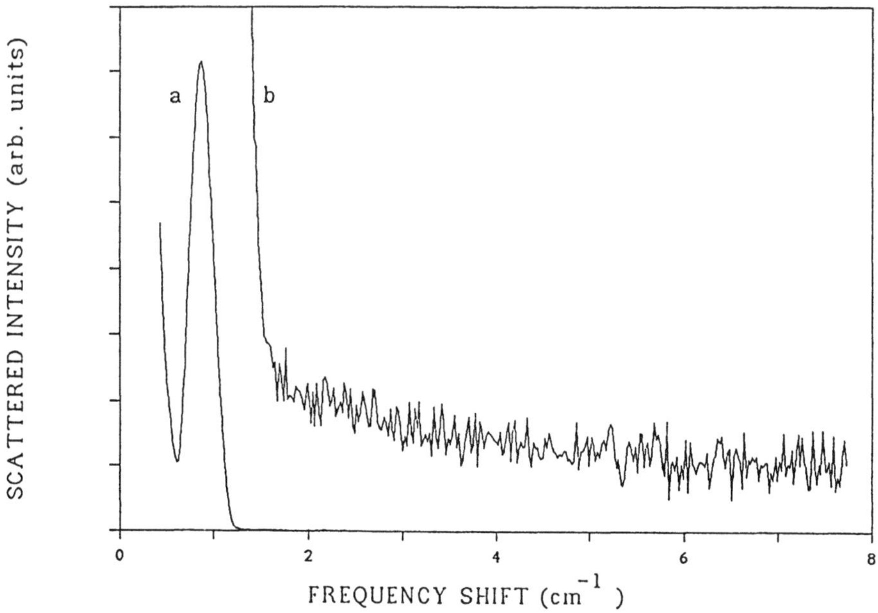

Figure 2 a) Brillouin and Raman scattering on the same scale; b) the same datum of a) with the y scale changed 1000 times.

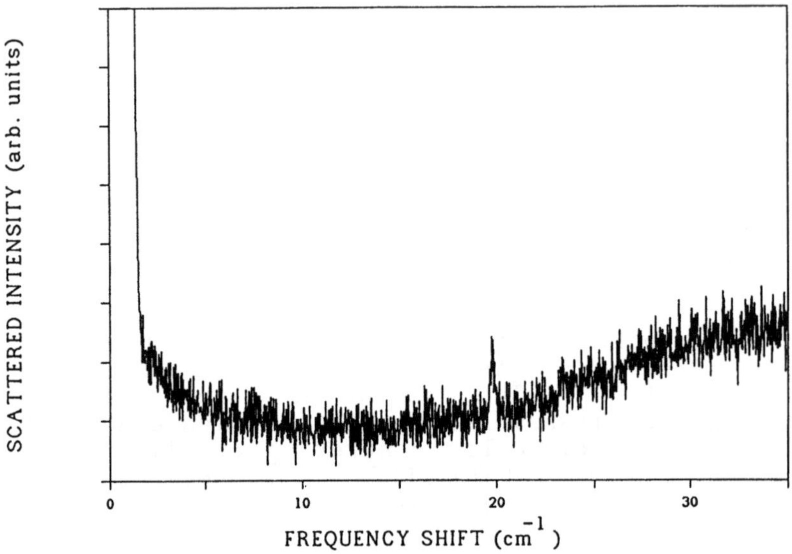

Figure 3 Brillouin and Raman scattering from 0.10 to 35.0 cm^{-1}. Brillouin is not in scale.

In conclusion we studied the light scattering intensity of a solid from about 0.40 to 8.0 cm^{-1}, demonstrating that the L.S.E. is not due to a spurious elastic component. From the preliminary examination of these data, it results that the L.S.E. in amorphous silica, is not due to the tails of Brillouin lines, confirming the hypothesis made by Winterling in his previous work [5]. Having measured the L.S.E. shape, it is now possible to verify the developed theories with confidence: this work is in progress.

References

1. B.B.Mandelbrot: In The Fractal Geometry of Nature (ed by Freeman, S.Francisco, 1982)
2. A.Fontana, F.Rocca, and M.P.Fontana, Phys.Rev.Letters 58,503 (1987)
3. F.Rocca, and A.Fontana, Philosophical Magazine, umpublished
4. R.Zeyher, Z.Phys. B31, 127 (1978)
5. G.Winterling, Phys.Rev B12, 2432 (1975)
6. V.Mazzacurati, P.Benassi, and G.Ruocco, umpublished.

Longitudinal and Transverse Brillouin Intensities in Glasses: Experiments and Interaction Induced Contributions

V. Mazzacurati[1], *M. Nardone*[1], *G. Ruocco*[2], *and G. Signorelli*[2]

[1]Dipartimento di Fisica, Università di Roma La Sapienza, Piazzale A. Moro 2, I-00185 Roma, Italy

[2]Dipartimento di Fisica, Università di L'Aquila, L'Aquila, Italy

1.Introduction

We present in this work a microscopic theory of Brillouin scattering intensities in a harmonic disordered solid composed of units carrying a spherical point polarizability α. The theory includes dipolar propagation processes at all orders accounting for the full expression of the dipole propagator [1]. It follows closely the analogous derivation already made for rare gas crystals [2] and accounts for the disorder by expanding the various microscopic quantities in fluctuations about their average values. Neglecting the fluctuations and identifying averaged quantities with the equilibrium averaged ones, it yields expressions for the Brillouin intensities which are functions of the refractive index n of the medium.

This theory yields for the first time, non trivial expressions for the depolarized intensities of both transverse and longitudinal modes in disordered systems. A comparison with experiments on a number of real multicomponent glasses suggests that the theory is a reasonable approximation also for these systems.

2.Basic Theory

The intensities in the various scattering configurations are given by the Fourier transforms of the correlation functions [2]

$$G_{\alpha\beta\gamma\delta}(t)=\sum_{ij} < \Pi^{i}_{\alpha\beta}(t)\, \Pi^{j}_{\gamma\delta}(0) \exp(i\vec{q}\,.\,\{\vec{r}_i(t)-\vec{r}_j(0)\}) > \tag{1}$$

$\vec{q}=\vec{k}_i-\vec{k}_s$ is the exchanged wavevector and α, β, γ, δ are cartesian indices.

The istantaneous atomic polarizability tensor components [3] $\Pi^{i}_{\alpha\beta}$ are shown to obey the self consistent equation [2]

$$\overline{\Pi}^{i}(t)=\alpha\,\bar{I}+\alpha\sum_{j}\overline{W}(\vec{r}_{ij}(t))\exp(i\vec{k}_i.\,\vec{r}_j(t)) \tag{2}$$

where $\vec{r}_{ij}=\vec{r}_i-\vec{r}_j$, $\bar{I}$ is the identity tensor and

$$W_{\alpha\beta}(\vec{R})=\left\{\nabla_\alpha\nabla_\beta+(\omega_o/c)^2(\hat{k})_\alpha(\hat{k})_\beta\right\}\exp\left[i(\omega_o/c)|\vec{R}|\right]|\vec{R}|^{-1} \tag{3}$$

is the modified dipole propagator resulting from the resummation of boundary effects into the driving (Maxwell) field [4]. ω_0 is the frequency of the incident radiation and ^ denotes unit vectors. When $\overline{W}(\vec{r}_{ij})$ is expanded in atomic displacements from the equilibrium position $\vec{u}_i(t) = \vec{r}_i(t) - \vec{x}_i$, trivial algebra yields:

$$\sum_i \overline{\Pi}^i(t) \exp\left\{ i\vec{q}\cdot\vec{r}(t)\right\} \cong \sum_i \sum_\mu \exp\left(i\vec{q}\cdot\vec{x}_i\right) \overline{C}^i_\mu u^i_\mu(t) \tag{4}$$

where

$$\overline{C}^i_\mu = q_\mu \left(\overline{\Pi}^i\right)_{eq} + \sum_l \left(\partial\, \overline{\Pi}^i / \partial\, u^l_\mu\right)_{eq} \exp\left\{ -i\vec{q}\cdot\left(\vec{x}_l - \vec{x}_i\right)\right\} \tag{5}$$

This tensor is an "electrical" property which varies from site to site owing to the disordered structure of the glass. With obvious definitions we further separate this quantity into a configurationally averaged value

$$\overline{C}_\mu = (1/N) \sum_i \overline{C}^i_\mu = q_\mu \left(\overline{\Pi}\right)_{eq} - i\, \overline{Q}^{eq}_\mu \tag{6a}$$

and into local fluctuations

$$\delta \overline{C}^i_\mu = \overline{C}^i_\mu - \overline{C}_\mu \tag{6b}$$

which we assume to be uncorrelated over dimensions smaller than $\left|\vec{q}\right|^{-1}$. While the former gives rise to the appearance of $\vec{k} = \vec{q}$ wavector modes (much like in a perfect crystal), the latter is generally responsible for a broad underlying contribution, generally referred to as disorder induced one-phonon scattering, which will be hereafter neglected [5].

However, in attempting to solve the self consistent equation (2) for $\overline{\Pi}^{eq}$ and $\overline{Q}^{eq}_\mu$ one finds that these equilibrium averaged quantities are coupled to their fluctuations through the propagator $\overline{W}(\vec{R})$. As a first step we have solved these equations in the zero order approximation (i.e. neglecting all fluctuation terms) obtaining

$$^{\circ}\overline{\Pi}^{eq}\left(\vec{k}_i\right) = \alpha \left[I + \alpha\, \overline{\tau}\left(\vec{k}_i\right)\right]^{-1} \tag{7}$$

$$^{\circ}\overline{Q}^{eq}_\mu\left(\vec{k}_i, \vec{k}_s\right) = -{}^{\circ}\overline{\Pi}^{eq}\left(\vec{k}_s\right) \cdot \overline{\Omega}_\mu\left(\vec{k}_i, \vec{k}_s\right) \cdot {}^{\circ}\overline{\Pi}^{eq}\left(\vec{k}_i\right) \tag{8}$$

where

$$\overline{\tau}\left(k_i\right) = \left(1/N\right) \sum_{ij} \overline{W}\left(\vec{x}_i - \vec{x}_j\right) \exp\left\{ -i\vec{k}_i \cdot \left(\vec{x}_i - \vec{x}_j\right)\right\} \tag{9a}$$

$$\overline{\Omega}_\mu\left(\vec{k}_i, \vec{k}_s\right) = \left(1/N\right) \sum_{ij} \left\{ \nabla_\mu \overline{W}\left(\vec{x}_i - \vec{x}_j\right)\right\} \exp\left\{ -i\vec{q}\cdot\left(\vec{x}_i - \vec{x}_j\right)\right\} \tag{9b}$$

These quantities may be explicitly calculated once the bare polarizability α and the structure of the system is known. However if we assume

$$^{\circ}\overline{\Pi}^{\alpha q} \cong \overline{\Pi} = \left(\overline{\varepsilon} - \overline{I}\right)/4\pi\rho \tag{10}$$

we can evaluate directly the perpendicular part of $\overline{\Pi}$ from the knowledge of the refractive index. Since the Lorentz-Lorenz relationship follows directly from eqs. (10) and (7), we can also obtain a value for α and hence for $^{\circ}\overline{Q}_{\mu}$.

The final expressions for the Brillouin scattering intensities are obtained performing a normal mode expansion of the atomic displacements. In the classical limit ($h\,\omega/2\pi << kT$) and describing the eigenmodes within the framework of the elastic theory of continuous media [6], we obtain

$$\begin{aligned} J_{VV} &= \left(2K_BT/\rho C_{11}\right)\left[(\varepsilon - 1)/4\pi\right]^2\left[1 + (\varepsilon - 1)/5\right]^2 \\ J_{HH} &= \left(2K_BT/\rho C_{11}\right)\left[(\varepsilon - 1)/4\pi\right]^2\left[(\varepsilon - 1)/5\right]^2 \\ J_{HV} &= \left(2K_BT/\rho C_{44}\right)\left[(\varepsilon - 1)/4\pi\right]^2\left[(\varepsilon - 1)/5\right]^2 \end{aligned} \tag{11}$$

Cij being the usual elastic constants. Connection with the phenomenological theory [7],is readily made in terms of the elasto optic coefficients Pij for which we obtain

$$\begin{aligned} P_{12} &= \left[(\varepsilon - 1)/\varepsilon\right]\left[1 + (\varepsilon - 1)/5\right] \\ P_{11} &= \left[(\varepsilon - 1)/\varepsilon\right]\left[1 + 3(\varepsilon - 1)/5\right] \\ P_{44} &= \left(P_{11} - P_{12}\right)/2 \end{aligned} \tag{12}$$

Notice that P_{44} is non vanishing and fulfils the isotropy condition.

It is also worthwhile to notice that, if we had used only the short range limit of the dipole propagator (usually encountered in DID theories instead of $\overline{W}(\vec{R})$) we would have obtained

$$P_{11} = P_{12};\; P_{44} = 0 \tag{13}$$

which implies the absence of depolarized Brillouin scattering.

3. Experiments and Comparisons

Brillouin frequency shifts and intensities have been measured using a 514.5 nm laser in single TEM $_{00}$ mode and a high resolution, high luminosity, high contrast monochromator recently developed by SOPRA [8]. Scattered photons are detected by a single photon counting device connected to an on line computer and corrected for dark counts. Typical resolution is 0.1 +0.05 cm^{-1}.

The following procedure has been found to give the more reliable and reproducible results:

i) 90° scattering measurements are performed with the incoming radiation polarized at 45°with respect to the scattering plane. This allows to measure $(J_{VH}+J_{HH})/2$ or $(J_{VV}+J_{HV})/2$

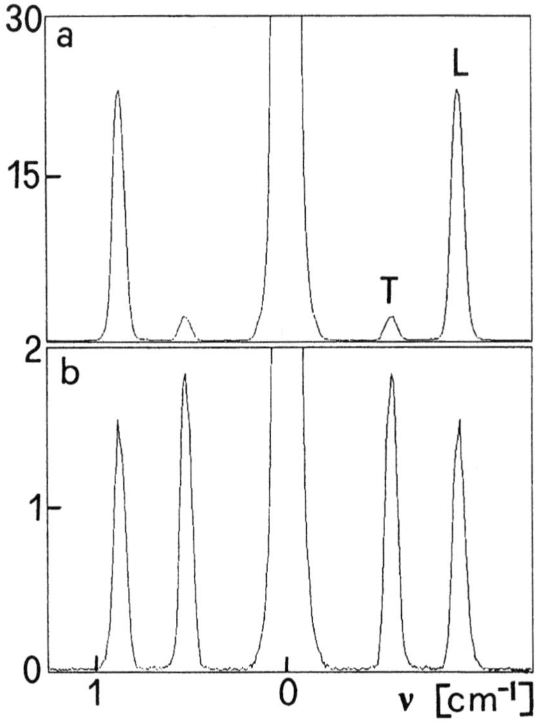

Fig. 1 Typical spectra of room temperature BK7 measured with a resolution of .07 cm . a) HV+VV configuration; b) VH+HH configuration

configurations depending on the selected polarization of the scattered radiation (see Fig.1). After having verified, from the former configuration, that for BK7,

$$J_{HH}/J_{VH}=2\,C_{44}/C_{11}=2\left(v_T/v_l\right)^2$$

within 5 x 10^{-4}, the latter has been used to obtain P_{44}/P_{12}.

ii) Measurements of absolute intensities, using toluene as external standard, are taken without focusing the incoming laser beam.

iii) Laser beam intensity has been monitored after the sample so to encompass the same reflectivity corrections as the scattered radiation.

In some glasses (namely SF6, SF14 and SF58) the intensity of the VH transverse modes is so weak that they are submerged by the tails of the longitudinal VV lines and the procedure described in (i) is no longer satisfactory. In these cases an accurate analysis of all VH, HV, HH, and VV spectra is necessary to account for unwanted polarization filtration and to obtain reliable values for P_{44}/P_{12}. (see fig. 2).

Measurements performed on a variety of commercial glasses are reported in Table 1 . In column 3 and 4 we report the values of P_{12}. as measured experimentally and as calculated using our zero order theoretical expression with the refractive index values reported in column 1 [9]. The two sets of numbers, exception made for SF6 and BaSF52, all lie within 30% of the predicted value. Something qualitatively similar occours for the P_{44} values reported in column 5 and 6, the deviations being considerably larger. It should be noted that although the uncertainty in our intensity measurements is only 2%, the absolute values of P_{12} are affected

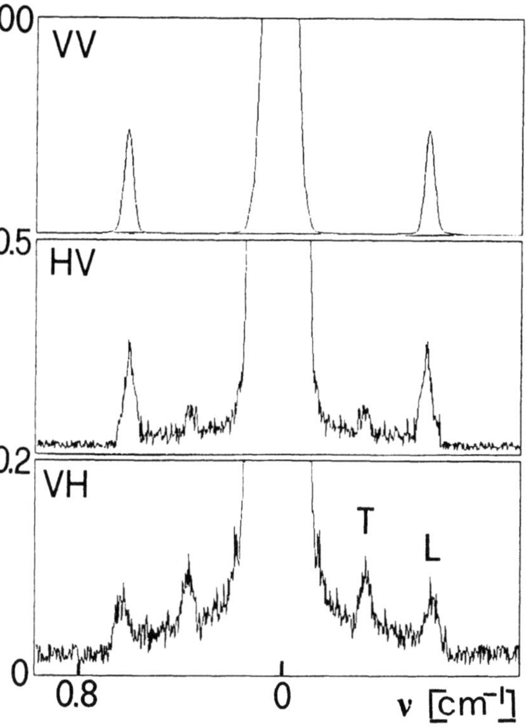

Fig. 2 Room temperature VV, HV and VH spectra of SF6 showing the longitudinal VV component leaking in the depolarized configurations.

also by the unknown uncertainity in the P_{12}. of toluene. However the values of P_{44}/P_{12}., being derived from the same spectrum, are reliable to better than 10^{-2}, indicating that our theory yields good predictions for BK7, LF5, F9, F1 and BaSF52.

The deviations of P_{44} from the predicted values seem particularly significant, since these deviations increase monotonically with mass density (as well as with the refractive index). The origin of these deviations may be due to the fact that our zero order theory accounts only

Table I Relevant parameters and elasto-optic coefficients for the glasses measured in this work.

		ρ	P_{12}		P_{44}		P_{44}/P_{12}	
Glass	n	(g/cm)	exp	th	exp	th	exp	th
BK7	1.517	2.51	.226	.310	.058	.064	.26	.21
LF5	1.581	3.22	.242	.312	.038	.073	.16	.23
F9	1.621	3.56	.248	.312	.035	.077	.14	.25
F1	1.626	3.65	.256	.312	.033	.077	.13	.25
BaSF52	1.702	3.96	.188	.312	.035	.086	.19	.28
SF14	1.762	4.54	.249	.310	.018	.092	.07	.30
SF6	1.805	5.18	.087	.309	.002	.096	.02	.31
SF58	1.918	5.95	.351	.304	.011	.106	.03	.35

partially (through the $^{\circ}\overline{\Pi}^{eq} \rightarrow \overline{\Pi}$ substitution) for the effects of local structural fluctuations and the presence of differently polarizable species on the Brillouin intensities.

Both these effects, hovewer, produce "fluctuations" in the microscopic electrical properties which should contribute also to the broad disorder induced component underlying our spectra [5] and partially visible in Fig. 2. It should therefore be usefull, for a better understanding of the approximations involved in our theory to see wether the depolarized Brillouin intensities are correlated to those of this disorder induced scattering. Work is in progress in our laboratory in this direction.

References

1. V. Mazzacurati, G. Signorelli and G. Ruocco Europhys. Lett. 2 (11) pp. 877- 881 (1986)
2. V. Mazzacurati and G. Ruocco, Mol. Phys. 61 1391, 1987.
3. A.D.Buckingham, Adv.Chem.Phys. 12, 107, 1967
4. See for example: B.U. Felderhof, Physica 76, 486, 1974.
5 V. Mazzacurati, M. Nardone, G. Ruocco and G. Signorelli on "Phenomena induced by intermolecular interactions" pag. 567. Edited by G. Birnbaum (Pfenum Press, New York (1985)) - A. J. Martin and W. Brening, Phys. Status Solid (b) 64, 163, 1974.
6. C. Kittel "Introduction to Solid State Physics", J. Wiley e Sons, New York, 1968.
7. R. Vacher and L. Boyer, Phys. Rev. B6, 639, 1970 L. Landau and E. M. Lifshitz, "Electrodynamics of continuous Media" (Pergamon, Oxford 1960)
8. V. Mazzacurati, P. Benassi and G. Ruocco J. of Physics E, Sci. Instrum. 21 pp. 798-804 (1988)
9. Commercial catalogue of Jena^{er} Glaswerk - Schott & Gen., Mainz.
10. D.Heimann, D.S.Hamilton and R.W.Hellwarth, Phys.Rev B19, 6583, 1979.

Phonons in Glasses

W. Schirmacher and M. Wagener

Physik-Department E13, Technische Universität München,
D-8046 Garching, Fed. Rep. of Germany

1. Introduction

It is well known that the vibrational excitation spectrum of amorphous solids and physical properties related to it differ appreciably from that of a crystal [1,2]. Above a certain frequency the observed phonon density of states $N(\omega)$ starts to deviate from the low frequency ω^2-law [3,4]. This anomaly, which appears also in the temperature dependence of the specific heat and thermal conductivity [1], has been interpreted in terms of a crossover from excitations in an isotropic medium ("phonons") to those in a medium with fractal topology ("fractons") [5]. This interpretation has been substantiated by effective medium calculations on a percolating lattice [6]. Nevertheless, it is not obvious how a glass can be related to a system with a fractal structure since in most cases its structure does not exhibit self-similarity [7].

In the present contribution we develop an effective medium theory for phonons in glasses. Like the percolating network approach [6] the theory is based on the mathematical analogy [8] between single particle hopping in a random medium and the phonon problem. Using the version of the effective medium approximation (EMA) suitable for hopping in 3-dimensional amorphous systems [9-12] we show that the anomalous behaviour of the density of states and the specific heat in glasses is the direct consequence of disorder, which is represented by a continuous pair distribution function $g(r)$. Within the present approach the frequency anomaly in $N(\omega)$ is accompanied by a sharp crossover from free wave propagation to a regime where the waves are strongly damped.

2. The Model

We consider a hypothetical one-component amorphous solid in which the total potential energy can be decomposed as the sum of pairwise interatomic potentials $\Phi(r_{ij})$ which depend only on the distance r_{ij} between the atoms. To be specific we take $\Phi(r) \sim \frac{1}{r}\exp(-\beta r)$ which corresponds to a screened Coulomb interaction. This applies explicitly to ionic and metallic glasses. As a further simplification we do not care about the polarization of the displacements $\vec{u}_i$ of the atoms in the solid, but rather treat them as scalar quantities u_i. (The full problem can be treated also within the present formalism as will be demonstrated in a future publication.)

With these model assumptions our starting point is the following set of equations of coupled harmonic oscillators of mass m:

$$m\frac{d^2}{dt^2}u_i = \sum_j K_{ij}u_j \tag{1}$$

where the force constants $K_{ij} = K(r_{ij})$ can be derived from the interionic potential $\Phi(r_{ij})$ by

$$K_{ij} = \left\langle \frac{\partial^2\Phi(r_{ij})}{\partial v_\nu \partial v_\nu} \right\rangle_{(\vartheta,\phi)} = \frac{1}{3}(\Phi''(r_{ij}) + 2\frac{\Phi'(r_{ij})}{r_{ij}}) \quad \text{with} \quad v_\nu = (\vec{u}_i - \vec{u}_j)_\nu \,. \tag{2}$$

$\langle ... \rangle_{(\vartheta,\phi)}$ denotes averaging over the angles (ϑ, ϕ). Inserting $\Phi(r) \sim \frac{1}{r}\exp(-\beta r)$ in (2) gives

$$K(r) = K_0(r)\exp(-\beta r) \tag{3}$$

with $K_0(r) \sim 1/r$. As it turns out, however, the qualitative features of our results do not depend on the details of $K_0(r)$ so that it is permissible for reasons of simplicity to treat $K_0(r)$ as a constant. (In the equivalent hopping problem (3) is then the jump rate for isoenergetic hopping in an impurity band ("r-hopping") [13].)

The averaged propagator corresponding to the equations of motion (1) is given by [10-12]

$$G(\vec{k},s) = \sum_{ij} e^{i\vec{k}\vec{r}_{ij}} \langle (s\hat{1} - \hat{\mathcal{K}})^{-1}_{ij} \rangle \tag{4}$$

where s is the Laplace variable with $s = -\omega^2 + i\epsilon$ for the phonons and $s = i\omega + \epsilon$ for the hopping problem ($\epsilon \searrow 0$). $\hat{1}$ is the unit matrix and $\hat{\mathcal{K}}$ has diagonal elements $m(\hat{\mathcal{K}})_{ii} = K_{ii} = -\sum_{j\neq i} K_{ij}$ and off-diagonal elements $m(\hat{\mathcal{K}})_{ij} = K_{ij}$. $\langle ... \rangle$ denotes the configurational average. The sum rule $K_{ii} = -\sum_{j\neq i} K_{ij}$ expresses infinitesimal translational invariance of the system in the phonon problem whereas it reflects particle conservation in the equivalent hopping problem.

The dynamical structure factor $S(\vec{k},\omega)$, which can be measured by inelastic neutron scattering, is given by $S(\vec{k},\omega) = -2\,\mathrm{Im}G(\vec{k},-\omega^2+i\epsilon)$. The phonon density of states $N(\omega)$ and the specific heat $C(T)$ can be calculated from $G(\vec{k},s)$ using

$$N(\omega) = -\frac{\omega}{\pi^3}\,\mathrm{Im}\int_0^\infty dk\,k^2 G(\vec{k},-\omega^2+i\epsilon) \tag{5}$$

and

$$C(T) = k_B \int_0^\infty d\omega \left(\frac{\hbar\omega}{k_B T}\right)^2 \frac{\exp(\frac{\hbar\omega}{k_B T})}{(\exp(\frac{\hbar\omega}{k_B T})-1)^2} N(\omega). \tag{6}$$

3. Effective-Medium-Approximation (EMA)

For performing the configurational average in (4) we use the two-site-EMA [9-12] derived for the hopping problem. In the hydrodynamic regime ($k \to 0$) the propagator is given by [11]

$$G_{hyd}(k,s) = (s + D(s)k^2)^{-1}. \tag{7}$$

The frequency dependent diffusivity $D(s)$ of the equivalent hopping problem, which is the square of the frequency dependent sound velocity $c(s)$, is given by

$$D(s) = c^2(s) = \frac{2\pi}{3} n \int_0^\infty dr\,r^4 g(r) \frac{1}{(\nu(s)+s)^{-1} + mK(r)^{-1}} \tag{8}$$

where n is the number density and $g(r)$ is the radial pair distribution function of the atoms. The self energy $\nu(s)$ obeys the EMA equation

$$\nu(s) = 4\pi a_p n \int_0^\infty dr\,r^2 g(r) \frac{1}{(\nu(s)+s)^{-1} + mK(r)^{-1}} \tag{9}$$

($a_p = \exp(-1)$ is a correction for double counting). Calculations of $D(s)$ for the hopping problem have shown to be almost indistinguishable from those of computer simulations [9,14].

As a further simplification we perform a generalized Debye approximation [6]

$$G(k,s) = \begin{cases} G_{hyd}(k,s) & , k \le k_D \\ 0 & , k > k_D \end{cases} \tag{10}$$

where $k_D = (6\pi^2 n)^{1/3}$ is the Debye wave number. Inserting (10) into (5) we obtain the following explicit formula for the phonon density of states

$$N(\omega) = -\frac{1}{\pi^3}\mathrm{Im}\left\{\frac{\omega k_D}{c^2(s)} + \frac{\omega^2}{2c^3(s)}\log\frac{\omega - c(s)k_D - i\epsilon}{\omega + c(s)k_D - i\epsilon}\right\}\Bigg|_{s=-\omega^2+i\epsilon}. \tag{11}$$

4. Results and Discussion

We have performed numerical calculations for a simplified $g(r) = \Theta(r-d)$ where d is a hard core diameter of the atoms. To be able to see the influence of the excluded volume we took both $dk_D = 3.96$ (which corresponds to a packing fraction of $\frac{\pi}{6}nd^3 = 0.55$) and $d = 0$. The latter "model" is insofar interesting as it represents the litteral translation of the "r-hopping"-model [13]. We measure all lengths in units of k_D^{-1}. The screening parameter β is chosen to be $\beta k_D^{-1} = 1.1$, which is a typical Thomas-Fermi-parameter for amorphous metals. The squared frequency ω^2 and the squared sound velocity $c^2(s)$ are given in units of $\frac{K_0}{m} = \omega_D^2$ and $\omega_D^2 k_D^{-2}$, respectively. The specific heat $C(T)$ and the temperature T are measured in units of k_B and $\Theta_D = \hbar\omega_D/k_B$.

Solving the self consistent equation (9) numerically and then inserting $\nu(s)$ into (8) we obtain a frequency dependence of the real and imaginary part of $c(-\omega^2+i\epsilon)$ as depicted in fig.1. (In figs.1, 3 and 4 straight lines correspond to $dk_D = 3.96$, dashed lines to $d = 0$.) Below a critical frequency ω_1 $c(-\omega^2)$ is a real quantity. Thus, in this frequency regime the only effect of the disorder is a reduction of the sound velocity compared to the ordered case ($c = 1$). In the frequency interval between ω_1 and ω_2 $c(-\omega^2)$ has a non-vanishing imaginary part which leads to a damping of the sound waves. This can be described by a frequency dependent damping constant $\alpha(\omega) = l^{-1}(\omega)$ with a characteristic length $l(\omega)$ given by [6]

$$l(\omega) = \frac{|\,c(-\omega^2+i\epsilon)\,|^2}{2\omega\,\mathrm{Im}\,c(-\omega^2+i\epsilon)}. \tag{12}$$

Near the maximum of Im$\{c\}$ we have for both models $kl \approx 2$, which means — applying the Ioffe-Regel-criterion — that the system is at the borderline of localization.

The most striking feature of our model is the sharp crossover from free wave propagation to strongly damped motion as ω is increased. This crossover corresponds to the transition from free diffusion (Gaussian transport) to anomalous diffusion (dispersive transport) in the equivalent hopping model [11,13]. As in the hopping system the crossover frequency strongly depends on the parameter β. This dependence will be explored elsewhere.

In fig.2 we have plotted the dispersion law $\omega = \omega(k)$ for $dk_D = 3.96$. The almost straight line in the regime $\omega < \omega_1$ corresponds to an undamped sound wave with a reduced sound velocity of 0.5 in our units. The corresponding dynamical structure factor $S(\vec{k},\omega)$ exhibits a Brillouin peak with corresponding dispersion. In the damped region $\omega_1 < \omega < \omega_2$ dots denote the maxima of $-\omega\,\mathrm{Im}G(k,-\omega^2+i\epsilon)$ and bars full width of this function at half maximum. It is interesting to note that in this frequency regime excitations exist throughout k-space. Especially the existence of Einstein-like modes with long wavelengths is remarkable [15].

In fig.3 we have plotted $N(\omega)$ for both models. The dash-dotted line corresponds to a Debye model with $c = 1$. The crossover from the free to the damped region at ω_1 is reflected by an infinite slope in $N(\omega)$, which leads to an "overshooting" of the curve as it is observed in some glasses [16]. Such a feature is also predicted by the percolating network model [6]. The reason, however, is a different one: whereas in the percolating network one has a crossover to an $\omega^{1/3}$-law as predicted by the Alexander-Orbach-conjecture [5] for self-similar structures, in our model with $d = 0$ the hump in $N(\omega)$ corresponds to an onset of Im$\{c^2(-\omega^2)\}$ which behaves essentially as $x\log^4 x$ with $x = \omega^2 - \omega_1^2 - \mathrm{Re}\{\nu(-\omega^2) - \nu(-\omega_1^2)\}$, $x > 0$. For finite d the behaviour is similar as can be seen from fig.1. This form of the onset of damping is directly related to the behaviour of $K(r)$ for large r; correspondingly it will be similar for any model with exponentially decaying potentials but different for algebraically decaying ones.

The crossover from the undamped to the damped region is also reflected by an interesting feature of the temperature dependence of the specific heat. In fig.4 we have plotted $C(T)/T^3$ against temperature T. This function starts as a constant, corresponding to a Debye T^3-law. At

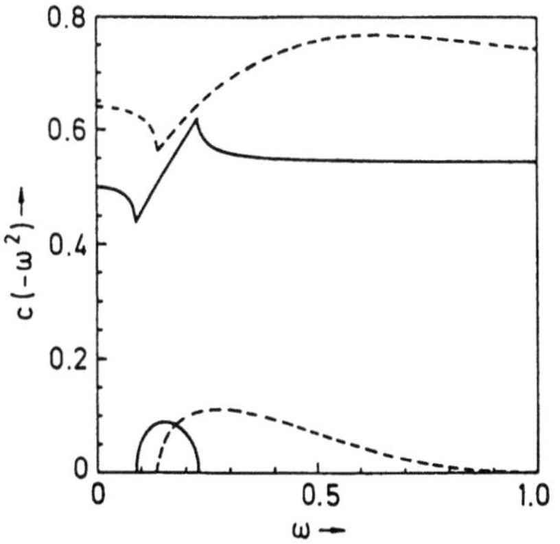

Fig.1 Frequency dependent sound velocity, upper curves: Re$\{c\}$, lower curves: Im$\{c\}$

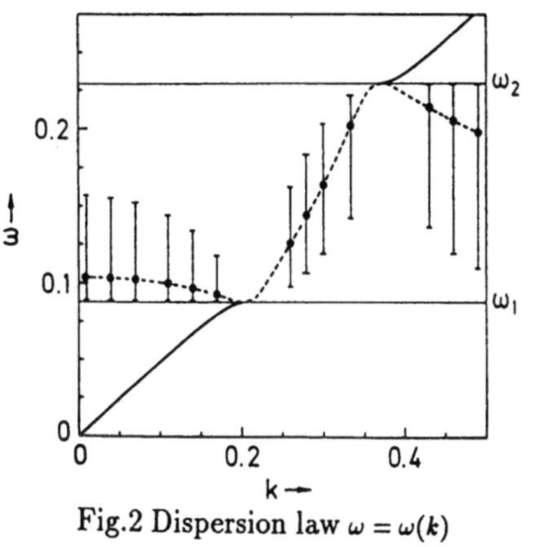

Fig.2 Dispersion law $\omega = \omega(k)$

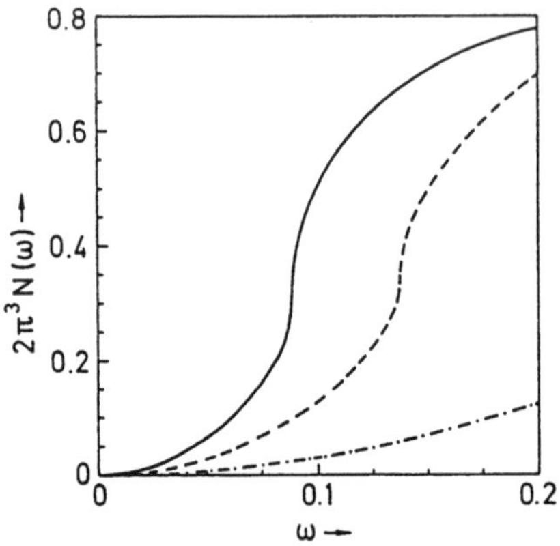

Fig.3 Phonon density of states $N(\omega)$

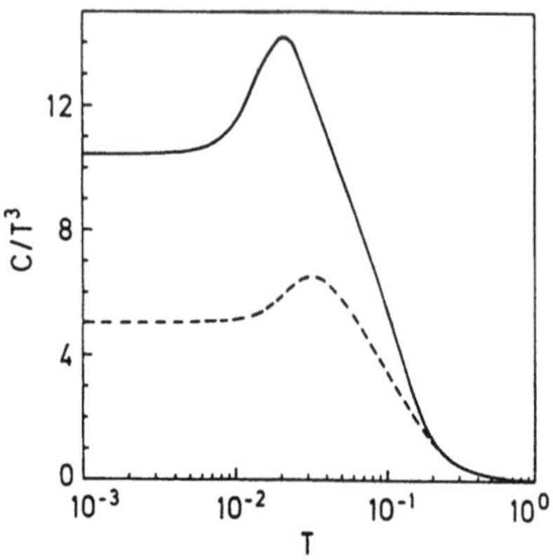

Fig.4 Specific heat $C(T)$ over T^3 vs. T

approximately $k_BT \approx 0.1\hbar\omega_1$ a characteristic hump appears which is absent in the Debye law. Such a hump has indeed been observed frequently [1,15].

Since our model is devised for small wavenumbers k ("generalized Debye approximation") it cannot account for the interesting features in $S(\vec{k},\omega)$ at large k [16].

5. Conclusions

Using the EMA we have shown that vibrational excitations in our model glass undergo a transition from free wave propagation to strongly damped motion as the frequency is increased beyond a certain critical frequency ω_1. This crossover, which is accompanied by characteristic anomalies in the density of states and the specific heat occurs as a direct consequence of disorder which is given by a continuous distribution of interatomic separations. We suggest that the observed anomalies in glasses have an origin related to this crossover, which is analogous to the transition from Gaussian to dispersive transport in hopping systems.

6. References

1. R.C.Zeller and R.O.Pohl, Phys.Rev.B4, 2029 (1971)
2. W.A.Phillips (Editor), Amorphous Solids - Low Temperature Properties, Vol.24 of Topics in Current Physics (Springer-Verlag, Berlin) 1981
3. J.B.Suck, H.Rudin, H.J.Günteroth, H.Beck, J.Daubert, and W.Gläser, J.Phys.C13, L167 (1980)
4. J.Hafner, J.Phys.C16, 5773 (1983)
5. S.Alexander and R.Orbach, J.Physique Lett.43, L625 (1982)
6. O.Entin-Wohlman, S.Alexander, R.Orbach and K.Yu, Phys.Rev.B29, 4588 (1984)
   B.Derrida, R.Orbach, and K.Yu, Phys.Rev.B29, 6645 (1984)
7. J.J.Freeman and A.C.Anderson, Phys.Rev.B34, 5684 (1986)
   J.A.Krumhansl, Phys.Rev.Lett.56, 2696 (1986)
   J.Maddox, Nature 322, 303 (1986)
8. S.Alexander, J.Bernasconi, W.R.Schneider, and R.Orbach, Rev.Mod.Phys.53, 175 (1981)
9. B.Movaghar, B.Pohlmann, G.Sauer, phys.stat.sol.(b)97, 533 (1980)
10. B.Movaghar and W.Schirmacher, J.Phys.C14, 859 (1980)
11. W.Schirmacher, Sol.State Comm.39, 893 (1981)
12. B.Movaghar, M.Grünewald, B.Pohlmann, D.Würtz, and W.Schirmacher, J.Statist.Phys.30, 315 (1983)
13. H.Böttger and V.V.Bryksin, Hopping Conduction in Solids, Akademie Verlag, Berlin 1985
14. J.A.McInnes, P.N.Butcher, and J.C.Clark, Phil.Mag.B41, 1 (1980)
15. C.C.Yu and J.J.Freeman, Phys.Rev.B36, 7620 (1987)
16. U.Buchenau, N.Nücker, and A.J.Dianoux, Phys.Rev.Lett. 53, 2316 (1984)
    U.Buchenau, M.Prager, N.Nücker, A.J.Dianoux, N.Ahmad, and W.A.Phillips, Phys.Rev.B34, 5665 (1986)

Vibrational States and Tunneling

H.R. Schober

Institut für Festkörperforschung, KFA Jülich,
D-5170 Jülich, Fed. Rep. of Germany

Many features of the low temperature glass anomalies, such as low frequency vibrations, relaxation processes and two level systems are also found in crystals containing interstitial atoms. There they originate from strong lattice distortions. Methods developed to study the properties of these defects can be of use to gain an understanding of the glassy state.

1. Structure and Dynamics of Interstitials

Let us consider a fcc metal. Small interstitial atoms, as e.g. hydrogen are found to occupy the octahedral interstitial site. Due to the local geometry their mass weighted vibrational spectrum is somewhat, but not dramatically, increased in the host frequency band /1/. With increasing "size" of the interstitial atom or of its interaction with the host one or more low frequency resonance (quasi localized) modes are formed simultaneously with high frequency localized ones. At a critical interaction strength the resonance frequency goes to zero and the mode becomes unstable /2/. The interstitial atom no longer sits at the high-symmetry octahedral site but instead at a site displaced in (1,1,1) or (1,0,0) direction. In this new position the particle will again have low-frequency resonant modes and the new site can again become unstable with increasing interaction. The high symmetry lattice site is surrounded by a number of equivalent sites of the lower symmetry and the interstitial atom can move between these sites like in a "cage" centered at the symmetry site. Fig. 1 shows three such possible cages. The third of these cages (c) is split into subcages where transitions within a subcage (square) are easier than between the subcages. This tilted dumbbell like configuration can be induced, e.g. for an interstitial atom normally in configuration (a), by minute changes in the electronic structure. Also agglomeration with a second defect can lead to such a tilting. Configurations with broken symmetry have been observed for numerous alloys both experimentally and numerically, see e.g. /3/ for a review.

The occurrence of resonant modes for interstitial atoms is a consequence of the large strains induced in the lattice by the defect. Large strains also cause the localized modes where they are not simply due to mass differences as for H or molecular binding effects. One therefore has to expect the simultane-

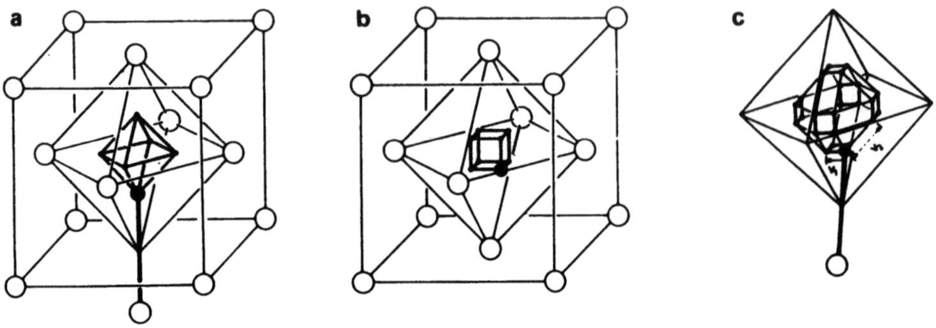

Fig.1 Cages formed by equivalent sites for a trapped interstitial: (a) octahedral cage of the mixed dumbbell [(1,0,0)-off-centre octahedral interstitial], (b) cubic cage of the (1,1,1)-off-centre octahedral interstitial, (c) split octahedral cage of a slightly tilted dumbbell interstitial /3/.

ous occurrence of both types of vibrational modes. A closer inspection of the resonant modes shows that the vibrational amplitudes are concentrated on a few atoms around the defect. The modes are therefore closely linked to the short-wavelength phonons. By changing the interstitial atom's mass one finds an effective mass of the resonant vibration, $m_{eff} = m(1 + x)$ where $x \approx 3$. In amplitude weighted average about four atoms participate in the resonance vibration. A convenient way to describe the dynamics of a defect is by its Green's function. In cases where the local vibrational spectrum of the defect has only little contribution from the eigenfrequencies of the ideal lattice this Green's function of the defect, $\mathbf{G}^{dd}$, can to a good approximation be described by the resonant and localized modes alone:

$$\mathbf{G}^{dd} = \sum_{\sigma} \frac{|\sigma\rangle\langle\sigma|}{m^{\sigma}\omega_{\sigma}^{2} - m^{\sigma}\omega^{2} - im^{\sigma}\gamma^{\sigma}(\omega)\omega} \qquad (1)$$

where $|\sigma\rangle$ represents the eigenvector of the mode σ in the defect subspace, ω_{σ} and γ^{σ} are the mode frequency and damping and m^{σ} its effective mass. At low frequencies $\mathbf{G}^{dd}$ is dominated by the lowest resonance modes.

Cage configurations exist also in the more complicated structures. Recently such a cage has been found in the hcp structure /4/. Soft vibrational modes are also found for interstitial atoms in silicon. It will be interesting to see their relationship to the soft vibrations found by Biswas et al. /5/ in amorphous silicon. The difficulties in finding an adequate interaction model accounting for covalent binding makes an investigation of symmetry breaking effect more difficult than in metals. It has, however, been stipulated that such interstitial configurations play a role in being the nucleus of two level systems /6/.

In contrast to the interstitial type defects, vacancies induce in metals only a softening of vibrations but no pronounced resonances since they cause no strong lattice strains.

2. Relaxation and Tunneling

The existence of equivalent configurations in the neighbourhood of a configuration of higher symmetry plus the low frequency vibrational modes make a motion between these sites at low temperatures possible. Depending on the details of the configuration and the experimental conditions this motion can be coherent tunneling, phonon assisted tunneling or classical hopping. All these can be estimated from the geometrical factors (jump distance) and the vibrational properties.

The relaxation rate in the classical jump process follows an Arrhenius law with an only weakly temperature dependent prefactor times Boltzmann factor

$$R = R_0 \exp(-E_a/kT) \tag{2}$$

The microscopic picture is that of an atom (interstitial or a neighbour atom to a vacancy) hopping over a potential barrier formed by the other atoms. The classical activation energy, E_a, is the minimal energy needed for that process. It corresponds to the energy necessary to move the atom adiabatically to the saddle point separating its initial and final equilibrium configurations. E_a is, therefore, the difference of the potential energies with the hopping particle in the saddle point and equilibrium configurations, respectively, and the rest of the lattice fully relaxed. E_a is often estimated in terms of elastic constants /7/. The local Green's function of the defect (1) allows one to relate E_a to lattice dynamics. The energy needed to displace an atom, *d*, from its equilibrium site with the other atoms relaxing adiabatically is given by

$$E(\mathbf{s}^d) = (1/2)\mathbf{s}^d[\mathbf{G}^{dd}(\omega=0)]^{-1}\mathbf{s}^d \tag{3}$$

Assuming a smooth, *sin*-like, shape of the potential surface along the "reaction" coordinate to the saddle point (4.2) can be extrapolated to give

$$E_a \simeq (2/\pi^2)\mathbf{d}[\bar{\mathbf{G}}^{dd}(\omega=0)]^{-1}\mathbf{d} \tag{4}$$

where $\mathbf{d}$ is the distance vector to the saddle point and $\bar{\mathbf{G}} = (1/2)(\mathbf{G}_{equilibrium} - \mathbf{G}_{saddle})$. Tests of this relation with computer models gave an error of $< 1\%$. If the defect Green's function for the interstitial atom is determined by low lying resonance vibrations (4) reduces to a sum over these resonances

$$E_a \simeq 2/\pi^2 \sum_r m^r \omega_r^2 d_r^2 \tag{5}$$

where d_r is the projection of $\mathbf{d}$ onto the eigenvector of the resonance mode with effective mass m^r and frequency ω_r. (4.4) establishes the observed correlation between the occurrence of low lying resonance modes and low activation energies.

At very low temperatures one expects tunneling to contribute to particle motion. The magnitude of the tunnel splitting, ΔE_T, for a particle in a one-dimensional double well potential is determined essentially by an exponential factor $\exp(-gm\omega d^2/\hbar)$ where g is a geometrical factor of order unity, m the mass, $\frac{1}{2}\hbar\omega$ the ground state energy and $2d$ the well separation. Tunneling of an atom in a crystal cannot be described by a simple onedimensional double well potential. The atom experiences the potentials exerted by its neighbours which in turn move in the potential of their neighbours and so on. The potential in which the atom tunnels thus becomes N-dimensional ($N \to \infty$). In tunneling between the two minima the atom carries its displacement cloud with it - it is "dressed".

For heavy interstitial atoms with pronounced resonant vibrations it has been shown that similar to the classical activation energy also the tunnel splitting energy can be approximated by these resonances /8/:

$$\Delta E_T \sim \exp[-g \sum_r m^r \omega_r d_r^2/\hbar]. \tag{6}$$

In metals there is an additional electron dressing factor not discussed here /9/. Experimentally tunneling cage motion has been observed in $Al : Zn$ /10/.

With increasing temperature lattice vibrations are excited. This causes a reduction and broadening of ΔE_T. Transitions between sites are then more probable by incoherent phonon assisted tunneling. The transition probability is then found by the usual "golden rule expression"

$$W^{if} = \langle\langle w^{if} \rangle\rangle_m \tag{7}$$

$$w^{if} = (2\pi/\hbar)|\langle\Psi_i(m)|H - E_{im}|\Psi_f(n)\rangle|^2\delta(E_{im} - E_{fn}) \tag{8}$$

where Ψ_i and Ψ_f are the initial and final state many body wavefunctions, m and n denote the respective excitational states, $\langle\langle\ \rangle\rangle_m$ denotes thermal averaging over the initial states m and summation over the final states n and finally $E_{im} = \langle\Psi_i(m)H\Psi_i(m)\rangle$. Taking account of the localization of the important resonant modes the expectation value in (8) can be evaluated in an "embedded molecule" approximation /11/.

3. Conclusions

We have seen that lattice strains induce besides high-frequency localized modes also low frequency resonant ones. These latter ones in turn lead to a tendency towards displacements from configurations with high symmetry to ones with lower symmetry. The motion between these sites, wether coherent tunneling, phonon assisted tunneling or classical hopping, is dominated by the resonant modes. Their localization makes embedded molecule methods possible. We think that similar effects will occur in amorphous materials where such a scenario has been postulated by Buchenau et al. /12/

1. H. R. Schober and V. Lottner, Z. Phys. Chem. (NF) **114**, 203, (1979).
2. R. Zeller and P. H. Dederichs, Z. Phys. **B25**, 139, (1976).
3. P. Ehrhart, K. H. Robrock and H. R. Schober: in *Physics of Radiation Effects in Crystals*, (R. A. Johnson and A. N. Orlov eds.), North Holland (Amsterdam 1986), p 3.
4. Y. Yoshida, M. Menningen, R. Sielemann, G. Vogl, G. Weyer and K. Schroeder, Phys. Rev. Lett. **61**, 195, (1988).
5. R. Biswas, A. M. Bouchard, W. A. Kamitakahara, G. S. Grest and C. M. Soukoulis, Phys. Rev. Lett. **60**, 2280, (1988).
6. V. J. B. Torres, P. M. Masri and A. M. Stoneham, J. Phys. C **20**, L143, (1987).
7. C. P. Flynn, Phys. Rev. **171**, 682, (1968).
8. H. R. Schober and A. M. Stoneham, Phys. Rev. **B26**, 1819, (1982).
9. J. Kondo, Physica **125**, 279, (1984).
10. K. L. Hultman, J. Holder and A. V. Granato, J. Phys. (Paris) **42**, 753, (1981).
11. H. R. Schober and A. M. Stoneham, Phys. Rev. Lett. **60**, 2307, (1988).
12. U. Buchenau, M. Prager, N. Nücker, A. J. Dianoux, N. Ahmad and W. A. Phillips, Phys. Rev. B **34**, 5665, (1986).

Part IV

Spin Glasses and Orientational Glasses

Critical Dynamics of the Spin Glass Transition

J.J. Prejean

CRTBT-CNRS, BP 166, F-38042 Grenoble Cedex, France

Abstract : the parameters (T_c, critical exponents) which are deduced from static magnetization measurements provide a stringent test for theories of the critical dynamics. An example of both static and dynamic measurements performed on the same sample is given. The analysis of a.c. susceptibility data (10Hz-1MHz) of an insulating manganese fluorophosphate spin glass provides the temperature dependence of the upper edge $\tau_{max}(T)$ of the distribution of the relaxation times. The data can be accurately fitted by various laws : Fulsher law, power of $t = (T-T_c)/T$, Fisher and Huse law. But each fit is obtained for a different value of the critical temperature T_c. Only the measurements of the equilibrium non linear susceptibilities $a_3(T)$ and $a_5(T)$ allow to deduce the value of T_c and therefore to conclude for the dynamics. Finally we find that :

1) $a_3(T)$, $a_5(T)$, τ_{max} vary by orders of magnitude in the 1.06-1.2 T_c range. Each of these quantities is a power of the others in agreement with the standard scaling hypothesis.

2) $a_3(T)$, $a_5(T)$, τ_{max} can be only interpreted in terms of powers of $t = (T-T_c)/T$

3) The standard power law fits accurately our data of the dynamics with reasonable values of the parameters (T_c, τ_0) in contrast with the other predictions.

1-Introduction

The spin glass transition problem has been the most exciting and disputed point since the first moment a phase transition was predicted by a mean field theory [1] which corresponds to the $d \rightarrow \infty$ limit of the space dimensionality d. On another hand, the problem of the existence of a transition in real systems, i.e. at d=3 arose when people began to measure the a.c. susceptibility of spin glasses [2]. Indeed it was found that the in phase response χ' of the magnetization to a small a.c. field (frequency=10 Hz for instance) exhibits a sharp peak at a temperature T_M (see the insert in Fig.1). T_M appears to separate two very different temperature regimes : the properties of each are summarized in the following.

Below T_M, no spontaneous magnetization is found but large magnetic hysteresis effects are observed : after a change of field at low T, the magnetization (and the associated energy flow [3]) depends strongly on the

Springer Proceedings in Physics, Vol. 37 **Dynamics of Disordered Materials**
Editors: D. Richter · A.J. Dianoux · W. Petry · J. Teixeira

magneto-thermal history, the applied field, the temperature T and relaxes very slowly (over hours, days...). The non exponential relaxations are interpreted in terms of the existence of a broad distribution of energy barriers which delay the response of the magnetization. It has been shown that i) *thermal activated processes* govern the relaxations and ii) an elementary relaxation process involves a *large number* ($\simeq 10^3$) *of correlated spins* [4].

In contrast, above T_M, it seems, at first sight, that a spin glass behaves as a simple paramagnet. No hysteresis over long times is observed, and the susceptibility χ is similar to that of a paramagnet in the simplest cases : χ *obeys a Curie law* (χ = C/T) as it is shown in the figure 1. Moreover, *the magnitude of C is that of a non interacting paramagnet* : from C ($= N\mu^2/3k_B$), one deduces an effective moment μ per magnetic atom rather near to the expected single moment value.

Therefore, the existence of a transition at a temperature close to T_M became the subject of many theoretical and experimental works. Let us summarize some of the theoretical predictions. The mean field results are expected to hold at and above the *upper critical dimension* $d_U = 6$ (=4 in ferromagnets). Below d_U and above the *lower critical dimension* d_L , a phase transition is expected to occur at finite temperature and critical scaling within this region should also apply, but with critical exponents distinct from those predicted by the mean field theory. At $d=d_L$, $T_c = 0$. Below d_L, the results imply a viscous paramagnetic state. Early numerical results predicted no transition for d=2 Ising spin glasses : $d_L>2$ ($d_L=1$ in Ising ferromagnets).

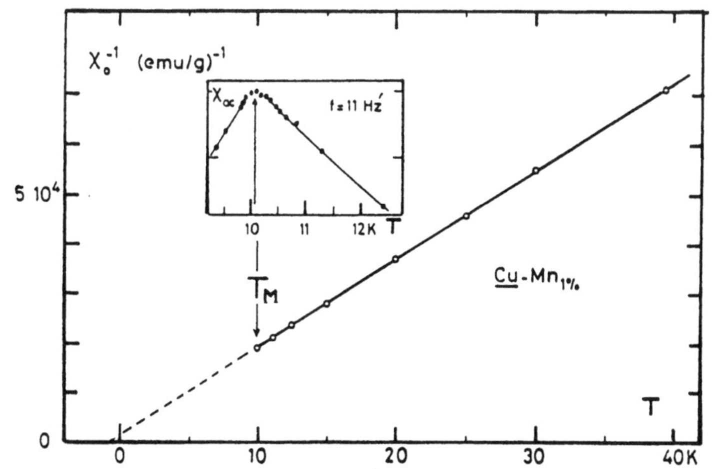

Fig. 1- *The spin glass* $Cu\text{-}Mn_{1\%}$ *susceptibility* χ *data are plotted in a* $1/\chi$ *vs. T diagram : this plot shows that* χ *follows a Curie law above the temperature* T_M *of the maximum of* $\chi_{a.c.}$ *measured at 10Hz (see the insert)* [15]

2-Theoretical predictions for the dynamics

The crucial question has been : what is the situation of our space dimensionality (d=3) with respect to d_L. If a transition exists, the spins are correlated over all the length scales up to the correlation length $\xi(T)$. At decreasing temperatures, $\xi(T)$ increases and diverges at T_c. In the following, only the behaviour expected or observed when T is lowered from the high temperatures is considered. The main point is : is it possible to measure quantities related to the correlation length $\xi(T)$? The characteristic relaxation time τ (T) of the dynamics is one of these quantities. In the early eighties, some clues in favor of a transition were given by the studies of the dynamics, but the definite answer has been provided by the study of the equilibrium magnetization. We shall come back on this point later. Let us focus on the dynamics first. The predicted behaviour of $\tau(T)$ depends on d_L.

2-1 At $d<d_L$ (no phase transition), $\tau(T)$ is expected to follow an Arrhenius law or to vary as :

$$\tau(T) \propto T^{-\Delta} \qquad (1)$$

following the early numerical results obtained in 2D Ising spin glasses [5].

2-2 At $d=d_L$ ($T_c = 0$), $\tau(T)$ and $\xi(T)$ are expected to vary as :

$$\tau(T) = \tau_0 \exp(A/T^{\sigma}) \qquad (2)$$

following Mc Millan [6] (σ =2 for ξ and 3 for τ).

2-3 At $d>d_L$ ($T_c > 0$), standard theories predict that $\xi(T)$ diverges as $t^{-\nu}$ where $t=(T-T_c)/T$. For the dynamics, different laws have been proposed :

2-3-1 Vogel-Fulsher law (VFL) : By analogy with the standard phenomenological analysis of glassy dynamics [7], the VFL was proposed for the frequency dependence of T_M :

$$\omega = \omega_0.\exp -(E/(T_M-T_c)), \qquad \omega_0=1/\tau_0 \qquad (3)$$

2-3-2 Power law : $\tau(T)$ is predicted to be a power of $\xi(T)$:

$$\tau = \tau_0 \,.\, \xi^z = \tau_0 \,.\, t^{-z\nu} \qquad (4)$$

It is the standard statement of dynamic scaling near a transition. The same expression is obtained in a simple fractal approach (see below).

2-3-3 Activated scaling : the relaxation times are given by the activated energy barrier W for overturning a correlation volume L^d

$(1 < L < \xi(T))$: $\tau = \tau_0 \exp [W(L)/T]$. In the frame of fractal models, two different predictions are given. The first one proposed by us [8] is an extension to the spin glasses of the results of [9] for percolation ferromagnetic clusters : $W = 2J\bar{d}.\mathrm{Log}(L)$ ($\bar{d}$ denotes the fractal dimension : $\bar{d}$=2.5). The other uses above T_c [10] the results obtained below T_c in the model of Fisher and Huse (FH) [11] : $W= J.L^{\theta}$. One finds respectively for $L = \xi$:

$$\tau = \tau_0 \, . \, \xi^{z'} = \tau_0 \, . \, t^{-z'\nu} \; , \; z' = 2J\bar{d}/T \qquad (5)$$

$$\tau = \tau_0 \, . \, \exp(A \, . \, \xi^{\theta} /T) = \tau_0 \, . \, \exp(A \, . \, t^{-\theta\nu}/T) \qquad (6)$$

which check the scale invariance: one finds respectively $\tau(2L)/\tau(L)$ and $W(2L)/W(L)$ independant of L.

i)*power law* : The only difference between the equation (5) and the standard one (4) is the temperature dependance of z'. We note that $z'=5(J/T_c)$ at T_C. Numerical studies provide $J/T_C \simeq 1$. Therefore our model predicts $z \simeq 5$ [8].

ii)*generalized VFL :* the Eq.(4) which allows to recover both the Arrhenius law (for $\theta = 0$) and the VFL (for $\theta\nu = 1$) can be considered as a generalized VFL.

2-3-4 Feigelmann and Ioffe (FI) predictions : they give a quantitative estimate of the involved parameters [12] in the case of long (but finite) range interactions. For Heisenberg spins, they predict a power law with $z\nu = 10.5$. F.I. note that a small amount of anisotropy leads to an Ising behavior :

$$\tau_{max} = \tau_0 \, t^{-z\nu}.\exp(A t^{-\theta\nu}) \text{ where } z\nu = 6, \; \theta\nu = 0.25.$$

2-4 A first check of a theoretical law is provided if the later fits the data with reasonable values of the parameter τ_0. The atomic time scale τ_0 has not to be significantly lower than the value of the inverse of the exchange frequency : 10^{-12} s. given roughly by $h/\tau_0 = JS(S+1)$, $J = kT_c$, for T_c=1K. We shall see that this stringent condition is not sufficient to give a definite answer.

3- Example of an experimental study of the dynamics

In standard studies, the relaxation time $\tau(T)$ which characterizes the dynamics is deduced from the measurement of the a.c. susceptibility χ. A wide range of measurement frequencies is necessary to obtain a wide range of values of $\tau(T)$ in order to check the validity of the theoretical laws. This can be done up to some MHz when the sample is insulating, i.e. with no eddy currents. Most of the data presented in the following have been obtained on

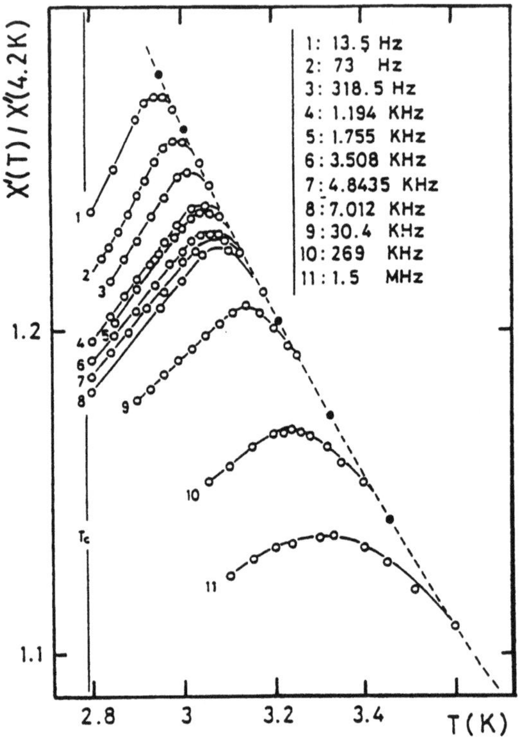

Fig. 2-*The in-phase response $\chi'(\omega,T)$ of the susceptibility of the insulating spin glass fluorophosphate of manganese is plotted vs. T at ω= const.(open circles). The solid circles represent the data $\chi(\omega=0)= \chi_{eq}$*

an insulating manganese fluorophosphate : its main properties are described elsewhere [8].

In the figure 2, the data are represented in a $\chi'(\omega,T)$ (i.e. the in-phase response of χ) vs. T diagram at ω= const.in the range 10 Hz-1 Mz (open circles). For technical reasons, only the data obtained at $\omega <$ 100 kHz are accurate enough to be used for a serious check the dynamics laws. The d.c. magnetization M is also measured in a low field SQUID magnetometer : the data M(T)/h obtained in 1 Oe (solid circles) are expected to give $\chi(\omega=0)$, i.e. the equilibrium susceptibility χ_{eq}. A Curie law (χ_{eq} =C/T) fits the data above 2.84 K with errors smaller than 1 mK. The value of C provides an effective moment μ (6.2 μ_B) per Mn rather near to the ionic value (5 μ_B).

3-1 Existence of a broad distribution P(τ) of times of relaxation τ, determination of a characteristic value of τ :

At any constant temperature, the data cannot be fitted by a simple Debye law : $\chi'(\omega) = \chi_{eq}/(1+\omega^2\tau^2)$ which implies a single value of τ. In contrast, the slow (pseudo logarithmic) frequency dependence of $\chi(\omega,T)$ [8] suggests the existence of a broad distribution P(τ) over a range of τ larger than the $1/\omega$ window : 10^{-6} s -10^{-1} s.

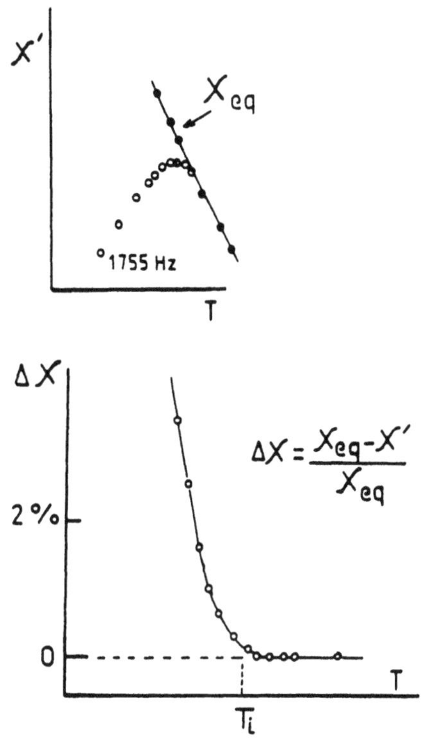

Fig. 3- *$\Delta\chi = (\chi_{eq}(T)-\chi'(\omega_i,T))/\chi_{eq}(T)$ is plotted vs. T for ω_i=1755 Hz*

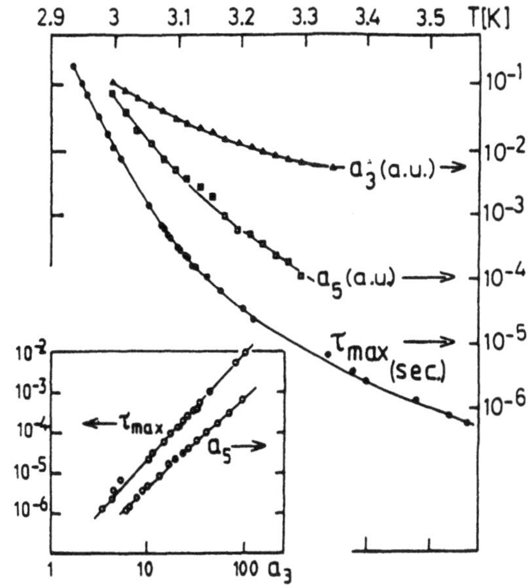

Fig. 4-*Temperature dependences of $\tau_{max.}(T)$, $a_3(T)$, $a_5(T)$*

At this step, the question is : how to characterize such a distribution? One can choose to study of either the average value $\tau_{av}(T)$ of τ [13] or the cut off of $P(\tau)$, i.e. the maximum relaxation time $\tau_{max}(T)$ [14]. The existence of the later is suggested by the data. Indeed, at ω = const.= ω_i, the normalised difference : $\Delta\chi = (\chi_{eq}(T)-\chi'(\omega_i,T))/\chi_{eq}(T)$ decreases rapidly down to zero when T is increased up to T_i (fig.3). So we assume that $\tau_{max}(T)$ decreases down to $1/\omega i$ when T reaches T_i : $\tau_{max}(T_i)=1/\omega i$. A set of measurement frequencies provides a set of values $\tau_{max}(T)$. The quantitative criterion of analysis of $\tau_{max}(T)$ and the condition of its stability are given with details in [14]. The temperature dependence of $\tau_{max.}(T)$ is shown in figure 4.

3-2 Experimental data versus theoretical laws.

The data $\tau_{max.}(T)$ are not consistent with the predictions for $d=3\leq d_L$: they can be fitted neither by an Arrhenius law nor by the expressions (1) and (2) (see the figure 5 where Log $\tau_{max.}$ is plotted vs. Log T and vs. $1/T^3$ respectively). But expression (2) fit the data for σ = 12.2 (see figure 6).

All the predictions for $d > d_L$ fit accurately the data. Here, only the data obtained with the best accuracy, i.e. at $\omega \leq 30$ kHz are used :

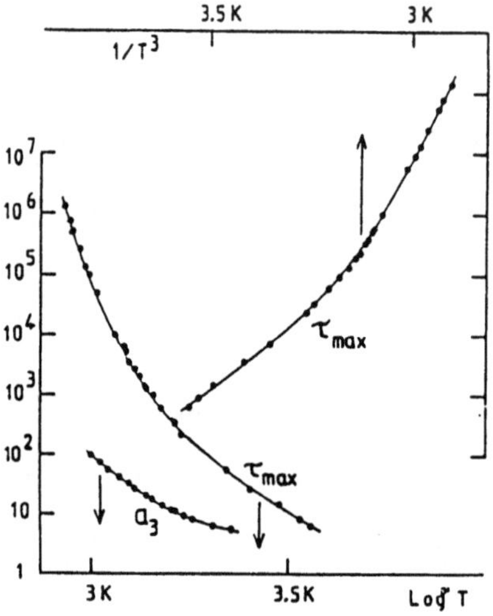

Fig. 5

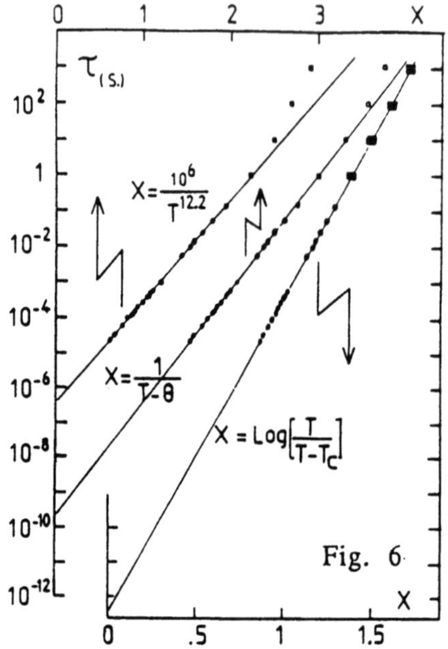

Fig. 6

Fig. 5-*Log* $\tau_{max.}$ *is plotted vs. Log T and vs.* $1/T^3$

Fig. 6-*Log* $\tau_{max.}$ *is plotted vs.* $1/T^{12.2}$, $1/(T-T_c)$ *for* $T_c=2.57K$, $Log(T/(T-T_c))$ *for* $T_c=2.79K$ *(the solid circles represent the data, the other symbols corresponds to extrapolated values : see text)*

i) The best fits by the Fulsher law (3) and the power law (4) are obtained for T_c =2.57K and 2.79 K respectively and yield a value of τ_0 nearly equal to that expected. For instance, in the later case, the computed linear fit of $Y = \log(\tau_{max})$ vs. $X = \log(t)$ yield a mean quadratic error E ($E^2=(1/N)\Sigma(Y_i-Y_{icalc.})^2$), which is minimum for T_c : 2.79K within 0.01 K (see the solid circles in the insert of Fig.7). It yields $\tau_0 = 10^{-12.5}$s. The slope of the straight line obtained in the fit $Log(\tau_{max})$ vs. Log(t) equals $z\nu = 9.0$ (see Fig.7). The data obtained at $\omega > 200$ kHz lie out but in the vicinity of the corresponding straight line showing the qualitative agreement of the fit in this range

ii) The best fit by the equation (6) is obtained for a set of couples of values (T_c, $\theta\nu$) but yield a value of τ_0 having a physical sense ($\tau_0 > 10^{-13}$ s.) only for $T_c < 2.78$K ($\theta\nu > 0.25$).

iii) Although the interactions are short ranged in our case, let us compare our results with the F.I. predictions (see section 2-3-4). For Heisenberg spins, their value $z\nu$ = 10.5 is slightly larger than our estimate : 9. Actually, in our system, the small amount of anisotropy due to the additional anisotropic dipolar interactions should induce an Ising behavior. The fit of our data with the corresponding law of F.I. is accurate but yields a too small value of τ_0 : 3.10^{-15} s. for $T_c \gtrsim 2.79$ K.

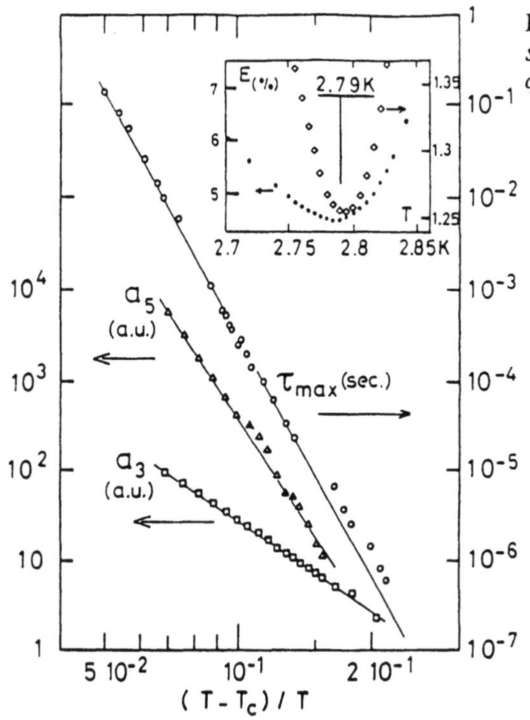

Fig. 7- *$\tau_{max.}(T)$, $a_3(T)$, $a_5(T)$ are shown to be powers of the same quantity $(T-T_c)/T$ with $T_c=2.79K$*

The main conclusion is that the same data can be fitted with the same accuracy by different laws, but each fit is obtained for a different value of Tc.

Of course, one may argue that the frequency range of the measurement is not large enough to give a definite answer. Let us try the following extrapolation. If we believe in the validity of the power law, we can deduce a set of coupled values (T, $\tau_{max.}(T)>0.1s$) corresponding to measurements which could be performed at very low frequencies (1, 10^{-1}, 10^{-2}, 10^{-3}Hz) : see the solid squares in figure 6. The later data, when they are reported in the other diagrams of the figure 6 (open squares), are observed to be accurately fitted by : i) the previous Fulsher law up to τ_{max} =100s and ii) expression (2) up to 10s. So to broaden the frequency range down to the very low values (which makes the experiment rather difficult) does not bring any help to us. Therefore another type of experiment is necessary to provide the value of Tc : once it is compared with these given by the previous fits, one can claim which law accounts for the critical dynamics.

4- Study of the equilibrium magnetization

A paramagnetic (or superparamagnetic) picture provides an equilibrium magnetization $M_{eq}(H,T)$ which can be developed in terms of odd powers of (H/T) :

$$M_{eq} = a_1(T)\chi_{eq}(T).H - a_3(T)(\mu H/kT)^3 + a_5(T)(\mu H/kT)^5 + ...$$

In a paramagnet, $a_1=a_2=a_3=..=1$. In ferromagnets, a_1, a_3, a_5,.., diverge at T_c. In spin glasses, $a_1=1$ but a_3, a_5, ..., are expected to diverge at T_c if a transition exists :

$$a_3 \propto t^{-\gamma} \quad \text{and} \quad a_5 \propto t^{-2\gamma-\beta} \qquad \text{for } T_c \text{ finite.}$$

Here γ and β denote the standard critical exponents of the transition in static studies. The determination of a_3, a_5 (called the non linear susceptibilities) is difficult but can be obtained by several accurate technics. The temperature dependence of $a_3(T)$ and $a_5(T)$ of the previous sample is represented in the figure 4 (see ref.[8], [14] for the experimental method).

The drastic temperature dependences of a_3, a_5, τ_{max} suggest the existence of a continuous phase transition which can be proved with no assumption for the temperature dependence of these parameters. Indeed, in agreement with the standard theory, a_3, a_5 and τ_{max} are powers of the same quantity, say the correlation length $\xi(T)$, since the data plotted in a a_5 and τ_{max} vs. a_3 log-log diagram are aligned on straight lines over several decades (see the insert of figure 4).

Among the different laws proposed at $d<d_L$ ($a_3(T) \propto T^{-\Delta}$: see Fig.5), $d=d_L$ ($a_3(T) \propto \exp(A/T^2)$), $d>d_L$, only the finite T_c predicted law fits the data $a_3(T)$: a_3 is a power of $t = (T-T_c)/T$. The computed linear fit of $Y = \log(a3)$ vs. $X = \log(t)$ yield a mean quadratic error E ($E^2=(1/N)\Sigma(Y_i-Y_{icalc.})^2$), which is minimum for T_c : 2.79K within 0.01 K (see the open rhombs in the insert of figure 7). In this figure, a_3 and a_5 are plotted vs. t in a Log-Log diagram for this value of T_c. The slopes of each straight line equal $\gamma=3.42$ and $2\gamma + \beta = 7.5$ (from which one deduces $\beta = 0.7$) respectively. A scaling analysis of M_{eq} very close to T_c (down to 1.01 T_c) gives the same results [14].

5- Conclusion

The value of T_c = 2.79K is in agreement only with that given by the power law (4) for the dynamics : we believe that our results support only the standard law for the critical dynamics. Let us conclude with two remarks.

The magnitude of $z\nu$ (= 9) appears to be surprisingly larger than that of the standard 3D ferromagnets ($z\nu$ = 1). Actually, only z is the relevant parameter for the dynamics ($\tau = \tau_0\xi^z$: eq.(4)) instead of $z\nu$. The exponent ν ($\xi(T) = t^{-\nu}$) is deduced separatly from γ and β ($d\nu = \gamma + 2\beta$), i.e. from the static study. It is found to be 3 times larger in our case (1.77) than in ferromagnets (0.6), the net result is a value of the exponent z only 3 times larger in spin glasses (z = 5) than in ferromagnets (z = 1.7). This difference is accounted by our fractal model (see above)

The value of ν is large enough to imply a huge number of spins in a correlation volume V_c even far from T_c : $V_c(T = 1.2\ T_c) = 10^4$ spins. In contrast, in 3D ferromagnets ($\nu = 0.6$), V_c=25 spins at 1.2 T_c and 10^4 spins at 1.006 T_c . V_c is the actual relevant parameter instead of the reduced temperature. This comparison (if it is correct) indicates that the high temperature corrections (i.e. at low values of ξ) should be small enough to insure the reliability of the above results obtained up to 1.2T_c.

REFERENCES

1. SHERRINGTON, D., KIRKPATRICK, S., Phys. Rev. Lett. **35** (1975) 1792
2. CANNELLA, V., MYDOSH, J.A., Phys. Rev. B **6** (1972) 4220
3. BERTON, A., CHAUSSY, J., ODIN, J., PREJEAN, J.J., RAMMAL, R., Solid State Comm.**49** (1984) 959 and references quoted therein
4. PREJEAN, J.J., SOULETIE, J., J. Physique France **41**(1980) 1335
5. BINDER, K., Z Phys. **B 48** (1982) 319
6. MC MILLAN,W.L., J. Phys. C **17** (1984) 3179
7. THOLENCE, J.L., Solid State Comm.**35** (1980) 113
8. BEAUVILLAIN, P., RENARD, J.P., MATECKI, M., PREJEAN, J.J., Europhysics Letters **2** (1986) 23, and
CARRE, E., PUECH, L., PREJEAN, J.J., BEAUVILLAIN, P., RENARD, J.P., Proc. of the Heidelberg Colloquium on glassy dynamics, Springer-Verlag, Lecture Notes in Physics (1986) 75
9. RAMMAL, R., J. Physique France **46** (1985) 1837 and
RAMMAL, R., BENOIT, A., J. Physique Lett. **46** (1985) L667
10. LEVY, L.P., OGIELSKI, A.T., Phys. Rev. Lett. **57** (1986) 3288
11. FISHER, D.S., HUSE, D., Phys. Rev. Lett. **56** (1986) 1601
12. FEIGELMAN, M.V., IOFFE, L.B., J. de Physique Lett. **46** (1985) L695
13. BONTEMPS, N.,RAJCHENBACH, J., CHAMBERLIN, R.V., ORBACH, R., Phys. Rev. B. **30** (1984) 6514
14. PREJEAN, J.J., CARRE, E., BEAUVILLAIN, P., RENARD, J.P., J. Physique France (1988) Proc. of ICM 88
15. OMARI, R., PREJEAN, J.J., SOULETIE, J., J. Physique France **44** (1983) 1069

Random Fields and Nonergodicity in the Orientational Glass $K(CN)_xBr_{1-x}$

K.H. Michel

Department of Physics, University of Antwerp UIA,
B-2610 Wilrijk, Belgium

Random strain fields due to substitutional disorder couple to translational and orientational dynamic modes in plastic crystals. A linear coupling of the random fields to the orientational order parameter leads to a continuous freezing. In strong random fields, the structural phase transition is suppressed. In weak random fields, and if the system comes close to a second order structural phase transitions, additional processes occur. Here restoring forces are weak and internal friction dominates. The resulting nonergodic instability marks the onset of a quadrupolar glass. It appears in addition to the weak continuous freezing induced by the linear coupling.

1. INTRODUCTION

Compositional disorder strongly affects cooperative phenomena in condensed matter. Examples are spin glasses [1] and structural glasses [2]. The existence of a quadrupolar glass state was first suggested in ortho-para hydrogen mixed crystals [3]. Here we study the alkalicyanides-halides mixed crystals $M(CN)_xX_{1-x}$ where M is a metal and X a halide ion as well as the related systems $(M_1)_x(M_2)_{1-x}CN$, where M_1 and M_2 are two different alkali ions [5]. The appearance of a narrow central peak in the neutron scattering spectrum [5] [6] suggests the existence of an orientational glass state [5] [7]. The glass state is characterized by a freezing in of orientational and translational correlations of T_{2g} symmetry (i.e. quadrupolar) without long range order [7].

Theoretical approaches have been based on the Potts glass [8] and on the uniaxial quadrupolar glass [8] [9]. Comparison of these models with experiments on $K(CN)_xBr_{1-x}$ is still an open question and currently this problem is investigated by Monte Carlo simulations [10].

Based on the experimental information that the nature of the substitutional halogens [11] (Cl or Br) is important we have proposed a different model which is not based on the random bond character of the CN-CN interaction but on the random strain fields which are generated by the substitutional atoms. These random strain fields interact with the remaining CN ions [12]. Molecular dynamics simulations give also evidence [13] for such a model.

2. MODEL

In KCN the bilinear coupling between dynamic acoustic lattice displacements $\vec{s}(\vec{q})$ and orientational modes $\vec{Y}(\vec{q})$ of T_{2g} symmetry leads to the translational rotation (TR) coupling

$$V^{TR} = \sum_{\vec{q}} v(\vec{q})\vec{s}(\vec{q})\vec{Y}(\vec{q}) \ , \qquad (1)$$

where $v(\vec{q})$ is the 3 x 3 coupling matrix [14]. This interaction is responsible for the ferroelastic phase transition, from a high temperature orientationally disordered phase of cubic symmetry to a low T phase with orientational order and

Springer Proceedings in Physics, Vol. 37 **Dynamics of Disordered Materials**
Editors: D. Richter · A.J. Dianoux · W. Petry · J. Teixeira © Springer-Verlag Berlin, Heidelberg 1989

orthorhombic symmetry. The elastic constants [15] c_{44} or the corresponding acoustic phonons become soft if T decreases towards T_c in the high symmetry phase. The static susceptibility $D^{-1}(\vec{q}) = \langle s^+(\vec{q})s(\vec{q})\rangle/T$ is related to the orientational susceptibility [14]

$$\chi(\vec{q}) = \langle Y^+(\vec{q})Y(q)\rangle/T \qquad (2a)$$

by

$$D(\vec{q}) = M(\vec{q})[1 + M^{-1}(\vec{q})v^+(\vec{q})\chi(\vec{q})v(\vec{q})]^{-1} \quad . \qquad (2b)$$

Here $\underset{\sim}{M}$ is the bare acoustic dynamical matrix, $\vec{q}$ is the wave vector. A divergence of $\chi(\vec{q}=o)$ at the transition temperature T_c leads to a vanishing of $D \propto q^2 c_{44}$.

In mixed crystals $M(CN)_x X_{1-x}$, the substitutional atoms generate static strains $\vec{s}^e(\vec{q})$ which couple to the orientational modes of CN by

$$V^{SR} = \sum_{\vec{q}} \vec{h}(\vec{q})\vec{Y}(-\vec{q}) \ , \qquad (3)$$

where $\vec{h} = v\, s^e$ is a random field with Gaussian mean $\overline{h^2}$]12]. If the random field is sufficiently strong, the divergence of χ is suppressed. Then a smooth maximum in $\chi(\vec{q}=o)$ corresponds to a minimum in the acoustic phonon frequency. This has been observed experimentally [16] [5] [6]. In addition, the interaction (3) leads to a static central peak in the neutron scattering law. The central peak intensity is given by [12]

$$\psi_h = x(1-x)\ \overline{h^2}(y/T)^2 \ , \qquad (4)$$

where y is weakly T dependent.

The random field model describes very well the experimental situation for $T \geq T_f$ [17]. Here T_f is the temperature where the phonon frequency is minimum. However, in $K(CN)_x Br_{1-x}$ near x = 0.5 and in $(K)_x(Na)_{1-x}CN$, x = 0.15, qualitative discrepancies remain. The central peak intensity increases more strongly than T^{-2} for $T \leq T_f$, [5] [6]. Low frequency experiments show that the turnover of the phonon frequency near T_f is a sudden phenomena [18]. The phonon dumping is much more pronounced [18] [19] [17] than is to be expected from theory [12]. This leads us to the conjecture that for $T<T_f$, dynamical processes are also relevant. Dynamic processes can lead to a nonergodic instability, as has first been shown for an electron moving in a random potential [20].

3. NONERGODIC INSTABILITY

The dynamic response function $\Phi^{SS}(\vec{q},z)$ satisfies the equation of motion [21]

$$[z^2 - D(\vec{q}) + z\Pi(\vec{q},z) + z\sigma(q,z)\ \Phi^{SS}(\vec{q},z) =$$
$$= [z + \Pi(\vec{q},z) + \sigma(\vec{q},z)]\ D^{-1}(\vec{q}) \ . \qquad (5)$$

Here z is the frequency, D the restoring force matrix, the memory kernels Π and σ account for phonon viscosity and orientational relaxation respectively. Anharmonic process give rise to scattering of orientational and translational modes with static random strains [21]

$$V^{RTS} = \sum_{\vec{p}\,\vec{k}\,\vec{q}} w(\vec{p},\vec{k},\vec{q})Y(\vec{p})s(\vec{q})\vec{s}^e(\vec{k}) \ , \qquad (6)$$

w is the anharmonic coupling. Using mode coupling theory one obtains contributions to the memory kernels which are in turn proportional to Φ^{SS} and Φ^{yy}:

$$\Pi(\vec{q},z) \equiv - \sum_{\vec{p}\,\vec{k}} |w(\vec{p},\vec{k},\vec{q})|^2 s^e(\vec{k}) s^e(-\vec{k}) \Phi^{YY}(\vec{p},z) \; . \qquad (7)$$

A similar expression $\propto \Phi^{SS}$ is obtained for $\sigma(\vec{q},z)$. In the low frequency limit, $\Phi^{SS} \propto \Phi^{YY}$. A nonergodic instability leads to a time independent contribution to the dynamic correlation function [7] or equivalently to a (1/z) singularity in the response function [20]:

$$^{XX}(q,z) = \frac{1}{z} f^{XX}(q) + \; ^{XX}(q,z) \; , \qquad (8)$$

where X stands for Y or s. From (7) follows a similar separation for Π and σ,

$$\Pi(\vec{q},z) = - \frac{1}{z} \rho(\vec{q}) + \tilde{\Pi}(\vec{q},z) \; , \qquad (9)$$

with

$$\rho(\vec{q}) = \sum_{\vec{p}\,\vec{k}} |w(\vec{p},k,\vec{q})|^2 \; s^e(\vec{k}) s^e(-\vec{k}) f^{YY}(\vec{p}) \; . \qquad (10)$$

The quantity $f^{XX}(q)$ corresponds to the central peak intensity Ψ in [7]. Collecting the (1/z) terms in Eqs. (5) - (10), we arrive at a closed integral equation for $\varphi(\vec{q}) = \chi^{-1}(q) f^{YY}(\vec{q})$:

$$\varphi(q) \, [1 - \varphi(q)]^{-1} = \sum_{\vec{p}} I(\vec{p},\vec{q}) \, \varphi(\vec{p}) \; . \qquad (11)$$

Here $I(\vec{p},\vec{q}) \propto |w|^2 \chi(\vec{p}) \chi(\vec{q})$. Concomittantly, the phonon resonances are given by

$$\Delta(\vec{q}) = D(\vec{q}) + \rho(\vec{q}) + \xi(\vec{q}) \; . \qquad (12)$$

Here $\xi(\vec{q})$ is the nonergodic part of $\sigma(\vec{q},z)$. Eq. (11) admits a nonzero positive solution if the kernel I is of order unity [21]. This situation is realized if χ becomes sufficiently large, or equivalently if the system can evolve close towards a second order phase transition which would occur at $T_c(x)$. The nonergodic instability then occurs at $T_f(x) \gtrsim T_c(x)$, and $(\vec{q})$ becomes different from zero. For $T < T_f(x)$, we have to take into account the depletion of thermal fluctuations by multiplying the effective interaction in $\chi(\vec{q})$ by $(1 - \varphi(\vec{q}))$. We have numerically solved the system of coupled equations for $\varphi(\vec{q})$ and $\chi(\vec{q})$ [22]. $\varphi(\vec{q})$ becomes nonzero for $T<T_f(x)$ and increases continuously with decreasing T. We obtain $\varphi(\vec{q}_1) > \varphi(\vec{q}_2)$ for $q_1 < q_2$, but $T_f(x)$, the threshold temperature is independent of $\vec{q}$, see Ref. [22]. The $\vec{q}$ dependence of $\varphi(\vec{q})$ and hence of $f(\vec{q})$ indicates deviations from the $1/q^2$ dependence of Huang scattering in the diffuse scattering law. Such deviations are also suggested by experiments [23].

Two effects can prevent the occurence of the nonergodic instability. i) A first order phase transition can occur before $\chi(\vec{q})$ is sufficiently large to trigger the instability. ii) A strong random field also keeps $\chi(\vec{q})$ small. This seems to be the case in $(K)_{0.5}(Na)_{0.5}CN$ [6].

Expression (12) for Δ has been evaluated numerically in Ref. [22]. For $T>T_f$, $\Delta = D$ decreases with decreasing T. Below T_f, the contributions ρ and ξ become nonzero and lead to an increase of Δ at low T.

The central peak intensity is the sum of two contributions:

$$\Psi(\vec{q}) = \Psi_h + f^{XX}(\vec{q}) \; . \qquad (13)$$

The term Ψ_h is always present due to a random field, the term f occurs if the nonergodic instability takes place.

REFERENCES

1. See e.g. K. Binder and A.P. Young, Revs. Mod. Phys. 58, 801 (1986).
2. E. Courtens, Jap. J. Appl. Phys. 24, Suppl. 24-2, 70 (1985).
3. N.S. Sullivan, M. Devoret, B.P. Cowan, and C. Urbina, Phys. Rev. B 17, 5016 (1978); A. Brooks Harris and H. Meyer, Can. J. Phys. 63, 3 (1985).
4. For a review see F. Lüty, in Defects in Insulating Crystals, ed. V.M. Tuchkevich and K.K. Shvarts (Springer-Verlag, Heidelberg, New York, 1981), p. 69.
5. J.M. Rowe, J.J. Rush, D.J. Hinks, and S. Susman, Phys. Rev. Lett. 43, 1158 (1979).
6. A. Loidl, T. Schräder, R. Böhmer, K. Knorr, J.K. Kjems and R. Born, Phys. Rev. B 34, 1238 (1986).
7. K.H. Michel and J.M. Rowe, Phys. Rev. B 22, 1417 (1980).
8. D.J. Gross, I. Kanter and H. Sompolinski, Phys. Rev. Lett. 55, 304 (1985).
9. P.M. Goldbart and D. Sherrington, J. Phys. C: Solid State Phys., 18, 1923 (1985).
10. H.-O. Carmesin and K. Binder, Z. Phys. B - Cond. Mat. 68, 375; H.-O. Carmesin, preprint.
11. J.Z. Kwiecien, R.C. Leung, and C.W. Garland, Phys. Rev. B 23, 4419 (1981); and 25, 5818 (1982).
12. K.H. Michel, Phys. Rev. Lett. 57, 2188 (1986); idem Phys. Rev. B 35, 1405 and 1414 (1987).
13. L.J. Lewis and M.L. Klein, Phys. Rev. Lett. 57, 2698 (1986).
14. K.H. Michel and J. Naudts, Phys. Rev. Lett. 39, 212 (1977); idem J. Chem. Phys. 67, 547 (1977).
15. S. Haussühl, Solid State Commun. 13, 147 (1973); J.M. Rowe, J.J. Rush, N. Vagelatos, D.L. Price, D.G. Hinks and S. Susman, J. Chem. Phys. 62, 4551 (1975).
16. S.K. Satija and C.H. Wang, Solid State Commun. 28, 617 (1978).
17. J.O. Fossum and C.W. Garland, Phys. Rev. Lett. 60, 592, (1988); J.O. Fossum, A. Wells and C.W. Garland, Phys. Rev. B 38, 412 (1988).
18. K. Knorr, U. Volkmann and A. Loidl, Phys. Rev. Lett. 57, 2544 (1986).
19. J.J. Vanderwal, Z. Hu and D. Walton, Phys. Rev. B 33, 5782 (1986); idem B 37, 9072 (1988).
20. W. Götze, Solid State Commun. 27, 1393 (1978).
21. K.H. Michel, Z. Phys. B - Cond. Mat. 68, 259 (1987).
22. C. Bostoen and K.H. Michel, Z. Phys. B - Cond. Mat. 71, 369 (1988).
23. A. Loidl, K. Knorr, J.M. Rowe and G.J. McIntyre, Phys. Rev. B 37, 389 (1988).

Nature of the Glass Transition in $(NaCN)_{1-x}(KCN)_x$

T. Schräder, M. Müller, and A. Loidl

Institut für Physik, Johannes-Gutenberg-Universität, Postfach 3980, D-6500 Mainz, Fed. Rep. of Germany

Alkali cyanide-alkali halide mixed crystals are model systems for orientational glasses. They exhibit a freezing dynamics of the orientational degrees of freedom analogous to that observed in spin glasses[1,2] and show low temperature thermodynamic properties which are characteristic for amorphous systems[3]. In a wide concentration range the mixed cyanides show frozen-in orientational disorder of the CN^- molecules. In addition the strong rotation-translation coupling yields frozen-in local shear deformations [4]. Theoretically, in analogy to spin glasses, this low temperature state has been interpreted to originate from frustration effects due to the anisotropic interactions between site disordered quadrupoles yielding a distribution of quadrupolar interaction forces [5-8].

Recently K.H. MICHEL has proposed a microscopic model where the orientational degrees of freedom are coupled to the random strains which are generated by the substitutional disorder[9]. Within this model he was able to describe the temperature and concentration dependence of the static and dynamic quadrupolar susceptibility. For collective freezing processes a non-ergodic instability was predicted [10]. The low temperature state is determined by the balance between the random strain-rotation coupling (RSRC) and the translation -rotation coupling (TRC): if the TRC is strong and the RSRC is weak the system undergoes a structural phase transition characterized by a soft acoustic mode. On the other hand, if, due to a high concentration of substitutional disorder, the RSRC is strong and the TRC is weak the low temperature state is generated by single ion freezing processes of quadrupoles within a highly anisotropic crystal field. Systems with a delicate balance between the RSRC and the TRC exhibit non-ergodic instabilities: here the shear modes are almost zero and only the freezing-in of critical fluctuations prevents the system from undergoing a ferroelastic phase transition. The non-ergodic instability marks the onset of the quadrupolar glass state. As well in pure random strain systems as in crystals which undergo a non-ergodic instability the freezing temperature T_f is defined by frequency dependent cusps in the quadrupolar susceptibility. The adequate way to characterize the low temperature state in disordered systems is to study the temperature dependence of the order parameter. In orientational glasses the order parameter is given by the configurational average of the squared thermal expectation value of the orientational coordinate: $\psi \sim \langle Y_\alpha(n)\rangle^2$ [10]. It is this property that leads to the existence of a static central peak in addition to the dynamic scattering law. According to MICHEL'S theory, in the cyanide glasses two contributions to the order parameter have to be taken into account: $\psi = \psi_1 + \psi_2$. The first term ψ_1 solely results from the response to the static random strain field h and is a smooth function of temperature, namely $\psi_1 \sim h^2/T^2$. ψ_2 accounts for the non-ergodic transition at T_f and exists only for temperatures $T < T_f$.

So far only a few experiments were reported which elucidated the nature of the glass transition in the cyanide glasses[11,12]. In order to test the predictions of the random strain theory we performed a systematic investigation of the elastic properties and the glass order parameter in $(NaCN)_{1-x}(KCN)_x$ mixed crystals applying inelastic neutron scattering and ultrasonic techniques. For an intermediate concentration range $0.15 < x < 0.89$ this system shows a low temperature glass state[13,14]. Random strains are implanted by the substitutional disorder of different sized cations. In contrast to alkali cyanid-alkali halide mixtures the CN^- sublattice remains undiluted. Thus $(NaCN)_{1-x}(KCN)_x$ is an ideal system to test the role played by random strains in orientational glasses.

Springer Proceedings in Physics, Vol. 37 **Dynamics of Disordered Materials**
Editors: D. Richter · A.J. Dianoux · W. Petry · J. Teixeira

Single crystals with concentrations x=0.19, 0.44, 0.59, 0.85 and 0.89 were investigated. The concentrations x=0.19, 0.44, 0.59 and 0.85 cover the glassy regime of NaCN:KCN, with x=0.19 and x=0.85 close to the critical concentrations. $(NaCN)_{0.11}(KCN)_{0.89}$ undergoes a structural phase transition into a coexistence region of cubic and rhombohedral phases[15]. The inelastic neutron scattering experiments were performed at the triple axis spectrometer IN3 at the Institut Laue Langevin Grenoble (France) and at the TAS7 at the Riso National Laboratory (Denmark). Here we show energy scans of transverse acoustic phonons as performed at constant momentum transfer Q=(2,0.2,0). Some typical raw data as observed in the 19% crystal are presented in fig.1. The squared frequency as determined by the center of the phonon side bands is proportional to the elastic constant c_{44}. On cooling from room temperature the phonon side bands soften, pass through a minimum and increase again towards lower temperatures. In addition a resolution limited central peak appears at zero energy transfer. Its intensity can be viewed as a direct measure of the order parameter of the glass state[9]. A summary of the results as observed in the glassy crystals with concentrations x=0.19, 0.59 and 0.85 is given in fig.2. The upper frame shows the temperature dependence of the elastic constant c_{44}; the lower frame documents the T-dependence of the central peak intensities. Obviously the different behaviour observed displays the different strength of the random fields:

x=0.85: near 110K c_{44} is almost zero. In this crystal we can expect a non-ergodic instability and the strong increase of the central peak intensity at temperatures below the minimum of the shear constant supports this view.

x=0.59: $c_{44}(T)$ is almost temperature independent. The central peak intensities exhibit a continous and very weak increase only. In the intermediate concentration regime of NaCN:KCN very strong random fields dominate the dynamics of the freezing process.

x=0.19: a broad and smeared out minimum in $c_{44}(T)$ is accompanied by a moderate increase of the central peak intensity. Although the crystal with x=0.19 is close to the critical concentration, $c_{44}(T)$ never becomes soft enough to support a non-ergodic instability.

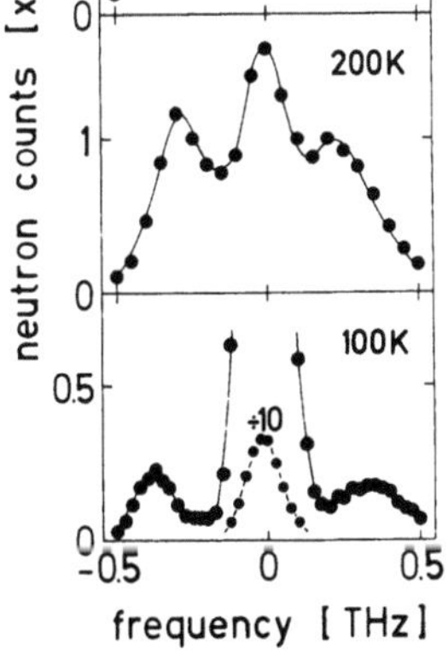

Fig.1

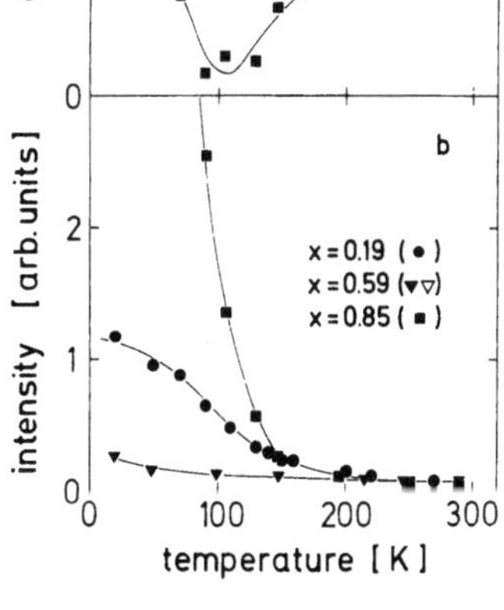

Fig.2

Fig.1: Constant Q scans at Q=(2,0.2,0) in $(NaCN)_{0.81}(KCN)_{0.19}$
Fig.2: $c_{44}(T)$ and central peak intensities for $(NaCN)_{1-x}(KCN)_x$

To get some insight into the dynamics of the quadrupolar freezing process we studied the temperature dependence of the elastic constants for the concentrations x=0.19, 0.59 and 0.85 by ultrasonic techniques. A conventional pulse echo overlap method with measuring frequencies of 10MHz and with bonding techniques as described by Feile et al.[16] was used. In the measurements of the elastic shear constant c_{44} the signal was lost near the minimum due to an extremely high sound absorption. However, we were able to follow $c_{11}(T)$ through the minimum which defines also the freezing temperature T_f of the elastic shear constant c_{44}. Figure 3 shows the elastic constants c_{44} as a function of temperature. The minima as determined by $c_{11}(T)$ are indicated as arrows. The ultrasonic results are compared with the elastic stiffness as measured by inelastic neutron scattering techniques at much higher frequencies. The data were scaled to a weighted average value at 300K. Above the freezing temperatures where both measuring frequencies are smaller than the inverse average relaxation time of the quadrupoles the neutron results agree well with the ultrasonic data. Significant shifts of the freezing temperatures appear only for the concentration x=0.19. In the following we present a more quantitative analysis of the data to demonstrate that a simple random strain model alone can not explain the experimental results in NaCN:KCN mixed crystals. According to K.H. MICHEL[9] $c_{44}(x,T)$ can be written in the form of a modified Curie-Weiss-law:

$$c_{44}(x,T) = c_{44}^{o} * \frac{(1 - T_s * \chi)}{(1 - T_c * \chi)} \tag{1}$$

χ (x,T) is the concentration dependent static susceptibility:

$$\chi(x,T) \sim \left(1 - \frac{T_f^2}{T^2} * \frac{1}{3}\right) * \frac{1}{T} \quad . \tag{2}$$

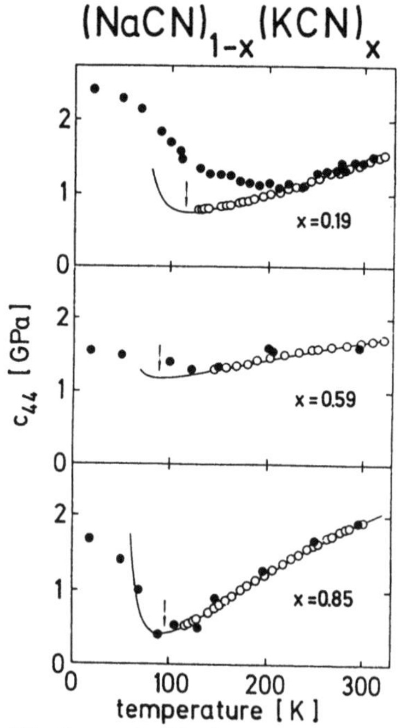

Fig.3

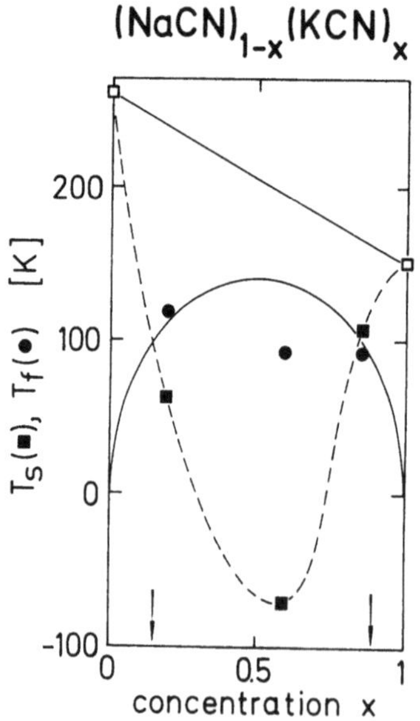

Fig.4

Fig.3: $c_{44}(x,T)$ determined by inelastic neutron scattering (closed symbols) and ultrasonic techniques (open symbols). The results of the random strain theory are represented by solid lines. Minima measured in $c_{11}(T)$ are indicated by arrows

Fig.4: $T_s(x)$ (closed squares) and $T_f(x)$ (circles) obtained from the data analysis of $c_{44}(T)$ for x=0.19, 0.59, 0.85. Open symbols are taken from ref.14. The solid line represents the concentration dependence of T_f, predicted by the random strain theory. The critical concentrations are indicated by arrows

The maximum of the susceptibility $\chi(x,T)$ defines the freezing temperature $T_f(x)$:

$$T_f(x) = \sqrt{3x(1-x)\xi} * h \qquad (3)$$

h is the strength of the static random strain and ξ a weakly temperature dependent function. Using these equations and fixing T_f to the minimum value of c_{11} we fitted $c_{44}(T)$ as observed in the ultrasonic experiments. The concentration dependence of T_s and of T_f is shown in fig.4. T_s deviates drastically from a linear interpolation based on the T_s values of the pure cyanides NaCN and KCN[14]. The problem that arise when $c_{44}(T)$ for x=0.59 is described within the random strain theory becomes apparent by the negative T_s obtained from the best fits. We interpret the large deviations from the linear interpolation by the fact, that strong random fields weaken the TRC, an effect that is not predicted by the random field model. Using eq.(3) the concentration dependence of $T_f(x)$ was fited to T_f(x=0.19 and 0.85). This defines a random field strength $h * \sqrt{3\xi}$ = 281 K. The concentration dependence is represented by the solid line in fig.4. T_f(x=0.59) clearly deviates from the calculated value. The theoretical model yields a much higher freezing temperature. Obviously, for the 59% crystal h/T is not a small quantity and the expansion used in the random field theory is not applicable.

To clarify the inconsistence of the elastic properties, especially of x=0.59, with the concentration dependend behaviour predicted by the random strain theory, we studied the glass order parameter in $(NaCN)_{1-x}(KCN)_x$ in more detail. Here we included results for crystals with concentrations x=0.89 and x=0.44. According to the theoretical approach the temperature dependence of the glass state order parameter follows a power law, namely $\psi_1 \sim h^2/T^2$. The order parameter for four different mixed crystals is shown in Fig.5 where the intensities of the central peak are plotted versus the temperature on a double logarithmic scale where power law behaviour is represented by straight lines. The striking different temperature dependence of the order parameter as obtained for x=0.19, 0.59, 0.85, and 0.89 documents the different regimes in the glass like phase of NaCN:KCN and nicely reflects the interplay of RSRC and TRC depending on the concentration : at high temperatures the crystals with x=0.19, 0.85, and 0.89 behave as pure random strain systems with a constant slope of approximately m=2, as predicted by the theory of MICHEL.

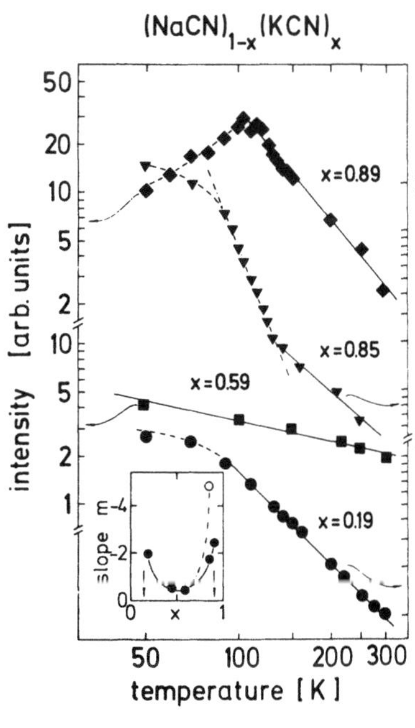

Fig.5: Log/log plot of central peak intensities versus the temperature for $(NaCN)_{1-x}(KCN)_x$ with $x_{c1} < x \leq x_{c2}$. The insert summarizes the concentration dependence of slope m for high temperatures (closed circles) and in the case of x=0.85 for temperatures below the glass transition (open circle)

For x=0.89, which coincides with the critical concentration, the anomaly in the T-dependence of the order parameter near 110K indicates the structural phase transition. In this crystal the molecular reorientations couple to the random field at high temperatures; however, the translation-rotation coupling is strong enough to overcome the RSRC and to establish an elastically ordered low temperature phase. For x=0.85, which is close to the critical concentration, the TRC drives the elastic shear mode c_{44} completely soft. However, the RSRC pins the dynamic modes and prevents the system from undergoing a structural phase transition. As a consequence of this non-ergodic instability an additional term $\psi_2(T)$ in the glass state order parameter occurs: the freezing-in of critical fluctuations results in a sudden change of slope in log(ψ) vs. log(T) from m=1.7 to m=4.8. This discontinuity is related to the minimum in $c_{44}(T)$ of Fig.3. At lower temperatures the strong increase of the central peak intensities is limited due to saturation effects. Apart from saturation effects observed at lower temperatures, no discontinuities could be observed for x=0.19. $\psi(T)$ fits well with the $1/T^2$ behaviour predicted by the random strain theory. Obviously, as can be seen from fig.3, the elastic shear mode becomes not soft enough to drive a non ergodic instability. The results for x=0.59 completety fall off from what we would expect from theory. $\psi(T)$ follows a power law $\psi(T) \sim T^{-m}$ from room temperature down to the lowest temperatures with a small exponent of m≈0.4 and no discontinuity occurs over the entire temperature range. As indicated in the insert of fig.5 similar results were obtained for x=0.44 which yield an exponent of m≈0.5. The deviation from the predicted $1/T^2$ behaviour of the order parameter is a result of the strong random strain fields. This demonstrates again that for intermediate concentrations in NaCN:KCN the high temperature expansion of the random field model is not applicable. The measured concentration dependence of the exponent m is indicated as a solid line in the insert of fig.5. The dotted-dashed line indicates deviations due to the appearance of a non-ergodic instability. NaCN:KCN can not be described by an unique behaviour of the glass state order parameter. The order parameter sensitively depends on the strength of the random fields: At high T for a low concentration of impurity ions the random strain theory as proposed by MICHEL [9] is valid. Here the order parameter is a smooth function of temperature and follows $\psi \sim h^2/T^2$. However even in these cases which seem to be well represented by the pure random strain theory, the TRC is considerably weaker as theoretically expected. Near the critical concentrations (x=0.85) a non-ergodic instability gives a considerably enhanced T-dependence of the order parameter below T_f. For intermediate concentrations however, in the regime of strong random strain fields the random strain model fails and higher order terms have to be taken into account.

1. J.M. Rowe, J.J. Rush, D.J. Hinks, S. Susman: Phys.Rev. Lett. 43, 1158 (1979)
2. A. Loidl, R. Feile, K. Knorr: Phys. Rev. Lett. 48, 1263 (1982)
3. for a review see:
   J.J. DeYoreo, W. Knaak, M. Meissner, R.O. Pohl: Phys.Rev.B 34, 8828 (1986)
4. K. Knorr, A. Loidl: Phys.Rev.Lett. 57, 460 (1986)
   A. Loidl, K. Knorr, J.M. Rowe, G.J. McIntyre: Phys.Rev. B37, 389, 1988
5. B. Fischer, M.W. Klein: Phys. Rev. Lett. 43, 289 (1979)
6. K.H. Michel, J.M. Rowe: Phys. Rev. B22, 1417 (1980)
7. P. Goldbart, D. Sherrington: J. Phys. C18, 1923 (1985);
   D.J. Gross, I. Kanter, H. Sompolinski: Phys. Rev. Lett. 55, 304 (1985)
8. H.-O. Carmesin, K. Binder: Z. Physik B68, 375 (1987)
9. K.H. Michel: Phys. Rev. Lett. 57, 2188 (1986);
   Phys. Rev. B35, 1405 (1987); Phys. Rev. B35, 1414 (1987)
10. K.H. Michel: Z. Physik B68, 259 (1987);
   C. Bostoen, K.H. Michel: Z. Physik B71, 369 (1988)
11. S. Elschner, J. Albers, A. Loidl, J.K. Kjems: Europhys. Lett. 4, 1139 (1987)
12. J.O. Fossum, C.W. Garland: Phys. Rev. Lett. 60, 592 (1988)
   J.O. Fossum, A. Wells, C.W. Garland: Phys. Rev. B38, 412 (1988)
13. F. Luty, J. Ortiz-Lopez: Phys. Rev. Lett. 50, 1289 (1983)
14. A. Loidl, T. Schräder, R. Böhmer, K. Knorr, J.K. Kjems, R. Born:
   Phys. Rev. B34, 1238 (1986)
15. T. Schräder, A. Loidl, T. Vogt: Physica B to be published
16. R. Feile, A. Loidl, K. Knorr: Phys. Rev. B26, 6875 (1982)

Correlation Between Glass Behaviour and an Intrinsic Limit of Metastability: A Time Resolved Investigation

M. Descamps

Laboratoire de Dynamique des Cristaux Moléculaires, Université de Lille I, F-59655 Villeneuve d'Ascq Cedex, France

1 Introduction

Quenching an orientationally disordered crystal so that the first order ordering transition is bypassed can result in a glassy crystal state [1] if the quench is deep enough. In this case one expects that only the orientational degrees of freedom are frozen on an otherwise unaffected lattice. In contrast to spin-glass molecular systems (such as $K\,Br_{1-x}\,(CN)_x$ [2] and $Rb_{1-x}\,(NH_4)_x\,H_2PO_4$ [3]) the glassy crystals are not intrinsically frustrated by a dilution. Instead they show essentially all the dynamic and thermodynamic features of a conventional glass.

In searches for the origin of the glass forming tendency, the necessity of clustering [4] or of a critical level of static local ordering [5] underlies a number of the older concepts as well as the most recently elaborated theories [6]. Owing to the average translational invariance, one can expect much more sensitivity to changes of local order for a glassy crystal than for the isotropic $S(|Q|)$ structure factor which is obtained with a glass of liquid.

In this paper we give an outline of some recent (partly unpublished) time resolved X-ray scattering measurements of the non-equilibrium kinetics associated with the order disorder transition in single crystals of adamantane derivatives. These provide clear structural signatures of both the ergodic to non-ergodic transition and the glass heat capacity anomaly.

At room temperature (R.T.) mixed crystals of cyano and chloro-adamantane ($Cl_x(CN)_{1-x}$ ADM) have isomorphous f.c.c. orientationally disordered structures, called I [7,8]. For $x \le 0.5$ they show the same tendency to form a glassy crystal on quenching [9]. Usual glass properties have been more extensively characterized on the x=0 compound. In particular : the dipolar freezing and Vogel Fülcher law [10] -frequencies of 33 mHz have been detected at 184 K- ; a β relaxation has been recorded [11] ; and upon reheating history dependent glass like jump in C_p is observed in the Tg range 165-185 K.

We have found it more convenient for the present study to use the compound x=.25 which has a simpler equilibrium phase diagram. Using D.S.C., the first order equilibrium transition was determined to occur at $T_t \equiv 237$ K, the glass transition being detected in the range 165-185 K [8]. As was shown for x=0 compounds [12,13], X-ray diffuse scattering studies proved that the quenched state is not in equilibrium with the slow-growth, even below Tg, of well defined spots at the X-boundary points of the B.Z. (Fig. 1 inset). This reveals the development, on a local scale, of an antiparallel ordering of the molecular dipoles which can be described by a near tetragonal structure (to be called III in conformity with [13]). The III-I first order transition could be reliably probed by the sudden disappearance of superstructure peaks at T_t in a heating process.

In the work reported here measurements were made after quenching the sample from an initial temperature $T_i > T_t$ to a final temperature $T_f < T_t$. The kinetics could be followed from T_t down to the glass

Springer Proceedings in Physics, Vol. 37 Dynamics of Disordered Materials
Editors: D. Richter · A.J. Dianoux · W. Petry · J. Teixeira © Springer-Verlag Berlin, Heidelberg 1989

state. This is exeptional for a system where the phase transformation is limited neither by the conservation of the order parameter nor by lattice diffusion [14]. The fundamental constraint which limits the intrinsic speed of the kinetics upon deep quenching is the strong steric hindrance to molecular reorientations. It is so high that even at R.T. residence times are longer than 10^{-7} s [10,11].

2 Evidence of an intrinsic transition between metastability and instability

Important information about the mechanism of ordering is given by the shape of the growth curves. Fig. 1 shows an isothermal (T_f = 177 K) growth curve at wave vector transfer 121 which is typical of a quench with 0 K < ΔT < 70 K ($\Delta T = T_t - T_f$). The quench is followed by an incubation time ($t_i \cong$ 1 hour at 177 K) before the onset of scattering. This indicates a metastable condition. However, the intensity level of the initial plateau is not zero. It corresponds to a low stationary peak which is quickly ($\tau \lesssim$ 3 min at 177 K) established after the quench. The curve measured above the plateau is sigmoïdal. The kinetics distinctly slow down after 4 hours and are very sluggish in the late stage.

The different regimes essentially reflect two types of fluctuations associated with the 1st order phase transformation which operate at the same $\vec{q}_x$ position in reciprocal space. Using the classification of Frenkel these are homophase and heterophase fluctuations [15,16] :

- The sigmoid is typical of a nucleation and growth mechanism of transformation (Avrami behaviour [17]). It occurs when heterophase fluctuations which form locally in the disordered metastable phase have reached a critical size. This is determined by the degree of super-cooling and the interface energy.

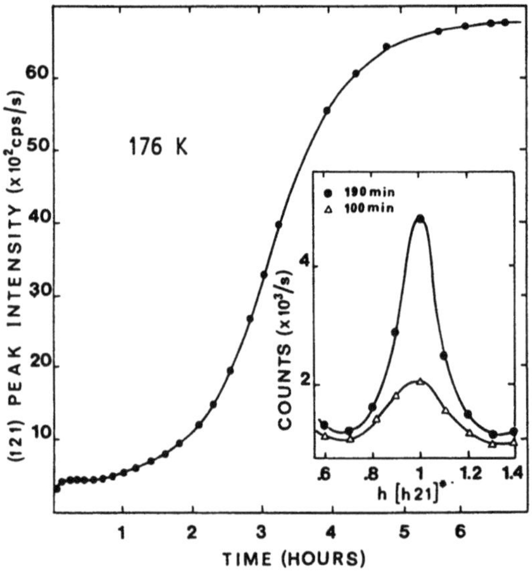

Fig. 1 : Typical sigmoïdal behaviour of the isothermal growth curve at the wave vector transfer $\vec{Q}$ = 121 when aging at $T_f \geq$ 165 K. The initial plateau points to a metastable condition. The sigmoïd is typical of a nucleation process. The intensity of the plateau (I_p) characterizes short range order. Inset : evolution of the peak profile.

- The initial stationary diffuse intensity $I_p(\vec{q}_x)$ is due to the short range order (S R O) which exists in the metastable phase during the life time of the latter. $I_p(\vec{q}_x)/T_f$ measures the order parameter static susceptibility $\chi(\vec{q}_x)$ associated with the homophase fluctuations.

The reciprocal of $\chi(\vec{q}_x)$ was found to be linear vs temperature for $T_f \geq 172$ K as is shown in Fig. 2. Extrapolation of this curve to low temperatures leads to a singularity in $I_p((\vec{q}_x)$ at an instability temperature $T_c \cong 165$ K which represents the intrinsic limit of metastability of the disordered phase. Such a diverging behaviour -of the form $\chi(\vec{q}_x) \cong (T-T_c)^{-1}$- would be expected only for a mean-field (M.F) condition. It thus appears that the system remains in a M.F. regime deeply below T_t which may occur by virtue of dipolar interactions. This approach of an instability was also observed in the progressive asymetrization of the sigmoids and the associated decrease of the time life of metastability. The universal scaling predicted by the Avrami law [17] is progressively violated on approaching T_c. Below T_c continuous ordering occurs immediately, but the kinetics are extremely slow (cf. Fig. 3). This is a way to probe activated reorientations on so large time scale -tens of hours- that they could not be seen otherwise.

3 Structural signature of the ergodicity breaking

The very beginning of the curve in Fig. 1 shows that, at 177 K, the time τ which is necessary to establish the stationary susceptibility after a quench becomes perceptible in a real-time experiment. In fact, by following the temperature evolution of $\chi(q_x)$ one observes a considerable increase of τ below 172 K. In a continuous cooling experiment experimental points distincly deviate from the initial M.F. straightline as is shown in the low temperature part of Fig. 2.

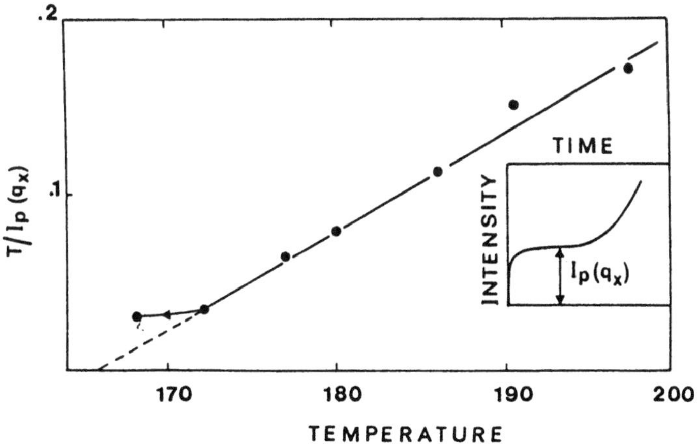

Fig.2 : Variation with temperature of the reciprocal of the measured S R O intensity (I_p) multiplied by T ($T/I_p = \chi^{-1}_{(q_x)}$). Deviation from the mean field expectation ($\chi_{(q_x)} \cong (T-T_c)^{-1}$) is observed in the neighbourhood of an intrinsic limit of metastability at about $T_c \cong 165$K. Ergodicity breaking appears in this locking of $\chi_{(q_x)}$ under continuous cooling.

In an isothermal aging in the intermediate temperature range 150 K $\leq T_f$ < 172 K the susceptibility then relaxes slowly. The quantity which is thus followed is :

$$\chi(q_x,t) = \sum_r < \vec{\mu}_o(t)\, \vec{\mu}_r(t) > \exp(i\, q_x.r)$$

It is the Fourier transform of the equal time orientational correlation function.

If we quench more deeply it appears that $\chi(q_x t)$ is locked in at the value it has when the system falls out of equilibrium. This value is a decreasing function of the quench rate with which the system crosses the temperature neighbourhood of the absolute limit of metastability. This implies that relaxations observed at $\vec{q}_x$, connected with molecular reorientations are slowed down and finally arrested when the system passes into the glassy crystal state.

4 Cluster reversion at the glass thermal anomaly

After an isothermal aging in the intermediate temperature range where a slow relaxation of $\chi(q_x,t)$ can be noticed we reheated the samples. The evolution of the system was followed by monitoring the superstructure peak as is shown in Fig. 3.

A 3 day aging at 157 K gives the relaxation seen in part [1] of the figure. It is typical of ordering kinetics in the instable region. A rapid increase in temperature ($\cong$ 6 K/min) induces rather suddenly the collapse of the peak at 180-183 K (part [2]). After the temperature step, reaging leads to a new peak increasing at exactly the q_x position with sigmoïdal kinetics standard at this temperature (part [3]). Upon a further reheating the equilibrium transition is finally observed by the definitive drop of the peak at T_t.

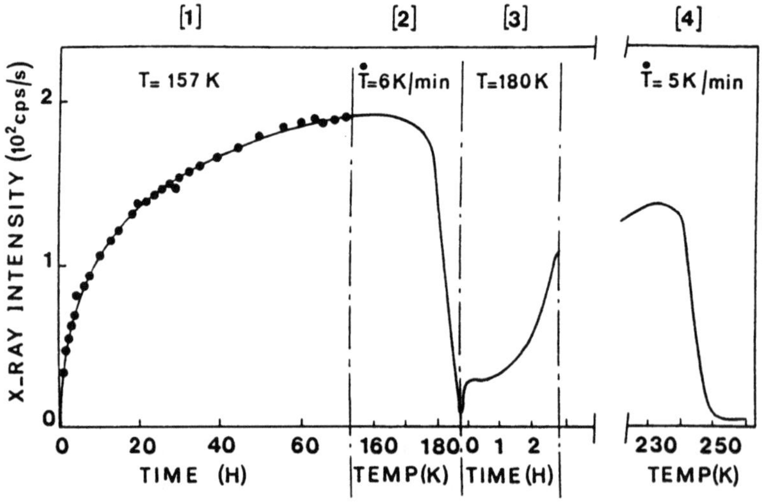

Fig. 3 : Evolution of X-ray intensity at the 121 position in response to a succession of aging [1], heating [2], reaging [3], reheating [4]. T_t is the equilibrium transition. The drop of intensity at 183 K marks the reversion of clusters occuring at the glass transition in a heating process.

The drop of the peak at 183 K marks the reversion of the ordered structure which has slowly developped at low temperature. This restoration of the metastable disordered phase is a purely non-equilibrium phenomenon which typically occurs 60 K below T_t. It is concomitant with the endothermic glass anomaly of C_p. It occurs at a temperature which is increased by a more prononced aging at low temperature.

5 Conclusion

The present study traces the structural manifestation of the glass transition in a glassy crystal.

- On cooling, the ergodic to non ergodic transition seems to be associated with a critical slowing down near the absolute metastability limit of the disordered phase.
- On heating the return to metastability involves the "dilution" of clusters of the ordered phase which had been formed during a lower temperature aging.

The observed hysteresis points to a stabilization effect by the clusters of the glassy state.

Acknowledgements

I thank C. Caucheteux, M. El Adib and G. Odou for their valuable help with X-ray experiments, N. Abraham for a reading of the manuscript. I have benefitted greatly by discussions with J.P. Simon.

References

1 - H. Suga in "Dynamic aspects of structural change in Liquid and glasses" C.A. Angell and M. Goldstein (Eds) N.Y. Acad of Sciences 484, 248 (1986).
2 - A. Loidl and B. Knorr, N.Y. Acad. of Sciences 484, 121 (1986).
3 - E. Courtens, F. Huard and R. Vacher, Phys. Rev. Lett. 55, 722 (1985).
4 - J.C. Phillips, N.Y. Acad of Sciences 484, 271 (1986).
5 - U. Bengtzelius and A. Sjölander, N.Y. Acad. of Sciences 484, 229 (1986).
6 - U. Bengtzelius, U.W. Götze and A. Sjölander, J. Phys. C 17, 5915 (1984).
7 - J.P. Amoureux, J.L. Sauvajol and M. Bée, Acta Cryst. A 37, 97 (1981).
8 - M. Foulon, Thèse doctorat d'état, Lille (1987).
9 - M. Foulon, J.P. Amoureux, J.L. Sauvajol, J. Lefebvre and M. Descamps J. Phys. C 16 L 265 (1983).
10 - J.P. Amoureux, G. Noyel, M. Foulon, M. Bée and L. Jorat, Mol. Phys. 52, 161 (1984).
11 - K. Pathmanathan and G.J. Johari, J. Phys. C : 18, 6535 (1985).
12 - M. Descamps, C. Caucheteux, G. Odou and J.L. Sauvajol, J. de Phys. Lett. 45, L 719 (1984).
13 - M. Descamps and C. Caucheteux, J. Phys. C 20, 5073 (1987).
14 - J.D. Axe, Japan. J. Appl. Phys. Suppl. 24, 46 (1985).
15 - J. Frenkel, Kinetic Theory of Liquids (Clarendon, Oxford) 374-390 (1946).
16 - H.E. Cook, M. Suezawa, T. Kajitani and L. Rivaud, J. de Phys. Colloque C7 suppl. au n° 212 38, 430 (1977).
17 - N. Hamaya, Y. Yamada, J.D. Axe, D.P. Belanger and S.M. Shapiro, Phys. Rev. B 33, 7770 (1986).

The Effect of Orientational and Composition Disorder on the Low Temperature/Low Energy Rotational Excitations of the NH_3 Groups in Calciumhexaamine $Ca(NH_3)_6$

P. Damay[1], *F. Leclercq*[1], *A.J. Dianoux*[2], *and W. Press*[3]

[1]LESI, UA 253 du CNRS, HEI, 13, rue de Toul, F-59046 Lille Cedex, France
[2]Institut Laue-Langevin, 156X Av. des Martyrs, F-38042 Grenoble Cedex, France
[3]Experimentalphysik, Universität Kiel, Olshausenstr. 40, D-2300 Kiel, Fed. Rep. of Germany

Introduction

Several alkaline earth metals form hexaammine complexes with ammonia. These complexes show some similarities with the transition metal salt hexaammines, but important differences are noticeable:

- the compounds are metallic with a low electron density (a tenth of that of Ca).
- the structure is very open. The volume expansion at formation is 45 percent.
- the complexes are formed with alkaline earth metals instead of transition metals; $Ca(NH_3)_6{}^{2+}$ does not exists in solution. The necessity of a rigid octahedral form of the complex is probably less stringent in the present case.

As seen from XRays [1] and early Neutron Scattering [2] the structure of the calcium compound is bcc with the ammonia group on the axis of the cell (00x). It has been shown more recently [3,4] that at high temperature the NH_3 groups undergo rotational diffusion with a short characteristic timescale (plastic phase). As temperature decreases, the diffusive motion freezes ($T \sim 30K$). A phase transition towards a lower symmetry structure was expected since the 3-fold symmetry of the NH_3 group is not compatible with the 4-fold symmetry of the site. But, even with a very slow cooling rate or annealing the structure remains bcc at least down to 1.5K. Then the NH_3 groups freeze in orientational disorder (molecular glass).

Proton tunneling is observed at low temperatures [3,4,5] ; but instead of observing well-defined transition energies, a broad peak is observed as a fingerprint of the proton disorder. This peak is explained by a continuous distribution of hindering potentials going from free rotation (no barrier) to barriers heights of several meV [5].

Beside the proton disorder, a detailed structural investigation [6] revealed a position disorder of the ND_3 group in a deuterated sample, $Ca(ND_3)_{5.7}$, at 6K. Indeed, only 80% of the ND_3 groups were found in the expected position (00x) while 20% were found in the 111 position (tetrahedral sites). Furthermore, it was found possible to remove ND_3 molecules from the $n = 6$ compound without breaking the structure. Moderate addition of NH_3 was also possible but with increasing values of n distortion of the lattice and then the precipitation of ammonia occur (n=6.5). The low temperature tunneling spectrum revealed itself to be very sensitive to departures from stoichiometry [5].

The purpose of the present work was to show how Inelastic Neutron Scattering (INS) could be used as a very sensitive local probe to study disordered materials. Mainly the $0 \rightarrow 1$ rotational tunneling transition observed with different samples (here samples with n = 5.5, 6.0 and 6.5) was used to

Springer Proceedings in Physics, Vol. 37 **Dynamics of Disordered Materials**
Editors: D. Richter · A.J. Dianoux · W. Petry · J. Teixeira

determine the density of states g(ω). The density of states used together with a realistic model for the shape of the hindering potential will be the appropriate tools to start the interpretation of the transitions observed at higher energy .

Experimental

The first indications found from the literature on $Me(NH_3)_6$ compounds (see for instance the review works of Mammano [7] or Glaunsinger [8]) is that preparation of good well-defined sample is a very difficult task. Homogeneity, powdering, knowledge of the exact composition and of course high vacuum conditions are the major points. The preparation techniques had to be improved considerably in the course of our study in order to obtain a better characterization of the samples. The rotational states turned out to be a very sensitive probe of the local environment, which changes with stoichiometry.

The sample preparation procedure has already been reported in detail [3]. As in previous works, the powdered sample covered the inner part of a ϕ20 mm niobium cylinder. The experiments have been performed on the IN6 spectrometer at the ILL in Grenoble. The chosen wavelength of 5.9Å provided an average resolution of 50 μeV in a Q-range extending from 0.2 to 1.75 $Å^{-1}$ ($\vec{Q} = \vec{k} - \vec{k}_0$, with $\vec{k}_0$ and $\vec{k}$ being respectively the incident and scattered wave-vectors)

Results and discussion

At low temperature (T=1.5K), several inelastic peaks are observed. As already said, they are very sensitive to the coordination number n. Most of the discussion will be devoted to the lower energy peak at E ≈ 0.7 meV; it corresponds to almost free quantum mechanical rotation (transition from rotational quantum number J = 0 to 1) of the uniaxial rotation of the NH_3 group in a 3-fold (or n×3fold) potential. In this low barrier limit the dependence of the observed transition energies with the barrier height V can be approximated by a Gaussian [9] $\hbar\omega = \hbar\omega_{free} \exp(-\alpha V^2)$. In ordered systems, tunneling transitions are seen at well defined positions with a linewidth equal to the spectrometer resolution. In the present case broad asymmetric lines are observed. Three examples are reported on figure 1 for for composition n = 6, 5.5 and 6.5.

Let us discuss the n = 6 compound first. The most striking feature is to find a cut-off of the inelastic intensity at the free rotor energy (0.73 meV as determined from our structural data [6]). That is probably the first proved evidence of a free uniaxial rotor at low temperature. Since the shape of the line for energy below -0.73 meV reproduces exactly the resolution of the spectrometer and since there is no temperature effect between 1.4 and 6K (except for detailed balance), one may consider that homogeneous broadening (for example coupling with conduction electrons) can play only a negligible role in this range and that inhomogeneous broadening due to local disorder must be invoked to explain the shape of the line.

From the observed spectrum, the real inelastic structure factor $S(\vec{Q},\omega)$ was obtained by removing the elastic peak and after deconvolution with the resolution. The structure factor $S(\vec{Q})$ was then obtained by integrating over ω. At $\vec{Q}$ = 1.5 $Å^{-1}$ one finds a SF which is 14.5 percent of the integrated intensity of the elastic peak. If all the NH_3 groups were involved in the motion, a ratio of 17 percent would have been found (see Heidemann[10]). Since any spurious scattering would increase the intensity of the elastic peak, one may only conclude that <u>at least</u> 85 percent of the NH_3 groups take part to the movement.

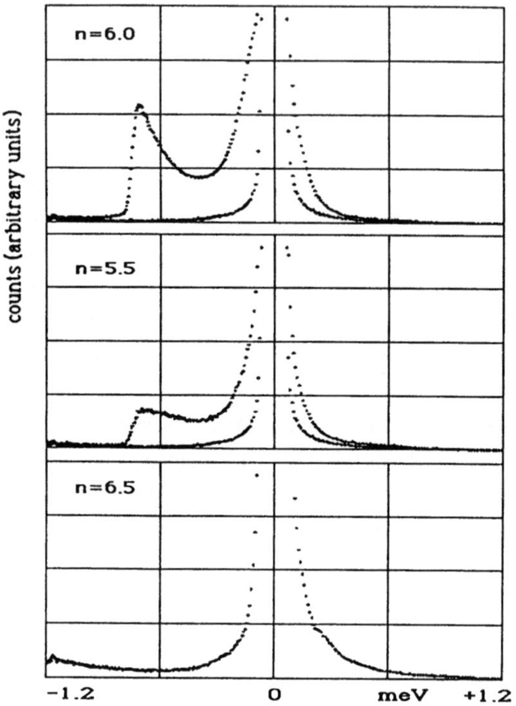

figure 1. First excitation (0→1 transition) of the uniaxial rotation of NH_3 groups in $Ca(NH_3)_6$ with n = 6.0, 5.5 and 6.5 respectively (T = 1.5K). Each spectrum was scaled at 5% of the elastic peak. The elastic line observed with a vanadium sample has been superimposed to indicate the spectrometer resolution. (Vector transfer $\vec{Q}$ = 1.5Å^{-1})

The density of state $g(\omega)$ may then be calculated from the expression:

$$\frac{d^2\sigma}{d\Omega\, d\omega} = \frac{k}{k_0} b_{inc}^2 \frac{Q^2}{6MN} e^{-2W} \frac{e^{\frac{\hbar\omega}{kT}}}{e^{\frac{\hbar\omega}{kT}} - 1} \frac{g(\omega)}{\omega} \qquad (1)$$

The results are reported in figure 2 . It is seen that for this compound most of the states have energies close to that of a free rotor.

An analysis of the 0 → 1 transition provides a complete description of the energy distribution but it is not sufficient to get a precise idea of what may be the shape of the hindering potential (from one observation, one obtains a one parameter function). The analysis of the 0 → 2 transition supplies new information. Indeed, the splitting of the energy level for the 0 → 2 transition increases with the barrier height in a 3-fold potential but decreases in a 6-fold one. In the first case, using the calculated $g(\omega)$, one expects a distribution extending from a minimum energy 4B to higher energies, while in a pure 6-fold potential, the distribution would extend from a maximum energy 4B to lower energy. Observation of the real spectrum (fig. 3)shows a distribution around E = 4 B = 2.92 meV. This indicates that at least a mixture of 3 and 6-fold potential is necessary to describe the system.

From all our observations it appeared that the n = 6 compound displayed more order than the others with $n \neq$ 6 (a majority of the states is close to the free rotor limit). Removing or adding ammonia in this system must be considered as an increase of the disorder. The characteristics of systems with n < 6 are described by a global decrease of the Inelastic Structure Factor and a shift of the distribution towards lower energy (i.e. an increase of the average barrier height). For a n = 5.5

compound (fig 1), the number of NH_3 groups involved in the motion is 60 percent of that of n = 6. The density near the free rotor limit remains rather high nevertheless.

The situation is very different for n > 6. As seen on figure 1 for n = 6.5, there is much less inelastic scattering for the 0 → 1 transition (35 % of that found with n = 6.0), and a peak near 1.2 meV shows up. The number of states near free rotation is then very low (fig. 4). A sharp peak is visible at 60 μeV and some structure appears in the distribution.

If for n < 6, NH_3 groups defects may perturb the local symmetry which leads to a high density of free rotors, for n > 6 additional NH_3 groups in interstitial positions seem to block most of the motion, giving rise to rather high hindering barriers.

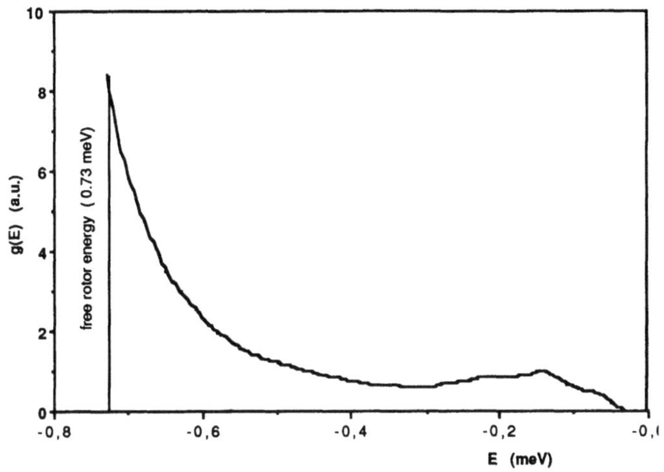

figure 2. Normalized density of states $g(\hbar\omega)$ in arbitrary units for the $Ca(NH_3)_{6.0}$ compound at 1.4K. Most of the states are located near free rotation $\Delta E = \hbar^2/2I = 0.73$ meV .

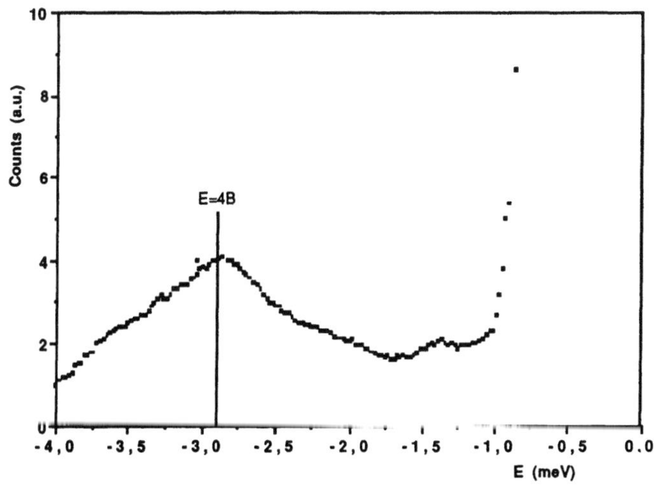

Figure 3 Transition 0→2 for the $Ca(NH_3)_{6.0}$ compound at T=1.5K (Y-scale is 0.75% of the elastic peak). As indicated in the text a mixture of 3-fold and 6-fold potential is necessary to explain the shape of the peak.

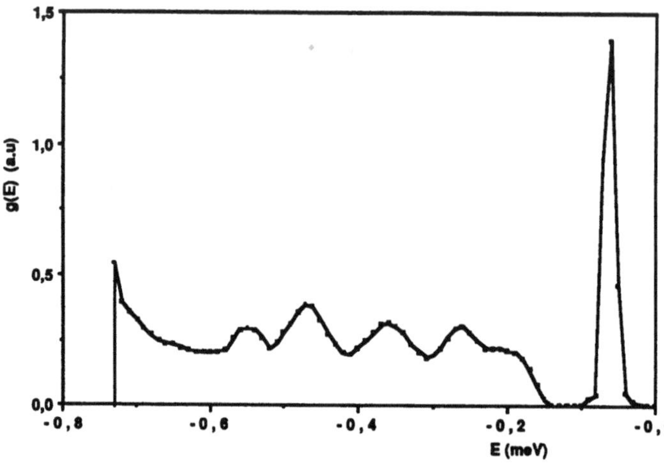

figure 4. Normalized density of states $g(\hbar\omega)$ in the same arbitrary units as in figure 2 for the $Ca(NH_3)_{6.5}$ compound at 1.5K.

Concluding remarks

The study of quantum excitations at low temperature thus gives a precise description of a disordered material. It is possible to determine the density of states g(ω) and a realistic shape of the potential. As temperature increases a transition to classical rotational diffusion is observed. In the classical regime, above the glass transition, the quasielastic spectrum was found to be insensitive to compositional disorder. One has still to find the appropriate perturbation of the quantum motion by phonons to explain the evolution of the spectra as temperature increases that is to say to describe the transition between the glassy state to the plastic phase. A quantitative explanation of the densities of states displayed on figures 2 and 3 is in progress.

Bibliography

1 - MAMMANO N. and SIENKO M.J., J.Solid State Chem., **1**, 534 (1970)

2 - VONDREELE R.B., GLAUNSINGER W.S., BOWMAN A.L. and YARNELL J.L, **79**, 2992 (1975)

3 - LECLERCQ F., Thèse de Doctorat d'Etat, LILLE (1985)

4 - LECLERCQ F., DAMAY P., CHIEUX P., **88**, 3886 (1984)

5 - LECLERCQ F., DAMAY P., DIANOUX A.J., PRESS W., Springer Proc.Phys., **17**, 103 (1987)

6 - PRESS W., DAMAY P., LECLERCQ F. CHIEUX P., to be published

7 - MAMMANO N., in " Metal Ammonia Solutions", Colloque Weyl II
J.J.LAGOWSKI and M.J. SIENKO ed., Butterworths, London (1970)

8 - GLAUNSINGER W.S., VON DREELE R.B., MARZKE R.F., HANSON R.C., CHIEUX P., DAMAY P. and CATTERALL R., J.Chem.Phys., **88**, 3860 (1984)

9 - GRONDEY S., PRAGER M., PRESS W., HEIDEMANN A., Springer Proc.Phys., **17**, 98 (1987)

10- HEIDEMANN A., ANDERSON I., JEFFRYES B. and ALEFELD B.,
Z.Phys. B - Condensed Matter , **49**, 123 (1982)

What Stabilizes the Orientational Glass Phase?

H.-O. Carmesin

Institut für Physik, Johannes-Gutenberg-Universität, Postfach 3980, D-6500 Mainz, Fed. Rep. of Germany

New address: Courant Institut of Mathematical Sciences, 251 Mercer Street, New York, NY 10012, USA

1. INTRODUCTION

There are mixtures of isotropic and anisotropic particles like $K\,Br_x(CN)_{1-x}$ /1/ or $Ar_x(N_2)_{1-x}$ /2/, which undergo a transition to an orientational glass phase. Such systems have been modelled as isotropic quadrupolar glasses /3/, but Monte Carlo simulations indicate /4/, /5/ that these undergo a transition to the quadrupolar glass state at zero temperature only in three dimensions.

It is shown by simulations that the quadrupolar glass phase is stabilized by cubic anisotropy. The similarity to the Potts glass /6/ is demonstrated.

The anisotropy due to an orthorhombic lattice /1/ yields a new saddle point in the mean field free energy /7/.

2. CUBIC ANISOTROPIC QUADRUPOLAR GLASS

Each particle is occupying a lattice site of a simple cubic lattice, and its orientation is given by a three-dimensional unit vector $\vec{S}_i$. There are nearest-neighbor symmetric Gaussian coupling constants. The Hamiltonians of the isotropic-(1a) and cubic anisotropic quadrupolar glass (1b) and of the Potts glass (1b) read:

$$H = \sum_{i<j} J_{ij} \Big(\sum_{\mu=1}^{3} S_i^{\mu} S_j^{\mu} \Big)^2 \qquad (1a)$$

$$H = \sum_{i<j} J_{ij} \sum_{\mu=1}^{3} (S_i^{\mu} S_j^{\mu})^2 \qquad (1b)$$

While the unit vector rotates freely in the case of the quadrupolar glasses, it aligns always parallel to one of the coordinate axes in the case of the Potts glass.

A Monte Carlo Simulation indicates that the cubic anisotropic quadrupolar glass undergoes a transition to the glass phase at $T_c \simeq 0.1$, Fig. 1 /7/. The same simulation shows, that the leading three moments of the one particle distribution function approach the values of the Potts glass with falling temperature, Fig. 2. From Fig. 2 the question arises, whether the quadrupoles will all align parallel to the coordinate axes at sufficiently slow temperature. A counterexample shows that this is not the case : Consider a particle whose neighbors are fixed due to large coupling constants. All coupling constants adjacent to the particle are one. To each axis, they align two of the neighbors. Thus, the particle will align along the spatial diagonal to minimize its energy.

Editors: D. Richter · A.J. Dianoux · W. Petry · J. Teixeira

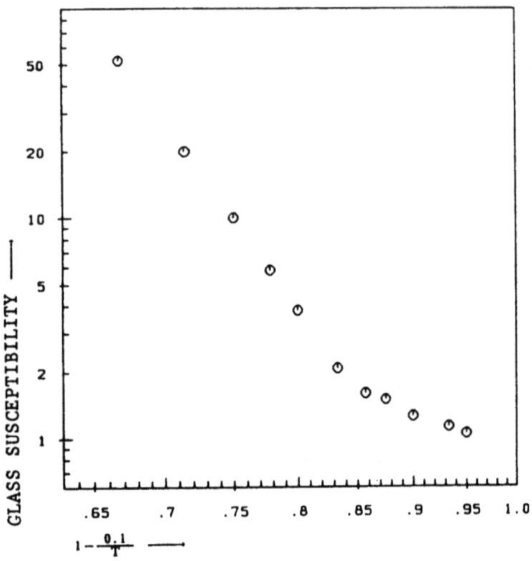

Fig. 1 : Log-Log plot for the glass susceptibility

$$\chi_G = \frac{1}{N} \sum_{i \neq j} \left[\langle \sum_{\mu=1}^{3} (S_i^{\mu} S_j^{\mu})_T^2 \rangle_T^2 \right]_{av}$$

Here $\langle\ \rangle_T$ denotes the average over the canonical ensemble, while $[\ ]_{av}$ denotes the average over the distribution of the bonds, which is chosen to be a gaussian with zero mean.

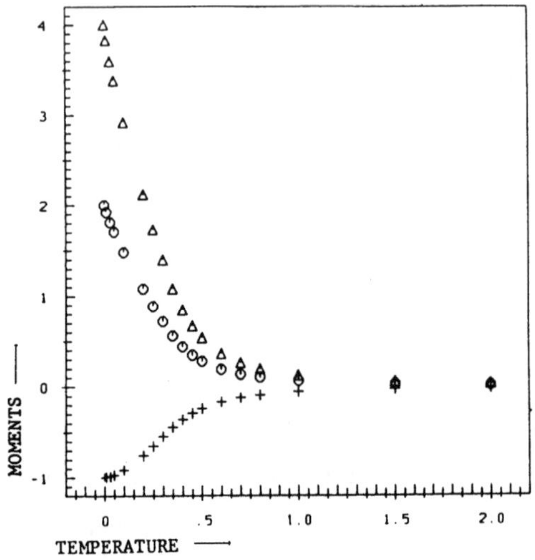

Fig. 2 : Moments of the one particle distribution function - at T = 0 : values for the Potts glass.

octogons : $5[\langle \sum_{\mu=1}^{3} (S^{\mu})^4 \rangle_T]_{av} - 3$

triangles : $7[\langle \sum_{\mu=1}^{3} (S^{\mu})^6 \rangle_T]_{av} - 3$

crosses : $105[\langle (S^1 S^2 S^3)^2 \rangle_T]_{av} - 1$

3. FERROMAGNETIC CORRELATIONS

Quadrupolar glasses and Potts glasses contain ferromagnetic correlations, which are of the same origin. This is demonstrated by the following simple algebra. Consider, for simplicity, coupling constants which are $\pm$ 1 with equal probability. Consider a chain of three particles, thus containing two bonds. First consider a two state Potts glass which is the Ising spin glass. Evaluate for the four possible configurations of bonds the ratios of probabilities of state one to state two at zero temperature:

first particle	1	1	1	1	1:0
first bond	+	−	+	−	
second particle	1	2	1	2	1:1
second bond	+	+	−	−	
third particle	1	2	2	1	1:1

Now evaluate for the three state Potts glass the ratios of probabilities of state one to state two to state three :

first particle	1	1	1	1	1:0:0
first bond	+	−	+	−	
sec. particle	1	2,3	1	2,3	2:1:1
second bond	+	+	−	−	
third particle	1	2,3	2,3	1,3,1,2	6:5:5

While in the case of the Ising spin glass, the ferromagnetic correlations are not propagated, they are propagated in the case of the three state Potts glass. These correlations are generalized to the continous degrees of freedom of the quadrupolar glasses in the spirit of section two. Furthermore, these correlations are the origin of the ferromagnetic phases in Potts glasses /8/ and quadrupolar glasses /3/,/5/. A classification of such correlations in multipolar glasses has been given /9/.

4. SADDLE POINTS

An orthorhombic lattice gives rise to a multiple anisotropic quadrupolar glass, described by the following Hamiltonian:

$$H = \sum_{i<j} J_{ij} \sum_{\mu=1}^{3} a \, (s_i^{\mu} s_j^{\mu})^2 \qquad (2)$$

It has been shown, that the conjugated mean field free energy contains a new type of saddle points /7/. This is due to one preferential direction. It is an open question, however, whether these saddle points lead to instabilities in realistic orientational glasses.

5. CONCLUSION

It has been shown with the accuracy of a Monte Carlo Simulation, that cubic anisotropy is sufficient to stabilize the quadrupolar glass state in three dimensions. The Potts glass is similar to the cubic anisotropic quadrupolar glass at low temperature. Ferromagnetic correlations in random bond systems can be understood by simple algebra. There exists a new type of saddle points in the mean field free energy of multiple anisotropic orientational glasses.

6. LITERATURE

1. K. Knorr: Physica Scripta T19, 531-536 (1987).
2. W. Press, B. Janik, H. Grimm: Z. phys. B49, 9-16 (1982).
3. P.M. Goldbart, D. Sherrington: J. Phys. C18, 1923 (1985).
4. H.-O. Carmesin, K. Binder: Z. phys. B68, 375-390 (1987).
5. D. Hammes, H.-O. Carmesin, K. Binder: to be published in Z. Phys. B (1988).
6. H.-O. Carmesin, K. Binder: accepted in J. Phys. A (1988).
7. H.-O. Carmesin: Modellierung von Orientierungsglaesern, PhD thesis, Mainz, unpublished (1988).
8. E. Elderfield, D. Sherrington: J. Phys. C16, L497 (1983).
9. H.-O. Carmesin: accepted in Z. Phys. (1988).

Neutron Scattering Study of a Crystal with Random Competing Interactions

H. Grimm

Institut für Festkörperforschung, KFA Jülich, D-5170 Jülich, Fed. Rep. of Germany

1. INTRODUCTION

Most of the research on freezing phenomena in strongly disordered materials is restricted to the time- or frequency (ω) domain [1,2]. The spatial aspect or the wave vector ($\underline{Q}$) dependence seems to play an inferior role in model approaches. This might be different for more realistic systems containing not merely translational degrees of freedom [3] and not random interactions of infinite range with even distribution functions [4]. It is hoped to gain some insight into this question from the experimental point of view by the investigation of disordered systems where only a few "freezing" degrees of freedom are embedded in an otherwise well ordered crystalline frame-work. The mixed crystals $Rb_{1-x}(NH_4)_xH_2PO_4$ (RADP) and their deuterated isomorphs belong to this class of systems [5]. These crystals exhibit randomly distributed ferro- and antiferro-electric local interactions. The competition acts primarily within the hydrogen bond network (acid protons) linking the phosphate groups (Fig.1). It is determined by the distribution of cations (Rb^+, NH_4^+) which favour two different acid-proton, and associated dipolar, orderings (the ferro- and antiferro-electric ones, respectively). The difference arises due to the formation of competitive hydrogen bonds (N-H...O) by the NH_4^+ cations. In the broad glass range, $0.22 \leq x \leq 0.74$, these long range orderings do not develope but the NH_4-orientations and acid proton displacements freeze randomly with short-range correlations, only.

These correlations are not just short-range versions of the ferro- or antiferro-electric fluctuations which grow at the Brillouin zone center (Γ-point) or the zone boundary (Z-point), respectively, for x outside of the glass range on approaching the phase transition temperature. They rather freeze-out for wave vectors $\underline{q}$ on the Σ-line which connects these two points and give rise to a cigar-shaped diffuse scattering [6-9].

In the time domain, the freezing is characterized by the developement of a broad spectrum of relaxation times on temperature lowering which culminates in a finite Vogel-Fulcher temperature. This spectral behaviour has been observed by dielectric measurements [5] which probe the system at the Γ-point and by the locally probing NMR-method [10].

In the following, some of the main results of neutron scattering experiments on RADP are reported. Among the many systems with substitutional disorder, RADP recommends itself

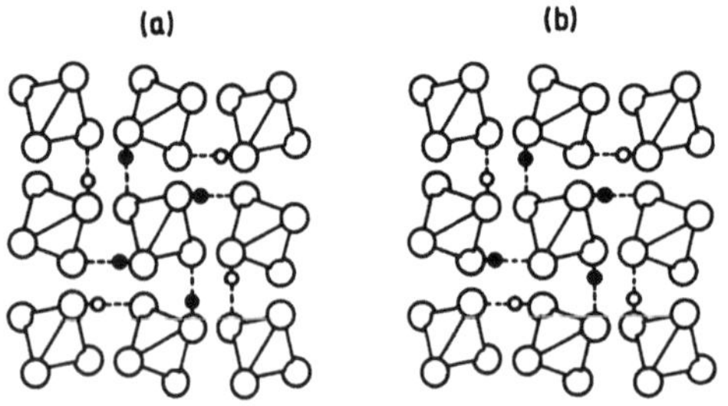

Fig.1 - Projection along the tetragonal axis of the hydrogen bond network linking the the PO_4-groups. The cations (Rb, NH_4), above and below them (not shown), favour the ferroelectric (a) and the antiferroelectric (b) acid proton arrangement.

Springer Proceedings in Physics, Vol. 37 Dynamics of Disordered Materials
Editors: D. Richter · A.J. Dianoux · W. Petry · J. Teixeira © Springer-Verlag Berlin, Heidelberg 1989

for this type of investigation by two advantages: (i) The *Rb*-ion matches well in size with the NH_4-ion and nearly strain free crystals of high quality and homogenous distribution may be grown. (ii) The freezing involves dominantly hydrogen atoms, i.e. the hindered rotation of the NH_4-group and the tunneling or jump within the $O-H...O$ hydrogen bonds. For the deuterated isomorphs, the transition and freezing temperatures are shifted to higher values but the phenomena, mentioned above, remain qualitatively the same. Due to these advantages, full use can be made of the capabilities of neutron scattering to explore the four dimensional dependence of correlation functions in the $(\underline{Q},\omega)$-space and to discriminate between self- and distinct correlation by H/D-exchange.

2. DISTINCT CORRELATION

Results for const-$\underline{Q}$-scans at the peak position $\underline{q}_m = 0.35\,\underline{a}^*$ (a = lattice constant) of the diffuse scattering on the Σ-line are summarized in Fig.2 for deuterated RADP with x=0.62 [12]. The incoherent "background" is always subtracted for the data, presented in this paragraph. This contribution, its elastic and its inelastic part, varies only smoothly in $\underline{Q}$-space and even less within a Brillouin-zone. Thus it can be easily identified by the asymptotic levels of scattering intensity obtained in the neighborhood of the Σ-line. One recognizes that both a quasielastic spectrum and a central component (CP) are associated with the diffuse scattering. The observed ω-width of the CP corresponds to the resolution of 35 GHz of the used three-axes-spectrometer. Note, that the CP is recognizable already at T=280 K. This value may be compared with the transition temperatures T=218, 237 K for x=0,1, respectively. As shown by the logarithmic scale, the CP grows rapidly on temperature lowering. It reaches its saturation value at T$\simeq$ 80 K. Thereby, the width of the quasielastic spectrum diminishes and vanishes within the resolution window at T$\simeq$ 110 K, as shown in the insert of Fig.2. Also displayed in the insert, is the temperature dependence of the spatial correlation perpendicular to the Σ-line by the lateral extensions of the cigar-shaped diffuse scattering. On recognizes the growth of these correlation lengths for decreasing temperature and its cessation between T=80 and 110 K. This saturation at about 20 Å corresponds to a suppression of an attempted transition to an incommensurate phase by random interactions and/or fields.

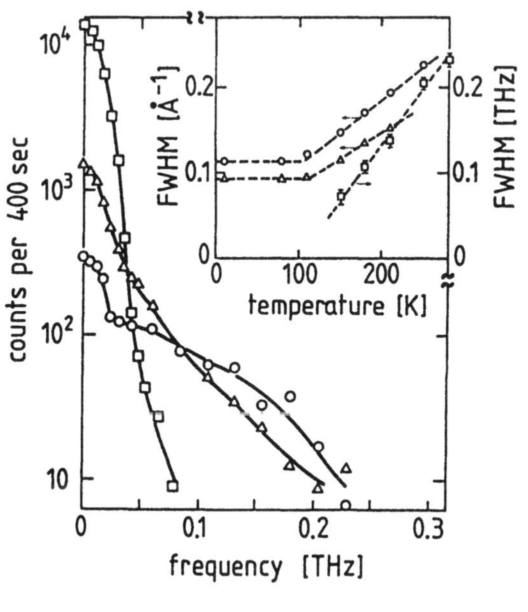

Fig.2 - Const-$\underline{Q}$-scans at the peak position of the diffuse scattering on the Σ-line for three temperatures T=80 (□), 210 (△), and 280 K (o). The energy gain and loss data have been averaged in order to improve the statistics. The insert shows the full width at half maximum values of the diffuse scattering: parallel (o), perpendicular (△) to the tetragonal axis, and (□) refers to the energy width of the quasielastic part.

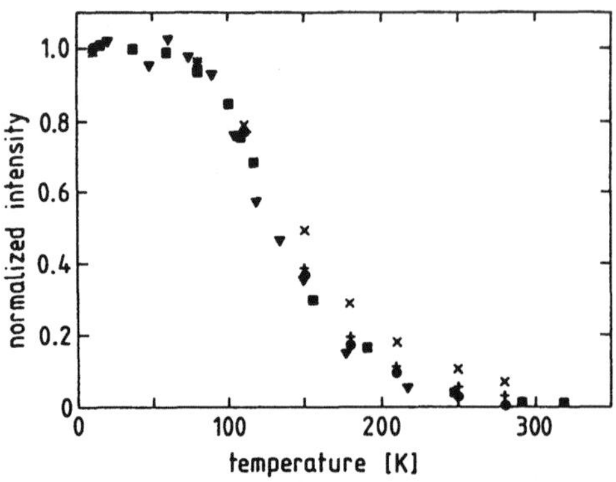

Fig.3 - Measured elastic intensity at $\underline{q}_m$ in units of the saturation value for x=0.62. The symbols refer to instrumental resolutions of 250 GHz (■), 35 GHz (•), and 2 GHz (▼). The symbols (×), (+) denote results of a resolution unfolding as explained in the text.

Does the CP represent a truly elastic signal, i.e. an order parameter for the freezing process? A hint to the answer is shown in Fig.3. The temperature variation of the measured peak intensity (frequency = 0 in Fig.2; normalized for T → 0) as obtained with the three resolutions 2, 35, and 250 GHz, practically collapses on a single curve, indicating the essentially elastic nature of the CP [11]. In order to achieve sufficient statistical accuracy, the $\underline{Q}$-resolution had to be relaxed to about $0.1 \mathring{A}^{-1}$ in the case of the 2GHz-data. The symbols (×) represent the volume (in $\underline{Q}, \omega$-space) of the diffuse scattering as obtained from a four-dimensional resolution analysis of the 35GHz-data [12]. If translated into the peak amplitude by dividing these values through the spatial correlation lengths (insert of Fig.2) the plus signs (+) result. Their near coincidence with the measured normalized peak intensity demonstrates the smallness of the effect of the quasielastic contribution (Fig.2).

It has been verified, that this temperature dependence of the CP cannot be explained by the high frequency extrapolation of the relaxation spectrum as observed by dielectric measurements and Brillouin scattering [12]. This is in contrast to the quasielastic spectrum whose width (insert of Fig.2) smoothly extrapolates to the spectral response obtained at lower probing frequencies which predict a Vogel-Fulcher temperature of T≃ 30 K [13]. Note, that this temperature is much smaller than e.g. the temperature at which the CP saturates and the spatial correlations stop to grow. The recent observation of line broadening of deuteron NMR [14] already at T≃ 230 K for x=0.44 has presumably the same phenomenological origin as the CP.

The wave vector dependence of the CP is demonstrated in Fig.4. These data refer to a crystal with x≃0.4, where the diffuse scattering peaks at the wave vector $\underline{q}_m \simeq 0.25\ \underline{a}^*$ [6]

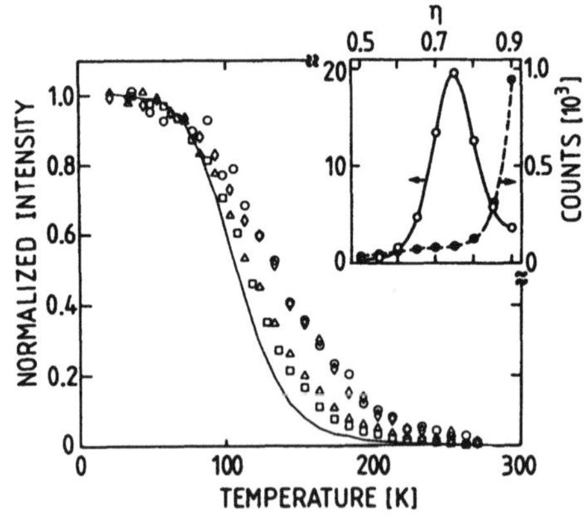

Fig.4 - CP variation at wave vectors (5, η, 0) for η = 0.55 (o), 0.65 (□), 0.75 (solid line interpolation), 0.85 (△), and 0.90 (◊). The asymptotic high and low temperature intensity levels were used for subtration and normalization, respectively. The insert shows the raw data for T=270 K (•) and T=30 K (o). The lines are guides to the eye.

rather than at $\underline{q}_m \simeq 0.35\ \underline{a}^*$ as observed for x=0.62 [8]. The relative increase of the CP with decreasing temperature is maximum for $\underline{q}_m$ and smaller for wave vectors being either closer to the Z- or the Γ-point. A deviation from $\underline{q}_m$ perpendicular to the Σ-line (not shown) causes about the same change of the CP variation as twice this step along the Σ-line. A similar wave vector dependence of the CP results from the measuremnts for x=0.62 [12].

3. SELF CORRELATION

The dominantly incoherent scattering due to hydrogen for non-deuterated RADP may be used to discriminate between the freezing of the NH_4-orientations and the acid proton displacements. This distinction is based on the different form factors associated with these two types of motion. The temperature dependence of the measured elastic incoherent intensity is shown in Fig.5 for a crystal with x=0.35. The experimental resolution is 2GHz and the data, normalized at T=8 K, are grouped within four ranges of $\underline{Q}$. One recognizes two discontinuities in the slope of the intensity.

The first discontinuity, near T≃180 K, corresponds to the slowing down of the NH_4-reorientations as verified by the determination of the incoherent formfactor [15]. At that temperature, this quasielastic spectrum collapses within the energy window of the instrument as indicated by the Arrhenius plot of its width in the insert of Fig.5. The continuation of this thermally activated process down to T≃ 120 K is known from NMR-experiments [10]. At that temperature the extrapolated linewidth is $\Gamma \simeq 1$ MHz. Thus, on the frequency scale of GHz, the ammonium orientations may be considered as frozen much above $T \simeq 60K$, the temperature at which a second discontinuity of the measured elastic incoherent intensity occurs.

Hence, this second discontinuity is being caused essentially by the freezing of the acid protons. This conclusion is corroborated both by Raman- [16] and X-ray [6] scattering experiments. The former method shows that the NH_4-librational mode appears first at T≃ 200 K and is saturated below T≃ 60 K, whereas the latter shows the subsequent appearance of the diffuse peaks on the Σ-line below this temperature. The relative (to neutrons) weakness of this signal for x-ray scattering results presumably from the indirect observation of the acid proton freezing by means of the associated PO_4-distortions. The latter give rise to a leakage of PO_4-vibrations into forbidden $I\bar{4}2d$-symmetries below T≃60 K [16]. It has been shown that the diffuse scattering transforms according to the Σ_{odd}-representation [8], i.e. the

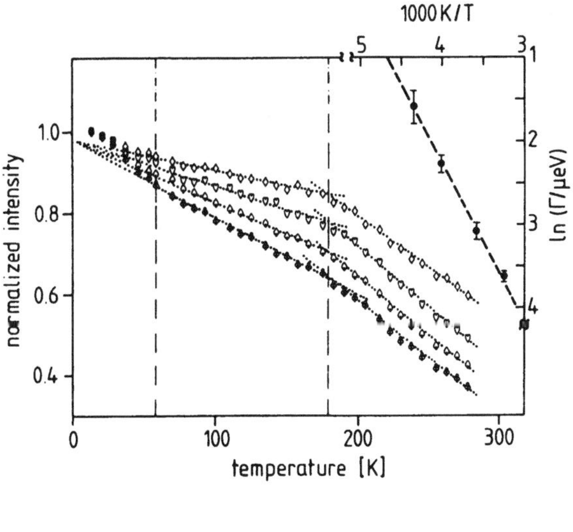

Fig.5 - Normalized (at T=8 K) elastic incoherent intensities, averaged over four ranges of scattering angles ($23^o \leq$ ◇ $\leq 47^o \leq$ ♡ $\leq 73^o \leq$ ♠ $\leq 98^o \leq$ ♣ $\leq 128^o$). The incident wave vector is $k_i = 2.82 Å^{-1}$. The dotted straight lines are guides to the eye. The Arrhenius plot for the observed relaxation rates of the NH_4-reorientations is shown in the right upper part of the figure with separate scales. The dashed line corresponds to $ln\Gamma = ln(18.0meV) - 1880K/T$.

two-fold rotation around the Σ-line is lost locally on freezing. Again, the Vogel-Fulcher (T≃ 9 K [13]) freezing occurs at a much lower temperature.

4. DISCUSSION

Which quantity is the suitable order parameter for the freezing process? One might extract an Edwards-Anderson type of order parameter [1,2] from the incoherent elastic intensity (Fig.5) by an extrapolation of the temperature dependence observed for 60 K$\leq$ T $\leq$ 180 K. However, using e.g. a Debye solid as reference for the extrapolation might be misleading in this case, since the Raman-experiment [16] reports a continuous increase of the NH_4 librational mode intensity in this temperature range. Thus, the slope of the measured incoherent intensity represents a mixture of slowing down and freezing of NH_4-reorientations together with contributions from the acid protons. In view of the obvious smallness of the additional increase of the incoherent signal, the separation between contributions from "normal" vibrational entropy loss and actual freezing of reorientations and jumps becomes extremely sensitive to the type of extrapolation.

The derivation of a generalized order parameter [17] from the coherent elastic intensity is less sensitive to systematical errors, since the relevant signal varies by more than an order of magnitude in the freezing regime (Figs.2,4). However, the details of these curves depend obviously on the wave vector (Fig.4). An analysis of the freezing process in terms of a Sherrington-Kirkpatrick model [4] including random fields [14] ignores e.g. this dependence.

A further basic difficulty is the understanding of the large difference between the Vogel-Fulcher temperature and the temperature at which the diffuse scattering and the spatial correlations reach their asymptotic low temperature values. The relaxational dynamics of the system behaves as if bifurcating into two, more or less spatially separated, components with different "routes" to the frozen state. This viewpoint is also suggested by molecular dynamics simulations [18].

The bifurcation might be understood qualitatively in the following way: The effective medium [19] would condense at the wave vector $\underline{q}_m$. More or less localized correlations whose wave vector representation are "closer" to $\underline{q}_m$ are somewhat slower than "neighboring" ones. Since the intrinsic anharmonicity of the acid proton displacements and the NH_4-reorientations results in a strong coupling of all types of modes, this difference is progressively amplified by the feedback mechanism [20] between relaxation time and "closeness" to the preferred wave vectors on the Σ-line. However, the increasing condensation of localized fluctuations in regions with "advantageous" cation distribution (growth of diffuse peaks) determines conflicting boundary conditions for the intermediate regions. This conflict further impedes their already slower condensation.

The conflicting boundary conditions arise because of the initially rather independent realizations of equally slow relaxation rates in the "advantageous" regions. Emphazising the wave vector representation, this degeneracy corresponds to the four arms of the star of $\underline{q}_m$. Emphazising the representation in direct space, it corresponds to the six local acid proton arrangements (two ferro- and four antiferro-types). With growing spatial correlation length, one of the many global compromizes for these choices stops the growth of the CP and its correlation lengths. This compromize may be thought of as a kind of very slowly relaxing network enclosing the still thermally actived (with respect to proton jumps) intermediate regions [18]. The final, essentially $\underline{Q}$-independent, slowing down of these regions to macroscopic time scales would then correspond to the Vogel-Fulcher temperature.

The elements for this bifurcation behaviour of freezing - preferred wave vectors, anharmonicity, feedback, and frustration - are not specific only to RADP.

ACKNOWLEDGEMENTS

This study has benefitted substantially by the collaboration with K.Parlinski and E.Courtens. All the experiments for this project was carried out together with other scientists. In this context I thank B.Dorner, J.Martinez, M.Monkenbusch, W.Schweika, J.Slak, and P.Xhonneux for cooperation and fruitful discussions. A prior condition for this endeavour were the excellent crystals kindly provided by H.Arend. The assistance of A.Heidemann and F.Douchin in the performance of the experiments at the Institute Laue-Langevin is gratefully acknowledged. Work at Brookhaven National Laboratory was supported by the Division of Material Sciences U. S. Department of Energy under Contract DE-AC02-76CH00016.

5 REFERENCES

1. J.Jäckle: Rep. Prog. Phys. 49, 171 (1986)
2. K.Binder, A.P.Young: Rev. Mod. Phys. 58, 801 (1986)
3. E.Leutheusser: J. Phys. C 15, 2801,2827 (1982)
4. D.Sherrington, S.Kirkpatrick: Phys. Rev. Lett. 35, 1792 (1975)
5. For a recent review see: E.Courtens: Ferroelectrics 72, 229 (1987)
6. E.Courtens,T.F.Rosenbaum,S.E.Nagler, P.M.Horn: Phys.Rev.B29, 515 (1984)
7. R.A.Cowley,T.Ryan, E.Courtens: J.Phys.C18, 2793 (1985)
8. H.Grimm,K.Parlinski,W.Schweika,E.Courtens, H.Arend: Phys.Rev.B33, 4969 (1986)
9. E.Courtens,R.A.Cowley,H.Grimm: Ferroelectrics78, 275 (1988)
10. J.Slak,R.Kind,R.Blinc,E.Courtens,S.Zumer: Phys.Rev.B30, 85 (1984)
11. P.Xhonneux,E.Courtens,H.Grimm: Phys.Rev., to be published
12. H.Grimm,J.Martinez: Z.Phys.B64, 13 (1986)
13. E.Courtens: Phys.Rev.B33, 2975 (1986)
14. R.Blinc,J.Dolinsek,V.H.Schmidt,D.C.Ailion: Europhys.Lett.6, 55 (1988)
15. H.Grimm,E.Courtens,B.Dorner,M.Monkenbusch: Physica B, to be published
16. E.Courtens, and H.Vogt: J.Chim.Phys.(Paris) 82,317 (1985)
17. T.Geszti: J. Phys. C 16, 5805 (1982)
18. K.Parlinski,H.Grimm: Phys.Rev. B33, 4868 (1986) and B37, 1925 (1988)
19. R.A.Cowley,T.W.Ryan,E.Courtens: Z.Phys.B65, 181 (1986)
20. T.R.Kirkpatrick: Phys.Rev. A31, 939 (1984)

Quasielastic Broadening of the Coherent Diffuse Scattering from the Orientational Glass $(KBr)_{1-x}(KCN)_x$

P. Wochner[1], *E. Burkel*[1], *J. Peisl*[1], *C.M.E. Zeyen*[2], *and W. Petry*[2]

[1]Sektion Physik, Ludwig-Maximilians-Universität München, Geschwister-Scholl Platz 1, D-8000 München 22, Fed. Rep. of Germany
[2]Institut Laue-Langevin, 156X Av. des Martyrs, F-38042 Grenoble Cedex, France

We report first measurements of the quasielastic broadening of the Huang diffuse scattering (HDS) from a $(KBr)_{0.8}(KCN)_{0.2}$ single crystal, which we carried out with different energy resolutions at D10 (88 μeV) and IN13 (8 μeV). Both experiments showed that in contradiction to earlier observations the HDS is quasielastically broadened in the vicinity of the freezing temperature T_f. The broadening shows a distinct **q**-dependence.
The detailed investigation of the temperature behaviour of the energy integrated (quasistatic) diffuse scattering indicates a feedback mechanism, similar to the one on which the mode coupling theory of the orientational glass instability according to MICHEL is based on.

1. INTRODUCTION

In the field of orientational glasses the mixed molecular crystal $(KBr)_{1-x}(KCN)_x$ is one of the best studied systems. Enormous experimental and theoretical efforts have been undertaken to understand the formation and the static and dynamical properties of the orientational glass state [1-8]. The system is cubic (NaCl) at room temperature with fast reorientations of the CN^- ions ($\tau_{CN}^{-1}=6.2$ THz for KCN). At high CN^- concentrations ($x > 0.6$) it shows ferroelastic phases with long-range orientational order of the linear CN^- molecules. For $x < 0.6$ the system remains cubic down to lowest temperatures [9]. For these concentrations there exists a glasslike state [2] characterized by short range orientational correlations [10]. The temperature T_f where the freezing of the orientations of the CN^- becomes effective depends on the measuring frequency indicating a relaxational process [6].

The coherent neutron scattering showed a strong increase of the scattering intensity at zero energy transfer in the vicinity of the Bragg peak on cooling the crystals below T_f [11]. In this early measurements an always resolution limited energy width of this intensity was observed, which was independent of temperature. MICHEL and ROWE [4], therefore, concluded that an ergodic-nonergodic transition leads to a freezing of the orientations of the CN^- ions and to the formation of an orientational glass state.

Recently [10] we showed that this coherent elastic diffuse intensity is Huang diffuse scattering caused by the lattice distortions around the CN^- defects which are at the lowest temperatures predominantly frozen in [111] orientations. In the orientational glass state the quasistatic orientational pair-correlation function of the CN^- dumb-bells reveals the evolution of a parallel (or antiparallel) alignment, growing with decreasing temperature. At higher temperatures the interpretation of the quasistatic (energy integrated) diffuse intensity requires the consideration of relaxation time effects of the distortion field due to the nonadiabatic rapid reorientation of the CN^- defects. The width of the energy distribution of the HDS should therefore contain informations about the lifetime of the defect correlations as well as the distortion field of the single defects [12]. For a crystal with $x=0.17$ this quasielastic broadening has been proved for the first time [13]. The coherent nature of the scattering process and at last the influence of the CN^- correlations should result in a certain **q**-dependence of the quasielastic intensity contributions, where **q** is the reduced scattering vector [13, 14]. The scope of the present article is to show that the HDS is quasielastically broadened, depending on **q** and temperature T. The possible mechanisms for this broadening will be discussed.

Springer Proceedings in Physics, Vol. 37 **Dynamics of Disordered Materials**
Editors: D. Richter · A.J. Dianoux · W. Petry · J. Teixeira

2. EXPERIMENTS and RESULTS

Most of the experiments were carried out at the triple axis spectrometer D10 and a first test experiment was performed at the back-scattering spectrometer IN13. Both instruments are located on a thermal neutron guide at the Institut Laue-Langevin in Grenoble.

Since the energy integrated HDS is proportional to $(Q/q)^2$ and the demanded quasielastic intensity contribution or the quasielastic energy-width increases strongly with q, this gives serious limitations for the instrumental $(\mathbf{Q},\omega)$-resolution. With triple axis spectrometers it is possible to achieve good **q**- and ω-resolution at high **Q**-values. But only with the D10, the high q side of the (220) reflection was accessible for measurement with an energy resolution of at best 88 μeV. A vertically focussing graphite monochromator and analyser was used with an incident energy of 4.8 meV and a 10-10-30 min collimation. In order to avoid higher order contaminations a liquid-nitrogen–cooled beryllium filter in the incoming beam was employed.

With the backscattering spectrometer IN13 Q-values up to $5.6Å^{-1}$ combined with an energy resolution of 8 μeV can be obtained. The great disadvantage of this spectrometer is the bad **Q**-resolution ($\approx 0.22Å^{-1}$ in the scattering plane) which is partly required to achieve sufficient counting rates. In our case, the pronounced anisotropy of the HDS in combination with the quasi two-dimensional **Q**-resolution volume which is due to the excellent energy resolution, nevertheless enables it to determine the strongly **q**-dependent quasielastic broadening. One reason for this is that we can align the resolution "plane" nearly perpendicular to the relevant **q**-vector. The great advantage of IN13, however, is not only the improvement in energy resolution but also the accessibility of the (420) reflection, which yields an increase of nearly a factor of 4 in diffuse intensity compared with (220).

The single crystal of $(KBr)_{0.8}(KCN)_{0.2}$ which was grown by the Crystal Growth Laboratory of the University of Utah, had a volume of approximately 7 cm^3 and was used in both experiments (D10, IN13).

In Fig. 1 a typical measurement of the energy integrated diffuse intensity is shown, which was performed on D10 with an energy resolution of 0.4 meV. The symmetrized intensity I_{sym} which is corrected for resolution and background demonstrates the $1/q^2$ dependence in the vicinity of the Bragg peak. A high resolution spectrum measured at D10 is shown in Fig.2 for (2, 2.25, 0) at a temperature of 60K. The inco-

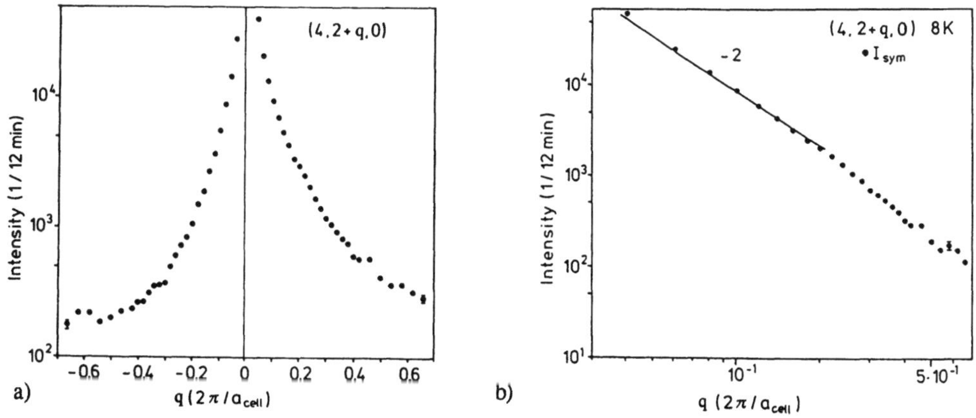

Fig.1 Distortion-induced diffuse scattering intensity at 8K close to (420) in the [010] direction a) the raw data on a logarithmic scale b) the symmetrized intensity, $I_{sym}(q) = 1/2\ (I(+q) + I(-q))$, on a double logarithmic scale

herent scattering has been determined at (1.65, 1.65, 0) where approximately no distortion scattering is to be expected and was found to be negligible. The spectrum in Fig.2 was fitted with a single gaussian plus a constant background only to reveal the quasielastic broadening of the HDS. The resulting FWHM of (109±9) μeV is the clear indication of a quasielastic broadening of the HDS. At 8K the FWHM is identical to the resolution width determined with the incoherent scattering of a vanadium sample. Fitting the data at 60K with a lorentzian convoluted with a gaussian resolution function of 88 μeV FWHM yields a lorentzian FWHM $\Gamma = (7.3 \pm 2.4)$ μeV. The q-dependence of the quasielastic broadening near the (220) reflection is demonstrated in Fig.3 for a temperature of 55K. Once more the spectra are fitted with a single gaussian. Obviously the FWHM is increasing with q. In Fig.4 the temperature dependence of the FWHM of the gaussian is shown for two different q-values. For the smaller q-value ξ=0.16 (x) the FWHM is practically unchanged by temperature variation, whereas at ξ=0.25 (•) a clear broadening is observed for T=55K and 60K.

A high resolution spectrum measured at IN13 is shown in Fig.5 for **Q**=(4.15, 2.13, 0) at a temperature of T=50K. Corrections for scattering from the empty cell have been made. The data was then fitted with a sin-

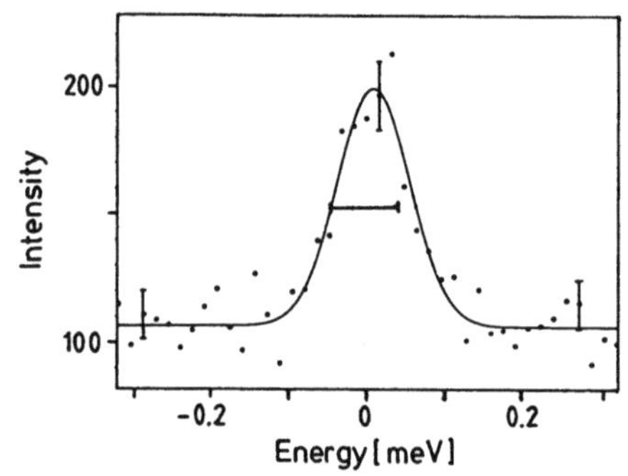

Fig.2 Const-**Q**-scan at (2, 2.25, 0) for the temperature T=60K measured at D10. The data is fitted with a single gaussian only to reveal the quasielastic broadening. This results in a FWHM Γ = (109±9) μeV.

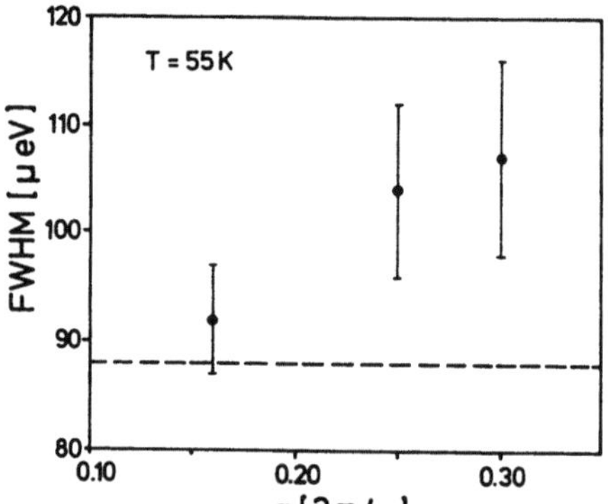

Fig.3 The q-dependence of the FWHM of the single gaussian fitted to the diffuse spectra at (220) in [010] direction. The dashed line marks the experimentally determined resolution width.

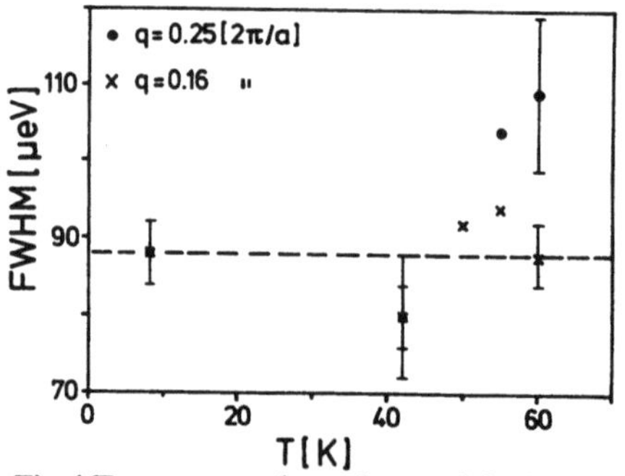

Fig.4 Temperature dependence of the FWHM of the single gaussian fitted to the diffuse spectra at (2, 2 + ξ, 0): 1) ξ=0.16 (x); 2) ξ=0.25 (•). The dashed line marks the experimentally determined resolution width.

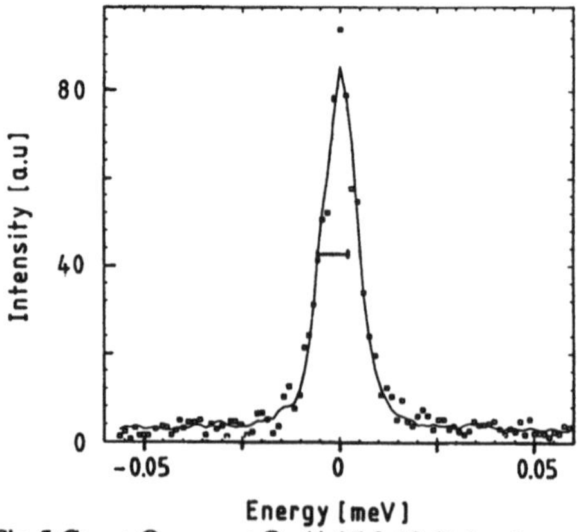

Fig.5 Const-**Q**-scan at **Q**=(4.15 2.13 0) for the temperature of T= 50K measured at IN13. The data is fitted with a single lorentzian convoluted with the measured resolution function. This results in a FWHM Γ = 1.8 μeV.

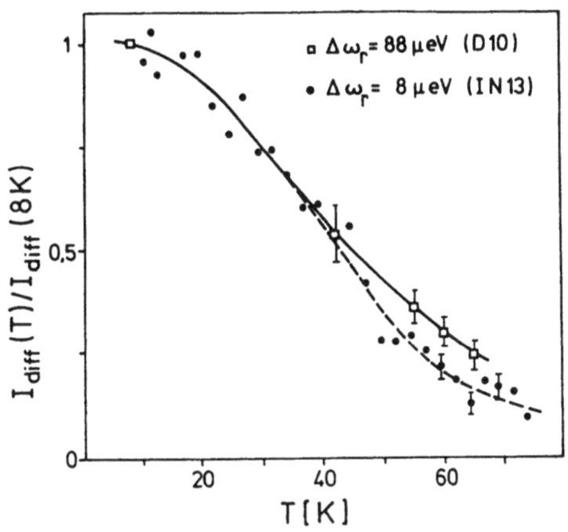

Fig.6 Energy integrated coherent diffuse intensity at $\mathbf{Q} = (2, 2.25, 0)$ measured at D10 (▫) compared with the elastic window measurement at $\mathbf{Q} = (4, 2.25, 0)$ measured at IN13 (•). Both intensities are normalized by their values at lowest temperatures.

gle lorentzian plus a flat background convoluted with the measured resolution function. This results in a FWHM $\Gamma = 1.8\,\mu eV$. Compared with the resolution of 8 μeV this is once more an obvious broadening but it seems to be smaller than the one measured at D10. This is supported by a comparison of the energy integrated intensity determined with D10 and the elastic window measurement of IN13 shown in Fig.6. In an elastic window scan only these neutrons are counted with an energy transfer smaller than the spectrometer resolution. This means that at temperatures above $T \approx 42K$ additional inelastic scattering falls out of the resolution volume of IN13.

3. DISCUSSION

In general the quasielastic broadening of the coherent, distortion induced diffuse scattering is a consequence of the limited lifetime of the distortion field of mobile defects. Roughly we can divide our models in two groups depending on whether the relaxation rate τ_l^{-1} of the lattice is fast compared with the defect (τ_d^{-1}) to follow this motion instantaneously or not. In the first case ($\tau_d^{-1} \ll \tau_l^{-1}$) the distortion field of a single defect is independent of reorientation. In the latter case ($\tau_d^{-1} > \tau_l^{-1}$) the strength and symmetry of the distortion field changes [15]. In case 1) the HDS can be written as [12,13]

$$S_{HDS}(\mathbf{Q}, \omega) = g_O(\mathbf{q}, \omega) \cdot S_O(\mathbf{Q}) + x(1-x)\, S_V(\mathbf{Q}) \cdot \delta(\omega) \tag{1}$$

$g_O(\mathbf{q}, \omega)$ is the Fourier transform of the orientational pair-correlation function of the CN^- defects; $S_O(\mathbf{Q})$ is the orientational part of the static structure factor of a single CN^- defect and $S_V(\mathbf{Q})$ the orientation independent part which is due to the volume mismatch between CN^- and Br^-. If the orientations are uncorrelated, we get in analogy to the convolution approximation of Vineyard [16]

$$g_O(\mathbf{q}, \omega) = x \cdot S_{inc}^{CN}(\mathbf{q}, \omega) \tag{2}$$

where $S_{inc}^{CN}(\mathbf{q}, \omega)$ is the incoherent scattering function of a single CN^- molecule performing reorientational jumps in the octahedral single particle potential. It consists of an elastic part as well as a lorentzian with q-independent width which is proportional to the reorientational frequency of CN^- [17]. But the relative intensity of the lorentzian increases with the reduced scattering vector q. If the orientations are correlated, a narrowing of the quasielastic width is expected in analogy to the DE GENNES narrowing [18] with a distinct **q**-dependence.

If case 2), which is most probably applying to the higher temperatures, is realized, the first term in (1) has to be modified [5,8].

$$S_{HDS}^{(1)}(\mathbf{Q}, \omega) = S_{q.e.}(\mathbf{Q}, \omega) + \{\, g_{n.e.}(\mathbf{q}) + g_{r.f.}(\mathbf{q}) \,\}\, S_O(\mathbf{Q}) \cdot \delta(\omega) \tag{3}$$

The elastic correlation term consists of two parts. The random field part $g_{r.f.}$ is due to the interaction between the orientation independent part of the displacement field of the dumb-bells and the trigonal shear distortions. The non-ergodic correlation factor $g_{n.e.}$ is a mode-coupling term which originates in a certain feedback mechanism [8]. Without mode-coupling theory the quasielastic part $S_{q.e.}(Q, \omega)$ can be approximated at not too high temperatures at which the quasielastic contribution and the phonons are well separated in energy, by a single lorentzian. Its FWHM Γ increases with temperature whereas the integrated intensity decreases [15]. The frequency dependence of $S_{q.e.}$ has not yet been determined by mode-coupling calculations. But if we consider the liquid-glass transition as a generalization of the orientational glass transition, we can adopt the results for the frequency dependence of $S_{q.e.}(Q, \omega)$ [19,20]. Therefore the quasielastic contribution $S_{q.e.}$ should follow a power law $\omega^{-\beta}$ $(0.5 < \beta < 1)$ in the vicinity of T_f which for our concentration is about 65K. Probably this is the adequate description since our quasistatic experiments [10] showed a similar behaviour of the KBr:KCN system like the one described by the mode-coupling theory of MICHEL [8]. In this quasistatic experiments we found that with increasing temperature the orientational correlations of the CN^- dumb-bells frozen at lowest temperatures decay, followed by a decrease of the shear distortion field of the individual CN^- defects. Since the interaction of the CN^- dumb-bells is mediated by lattice strains, this effect describes a dynamical feedback mechanism which is amplified by a softening of the elastic shear constants.

All the theoretically possible line shapes for our observed coherent diffuse intensity discussed here, show that there should always be an elastic $\delta(\omega)$ peak, not least resulting from the volume mismatch, and a quasielastic peak which is either a lorentzian or shows a power law behaviour. Qualitatively our results can be described by such a behaviour but still it is neither possible to distinguish between the different models nor to resolve definitely a $\delta(\omega)$ part. The q dependence of $S_{HDS}(Q, \omega)$ is qualitatively the one required by e.g. case 1) because the increase of the FWHM of the fitted gaussian can either be due to an increasing lorentzian intensity contribution or/and to an increasing lorentzian width. The differences between the IN13 and D10 measurements probably indicate a wide distribution of relaxation times similar to ordinary glasses. To solve the open questions especially the high resolution measurements at IN13 should be continued to higher temperatures and q-values including a wider energy range.

4. SUMMARY

With two different energy resolutions, 88 and 8 µeV, a quasielastic broadening of the HDS caused by the distortion field around reorienting CN^- dumb-bells was observed in the vicinity of the freezing temperature T_f. The observed increase of the quasielastic broadening with increasing reduced scattering vector **q** is a signature of its coherent origin. We showed that a detailed investigation of the quasielastic intensity distribution of the *coherent* diffuse scattering at the glassy crystal $(KBr)_{1-x}(KCN)_x$ together with quasistatic measurements should enable us to get an answer on many open questions concerning the nature of the glass transition perhaps not only in orientational glasses.

ACKNOWLEDGEMENTS

We are grateful to F.Lüty who supplied the sample.Stimulating and clarifying discussions with W. Fenzl are gratefully acknowledged. This work has been funded by the German Federal Minister for Research and Technology (BMFT) under the contract number 03PE1LMU-2-170308.

5. REFERENCES

1. Lüty F., in Defects in Insulating Crystals, ed. Turkevich V.M. and Shvarts K.K., Springer-Verlag,New York (1981), p. 69
2. De Yoreo J.J., Knaak W., Meissner M., Pohl R.O., Phys. Rev. B 34, 8828 (1986)
3. Sethna J.P., Chow K.S., Phase Transitions 5, 317 (1985)

4. Michel K.H., Rowe J.M., Phys. Rev. B 22, 1417 (1980)
5. Michel K.H., Phys. Rev. B 35, 1405 and 1414 (1987)
6. Loidl A., Feile R., Knorr K., Phys. Rev. Lett. 48, 1263 (1982)
7. Lüty F., Ortiz-Lopez J., Phys. Rev. Lett. 50, 1289 (1983)
8. Michel K.H., Z. Phys. B 68, 259 (1987)
9. Knorr K., Loidl A., Phys. Rev. B 31, 5387 (1985)
10. Wochner P., Burkel E., Peisl J., Eckold G., submitted to Phys. Rev. Lett.
11. Rowe J.M., Rush J.J., Hinks D.G., Susman S., Phys. Rev. Lett. 43, 1158 (1979)
12. Gillan M.J., Wolf D., Phys. Rev. Lett. 55, 1299 (1985)
13. Wochner P., Burkel E., Peisl J., Zeyen C.M.E., to be published
14. Wochner P., Ph.D. Thesis Ludwig-Maximilians-Universität München (1988), unpublished
15. Michel K.H., Naudts J., J. Chem. Phys. 68, 216 (1978)
16. Vineyard G.H., Phys. Rev. 110, 999 (1958)
17. see for example: Press W., Springer Tracts in Mod. Phys. 92, Springer Verlag Berlin (1981)
18. DeGennes P., Physica 25, 825 (1959)
19. Götze W., private comunication
20. Bengtzelius U., Götze W., Sjölander A., J. Phys C 17, 5915 (1984)

Part V

Fractal Dynamics

Dynamics of Tenuous Structures

O. Entin-Wohlman[1], *R. Orbach*[2], *and G. Polatsek*[3]

[1]School of Physics and Astronomy,
Raymond and Beverly Sackler Faculty of Exact Sciences,
Tel Aviv University, Tel Aviv 69978, Israel
[2]Department of Physics, Unversity of California, Los Angeles, CA 90024, USA
[3]The Racah Institute of Physics, The Hebrew University, Jerusalem 91904, Israel

The fracton picture for vibrational excitations in tenuous structures is reviewed. The implications of the fracton scaling model to the excitation dispersion law, density of states and lifetime are explained. A comparison is carried out with experimental data, explicit computations of fractal structures and effective medium approximations. The dynamic structure factor for scattering experiments is analysed in detail and compared with current measurements.

1. Introduction

Tenuous structures are systems that when observed on small length scales appear to be thin and dilute. Examples are aerogels, epoxy resins, aggregates of small particles. In those, the mass distribution, or the underlying bond network on small length scales is dilute and extremely inhomogeneous. One therefore expects the vibration dynamics to be different from that of homogeneous systems. Another example are randomly diluted magnetic systems. In these materials the spin-wave spectrum is different from that observed in homogeneous magnetic systems.

There is a considerable interest in the dynamics of such materials. They form highly disordered systems and hence the excitations belonging to higher frequencies are localized. They are also interesting because the understanding of their dynamics might be relevant to the explanation of the universal thermal properties of amorphous materials.

The common feature of the examples quoted above is a crossover behaviour. The material appears to be homogeneous above a certain length ξ; below it, the structure is tenuous and irregular. The classical theoretical model for such systems is the percolation network, in which ξ is the connectivity length. For lengths L greater than ξ, a percolation system is homogeneous, for $L<\xi$, its structue is fractal and the properties of the sample depend upon the length scale over which they are measured. How will this crossover be reflected in the dynamics? This question was first addressed to by Alexander and Orbach [1]. They considered the vibration spectrum of fractal objects, and proposed a scaling picture, in which the modes pertaining to short length scales (i.e., to high frequencies) are called "fractons". Fracton modes are localized. The excitation modes belonging to long length scales over which the system is homogeneous (i.e., to low frequencies) are extended. Thus, the fracton model describes the dynamics of structures where a mobility edge separates the excitation spectrum into extended modes and localized modes. In the following, the percolation network is used as a basis for analytic calculations within the fracton model. The conclusions drawn, however, may well have broader implications.

Springer Proceedings in Physics, Vol. 37 **Dynamics of Disordered Materials**
Editors: D. Richter · A.J. Dianoux · W. Petry · J. Teixeira

The next Section introduces the scaling approach to the fracton model and surveys some of the pertaining experimental data. Section 3 is devoted to the calculation of the dynamic structure factor, the quantity measured in scattering experiments. Section 4 includes some conclusions.

2. The Scaling Approach

In the fracton model, the characteristic length ξ corresponds to a characteristic frequency, ω_c, separating the spectrum into two portions: the low frequency part ($\omega<\omega_c$) corresponding to modes of long wavelengths, $\lambda>\xi$, and the high frequency ($\omega>\omega_c$), fracton part. The long wavelength modes belong to the homogeneous regime and one expects their spectrum to have the same features as that of phonons of an homogeneous, periodic system. These include a dispersion law linear in the wavevector k

$$\omega = c\,k \quad , \tag{1}$$

and a Debye-type law for the density of the states

$$N(\omega) \propto \omega^{d-1} \quad , \tag{2}$$

where d is the Euclidean dimensionality of the system. The fact that the system is tenuous and hence irregular over short length scales, manifests itself in the dependence of the sound velocity c upon ξ. The sound velocity tends to zero as ξ increases

$$c \sim \xi^{-\theta/2} \quad . \tag{3}$$

Here θ is the exponent characterizing the spatial decay of the force constant (this exponent was first introduced in the context of anomalous diffusion on fractals [2]). Similarly, the proportionality factor in the density of states expression (2) depends upon ξ [1, 3].

The phonon regime, $\omega<\omega_c$, is thus characterized by two lengths, the wavelength λ (λ=c/ω) and ξ. The scaling model for fractons postulates that the excitations in the high frequency regime, $\omega>\omega_c$, are characterized by a single length, λ_{fr}. These excitations belong to the short length scale regime over which the system is very irregular and disordered. The modes are therefore expected to be localized and λ_{fr} denotes their localization length. According to the scaling argument [3,4], λ_{fr} is strongly dependent upon the frequency,

$$\lambda_{fr} \sim \omega^{-(1+\theta/2)} \quad , \tag{4}$$

and decreases as the frequency increases [4]. The density of states of fracton modes is characterized by a novel exponent, $\bar{\bar{d}}$,

$$N(\omega) \propto \omega^{\bar{\bar{d}}-1} \quad , \tag{5}$$

which is called the fracton dimensionality [1], and is given in terms of the exponent θ and the Housdorff dimensionality D (describing the mass distribution on a fractal)

$$\bar{\bar{d}} = 2D/(2+\theta) \quad . \tag{6}$$

In all explicit examples investigated so far it was found that $\bar{\bar{d}}<2$. This is another indication for the localization of fracton modes: generalization [5] of Thouless argument for localization on a fractal object shows that $\bar{\bar{d}}$ replaces the Euclidean dimensionality, d, and should be less than two for localization, independent of D.

The fracton model has attracted considerable interest and has been subject to various numerical and experimental tests. Most straightforward are computer simulations of percolation clusters. Those revealed the crossover behaviour of the density of states. Carried out on a 2-dimensional system, for example [6], the computed density of states obeyed Debye law at frequencies smaller than ω_c and followed (5) at $\omega>\omega_c$, with $\bar{\bar{d}}=4/3$, thus verifying the "Alexander-Orbach conjecture" [1] for percolation clusters in $d\geq2$. But even more interesting are experiments carried out on physical systems. Much interest was devoted to spectroscopic measurements, used to determine the density of vibrational states of random systems [7,8,9]. Thus, for example, incoherent neutron scattering from DGEBA (epoxy resin) [9] showed a crossover in the density of states, with a break at about 1.2 meV and $\bar{\bar{d}}$ of the order of 1.5.

The fracton model is based upon the notion of scaling -- a single characteristic frequency which determines the scaling of the spectrum and the dynamics. It is thus desirable to test this assumption. This has been accomplished by Courtens and his collaborators [10] who have carried out extensive measurements on silica aerogels, using Brillouin scattering, small-angle and incoherent neutron scattering. Silica aerogels are porous materials which are self-similar at length scales between the grain size and a certain upper cutoff length, determined by the gel density. One can associate the percolation connectivity length ξ with the cutoff length. As the gel density decreases, ξ becomes larger. The gels are fractal for $L<\xi$ and form homogeneous systems at lengths larger than ξ, with Hausdorff dimensionality $D = 2.4$ [10]. The notion of single-frequency (or single-length) scaling was first established by analyzing Brillouin scattering data [10]. The position of the response function peak and the width at half maximum were measured over a series of samples of decreasing density, in the phonon regime ($\omega<\omega_c$), as a function of frequency and for a fixed wavevector. The peak location gave the sound velocity of the gel, $\omega = cq$, where q is the momentum transfer of the scattered light. The width at half maximum followed Rayleigh law. In order to use this data to examine the assumption of single-variable scaling, one needs an expression for the scattering width within the fracton model, in addition to (3) for the sound velocity.

The scaling prediction for the scattering width, $1/\tau$, is obtained as follows [11]. One writes down the form

$$1/\tau \;=\; \omega\, f\, (\omega/\omega_c) \quad , \tag{7}$$

where f is a scaling function. Then, Rayleigh scattering law, pertaining to phonons, implies that $f(x) \sim x^d$ for $x \ll 1$. In the fracton regime, according to the assumption of single-frequency scaling, there should be no dependence upon ω_c, and therefore $f(x) \rightarrow$ constant for $x \gg 1$. Thus, the scaling model yields

$$\frac{1}{\tau} \sim \omega_c^{-d} \quad \omega^{d+1} \quad , \quad \omega < \omega_c \quad , \tag{8a}$$

$$\frac{1}{\tau} \sim \omega \qquad\qquad , \quad \omega > \omega_c \quad . \tag{8b}$$

In particular, (8a) gives the scaling of Rayleigh law with the crossover frequency ω_c. This expression, in conjunction with the relation between ξ and ω_c [1]

$$\omega_c \;\sim\; \xi^{-(1+\theta/2)} \quad , \tag{9}$$

was used by Courtens et al. [10] for the Brillouin scattering data. It was indeed found that the curve of $(\tau\omega_c)^{-1}$ as a function of ω/ω_c is independent of the density (which determines ξ and ω_c), i.e., obeys single-frequency scaling. The analysis gave ω_c of the order 6 GHz, $\xi \sim 500$Å, $\theta \sim 1.51$ (see equation (3)) and $\bar{\bar{d}} \sim 1.3$. The last two values are in accordance with the predictions of scalar elasticity [12].

The fracton model prediction for the scattering width in the fracton regime, equation (8b), implies that fractons are localized according to the Ioffe-Regel criterion: dividing both sides of (8b) by the sound velocity yields a mean free path of the order of the wavelength. A relatively high-frequency regime in which such a condition holds is indeed found in many glasses and aggregates [13], though its interpretation in terms of the fracton model is still controversial. Analysis of thermal conductivity data of amorphous materials [13], using the kinetic heat-conduction formula, suggests that phonons of frequency roughly equal to the "plateau" temperature possess mean free paths comparable to their wavelengths. This occurs for λ of the order 20-50Å, a value larger than the structural unit length of these materials, of about 4Å, by roughly an order of magnitude. Thus vibrations in glasses may exhibit the strong scattering frequency regime [11] indicated by the Ioffe-Regel limit, over a large fraction of the excitation spectrum.

The spin-wave dynamics of antiferromagnets maps onto that of vibrational modes: both excitations have dispersion linear in the wavevector at long wavelengths. Thus, the excitation spectrum of randomly diluted antiferromagnets should follow the fracton model [4,14]. Inelastic neutron scattering off the site-diluted antiferromagnet $Mn_x\, Zn_{1-x}F_2$ (which forms a bcc percolation spin network) showed a crossover behaviour [15]. The percolation threshold concentration of this material is at x=0.25. Measurements taken at concentration x=0.5 exhibited a sharp spin-wave peak (as a function of frequency) when the neutron momentum transfer q was small. The peak broadened progressively with increasing q and the response at higher frequencies was only weakly dependent upon the wavevector.

The data was interpreted [15] as demonstrating a crossover from propagating spin-waves to localized or overdamped excitations. In particular, the peak position shifted rather slowly towards higher frequencies as q increased, while the width at half maximum increased very rapidly. This behaviour disappeared at concentration x=0.75, (i.e., very far from the percolation threshold).

While Brillouin scattering and incoherent neutron scattering measurements [10,15] generally support the fracton scaling model, it is clear that the data deserves comparison with a more detailed calculation which would yield the full line shape of the response function. In the next Section we describe such calculations for the dynamic structure factor.

3. The Dynamic Structure Factor

Inelastic neutron or light scattering measures the response function, i.e., the structure factor $S(q,\omega)$ as a function of the frequency and momentum transfer of the scattered particles. The determination of $S(q,\omega)$ necessitates the knowledge of the full Green's function of the vibrating structure,

$$S(q,\omega) = -\frac{1}{\pi}\,\mathrm{Im}\sum_{ij} e^{i\vec{q}\cdot(\vec{R}_i-\vec{R}_j)} \quad G_{ij}(\omega_+)\,. \tag{10}$$

Here, G_{ij} is the i,j- matrix element of the retarded Green's function in the site representation, $\omega_+=\omega+i\eta$, where η is an infinitesimal positive number and $\vec{R}_i$ denotes the coordinate of the i-th site.

In terms of the eigenmodes ϕ_α and eigenfrequencies ω_α, the structure factor takes the form

$$S(q,\omega) = \sum_{\alpha} \phi_\alpha(q)\,(\phi_\alpha(q))^*\,\delta(\omega^2-\omega_\alpha^{\,2})\,, \tag{11}$$

where $\phi_\alpha(q)$ is the Fourier transform of $\phi_\alpha(\vec{R})$. From this form it follows that $S(q,\omega)$ satisfies the sum-rule

$$\int d\omega\,\omega\,S(q,\omega) = 1\,. \tag{12}$$

Consider now a fractal object, for example, a percolating system at the percolation threshold. The assumption of a single-frequency scaling, in conjunction with (12), implies the form

$$S(q,\omega) = \omega^{-2} f\left(\omega/q^{(2+\theta)/2}\right)\,, \tag{13}$$

where f is a scaling function. To test this form, a computation of $S(q,\omega)$ was carried out for a 2-dimensional Sierpinski gasket [16]. An efficient algorithm for a successive construction of the Green's function was used, from which a recursion relation for $S(q,\omega)$ was derived. This enabled a computation for a large gasket (3^{10} sites). The results exhibited [16] a well-defined maximum of $S(q,\omega)$ as a function of q (or ω). The peak position, q_{max}, as a function of ω yielded $q_{max}\sim\omega^{2/(2+\theta)}$, and thus confirmed the fracton model prediction [1,4], with $2+\theta=\ell n5/\ell n2$. Moreover, rescaling the wavevector axis in terms of $q_{max}(\omega)$ lead to a collapse of $\omega^2S(q,\omega)$ (see equation (13)) of different frequencies on a single curve, thus confirming the idea of single-variable scaling.

This computation cannot describe the phonon-fracton crossover, as it is valid only at percolation, i.e., in the fracton regime. To consider the entire frequency range, the effective medium approximation (EMA) was exploited. The spirit of this model is to replace a random system by an effective homogeneous (and periodic) medium, which one can solve for the dynamics. The parameters of the effective medium depend, in a self-consistent way, upon the parameters of the random system.

Consider a system in which the isotropic elastic force constants obey a percolation distribution. In the EMA, the force constants are replaced by a uniform, frequency-dependent, complex force constant $W(\omega)$. This effective force constant is chosen in such a way that the scattering produced by one bond, for which the original force constant, K, is maintained, is zero on the average. This requirement determines the effective force constant $W(\omega)$.

Let us denote the Green's function of a system in which all the force constants are $W(\omega)$ by G, and the Green's function of the same system in which one force constant is allowed to fluctuate according to the percolation distribution by g. The two Green's functions are related by [17]

$$g = G + GTG , \tag{14}$$

where T is the T-matrix describing the scattering from that particular bond. One now requires that on the average (with respect to the percolation distribution) the T-matrix will vanish so that the two Green's functions will be equal. This requirement leads to the equation

$$\frac{1}{d} (W-K) (1-\omega^2 \sum_{\vec{q}} G(q,\omega)) - W + pK = 0 , \tag{15}$$

in which p is the percolation concentration. This is the self-consistency equation of the EMA for W, as G itself is given in terms of the effective force constant

$$G (q,\omega) = \left[\omega^2 - 2W \sum_{i=1}^{d} (1-\cos q_i)\right]^{-1} , \tag{16}$$

for a simple cubic lattice in d dimensions. The fact that $W(\omega)$ is a complex function of the frequency reflects the nature of the EMA: one forces a plane-wave solution upon the random system and consequently the wavevectors are complex.

Once the solution for the self-consistency equation is obtained, it is straightforward to solve for the dynamics, and in particular, for the dynamic structure factor [17]. The EMA yields a crossover frequency ω_c,

$$\omega_c \sim p - p_c \ , \tag{17}$$

where p_c is the percolation threshold (p_c=1/d within EMA) and connectivity length $\xi \sim (p-p_c)^{-1/2}$. For $q\xi < 1$, the structure factor as a function of ω has two peaks. The first, at lower frequencies, is high and narrow, centered around cq, where the sound velocity c scales as $(p-p_c)^{1/2}$. The width of this peak follows Rayleigh law. The second peak is located around ω_c and is much smaller. This peak represents the fracton contribution to $S(q,\omega)$. As q increases towards ξ^{-1}, the first peak moves towards ω_c and broadens, while the second peak rises but remains at ω_c. At $q\xi \sim 1$ the two peaks merge, and for $q\xi > 1$ the structure factor exhibits a very low and broad structure. This behaviour is in accordance with experimental data [10,15] and also with the predictions of the fracton scaling model, as the phononlike contribution to $S(q,\omega)$ is expected only for $q\xi < 1$ (i.e., long wavelengths).

The line shape of the response function, within EMA, can be written in terms of an effective sound velocity and line width, τ^{-1}, both functions of frequency [17]. One can then analyze their limiting behaviours, in the phonon ($\omega<\omega_c$) and in the fracton ($w>\omega_c$) regimes. By interpolating between these two limits and using the EMA line shape, Courtens et al. [10] were able to fit beautifully their Brillouin scattering data.

Within EMA, the sound velocity, up to ω_c, was found [17] to be independent of ω. Above ω_c it increased with ω. The scattering width τ^{-1} followed Rayleigh law in the phonon regime, $\tau^{-1} \propto \omega^{d+1}$, and became proportional to the frequency in the fracton regime, thus confirming the Ioffe-Regel limit for fractons, derived from scaling arguments (see equation (8b)). However, within EMA, the line width did not scale with ω_c. It is conceivable that this failure of scaling is connected with the approximation involved in the effective medium approach. However, there still exists the possibility of additional characteristic frequencies [11] in the fracton scaling model, though a Green's function perturbation calculation [18] found only a single characteristic frequency.

The EMA calculation described above pertains to the vibration spectrum in the scalar elasticity model [12]. The approach can be exploited for studying the excitation spectrum of other elastic networks [19], as well as for the spin-wave dynamics of diluted magnetic systems [14]. Motivated by the neutron scattering measurements of a diluted antiferromagnet [15], a detailed calculation of the dynamic structure factor was carried out [20] for the bond-percolation model of random antiferromagnets. The results are very similar to those quoted above for the vibrational spectrum, and are in quantitative agreement with the experimental data [15].

4. Discussion

We have reviewed the scaling approach of the phonon-fracton crossover for the dynamics of tenuous structures, and discussed experimental data pertaining to such materials. In particular, we have focused upon the dynamic structure factor measured in scattering experiments, and presented detailed results computed within the effective medium approximation.

The fracton model was first invoked in conjunction with thermal properties of epoxy resins [1,21] and other amorphous materials. The phonon-fracton crossover was proposed as an explanation for the excess specific heat over the Debye T^3-contribution, occurring at about the same temperature as the "plateau" in the thermal conductivity [13]. The concepts of the model are based upon the dynamics of fractal materials, but it was suggested [21] that a much broader class of amorphous materials might obey similar dynamics. The main feature of the fracton model is a crossover in the excitation spectrum, at ω_c, from extended to strongly localized modes [4], the latter being described by a single, frequency-dependent, length λ_{fr}. Using this picture, the thermal conductivity above the "plateau" temperature was considered [22]. It was shown that anharmonicity can generate hopping mechanism for localized vibrations, which leads to a contribution to the thermal conductivity linear in the temperature.

It would be of interest to explore further the implications of the fracton picture by calculating other properties associated with dynamics of tenuous and similar structures. These include high temperature transport or NMR relaxation rate for the diluted magnetic systems.

Acknowledgements

This work has been supported in part by the Fund for Basic Research administered by the Israel Academy of Sciences and Humanities, and by a grant from the National Science Foundation, DMR 84-12898.

References

1. S. Alexander and R. Orbach: J. de Physique Lett. 43, L625 (1982).
2. Y. Gefen, A. Aharony and S. Alexander: Phys. Rev. Lett. 50, 77 (1983).
3. A. Aharony, S. Alexander, O. Entin-Wohlman and R. Orbach: Phys. Rev. B31, 2565 (1985).
4. S. Alexander: Ann. Isr. Phys. Soc. 5, 144 (1983); R. Orbach: Science 231, 814 (1986); S. Alexander: Physica 140A, 397 (1986).
5. R. Ramal and G. Toulouse: J. de Physique Lett. 44, L13 (1983).
6. K. Yakubo and T. Nakayama: Phys. Rev. B36, 8933 (1987).
7. U. Buchenau, N. Nücker and A. J. Dianoux: Phys. Rev. Lett. 53, 2316 (1984).
8. T. Freltoft, J. Kjems and D. Richter: Phys. Rev. Lett. 59, 1212 (1987).
9. A. J. Dianoux, J. N. Page and H. M. Rosenberg: Phys. Rev. Lett. 58, 886 (1987).
10. E. Courtens, J. Pelous, J. Phalippou, R. Vacher and T. Woignier: Phys. Rev. Lett. 58, 128 (1987); R. Vacher, T. Woignier, J. Pelous and E. Courtens: Phys. Rev. B37, 6500 (1988); E. Courtens, R. Vacher, J. Pelous and T. Woignier: Europhys. Lett. 6, 245 (1988); Y. Tsujimi, E. Courtens, J. Pelous and R. Vacher: Phys. Rev. Lett. 60, 2757 (1988).

11. A. Aharony, S. Alexander, O. Entin-Wohlman and R. Orbach: Phys. Rev. Lett. 58, 132 (1987); O. Entin-Wohlman: in Condensed Matter Physics, ed. by R. Orbach (Springer-Verlag, 1987).
12. S. Alexander: J. de Physique 45, 1939 (1984).
13. R. C. Zeller and R. O. Pohl: Phys. Rev. B4, 2029 (1971); J. E. Graebner, B. Golding and J. C. Allen: Phys. Rev. B34, 5696; J. E. Graebner and B. Golding: Phys. Rev. B34, 5788 (1986).
14. K. W. Yu and R. Orbach: Phys. Rev. B30, 2760 (1984).
15. Y. J. Uemura and R. J. Birgeneau: Phys. Rev. B36, 7024 (1987).
16. U. Sivan, R. Blumenfeld, Y. Meir and O. Entin-Wohlman: Europhysics Lett., to be published.
17. G. Polatsek and O. Entin-Wohlman: Phys. Rev. B37, 7726 (1988).
18. A. Christou and R. B. Stinchombe: J. Phys. C19, 5717 (1986); A. Christou and R. B. Stinchombe: J. Phys. C19, 5895 (1986).
19. W. Tang and M. F. Thorpe: Phys. Rev. B37, 5539 (1988).
20. G. Polatsek, O. Entin-Wohlman and R. Orbach: Proc. of the International Conference on Magnetism, July, 1988, J. de Physique, to be published.
21. S. Alexander, C. Laermans, R. Orbach and H. M. Rosenberg: Phys. Rev. B28, 4615 (1983).
22. S. Alexander, O. Entin-Wohlman and R. Orbach: Phys. Rev. B34, 2726 (1986).

Dynamics of Silica Networks

J.K. Kjems[1], *T. Freltoft*[1], *D. Posselt*[1], *and D. Richter*[2]

[1]Risø National Laboratory, DK-4000 Roskilde, Danmark
[2]Institut Laue-Langevin, 156X Av. des Martyrs,
F-38042 Grenoble Cedex, France

Highly porous silica network materials like aerogels and smoke particle aggregates (Cab-O-Sil, Alfasil) have remarkable dynamic properties which can be studied through measurements of the density-of-states, the elastic properties and sound propagation as well as the thermal properties. The paper reviews some of the experimental studies which can be reconciled with a theoretical desription that involves long wavelength phonon modes at low frequencies, a crossover to localized fracton modes at intermediate frequencies and another crossover to bulk phonon-like modes at the highest frequencies. The interpretation of crossover frequencies and length scale will be discussed and compared to the corresponding parameters that are derived from structural studies along with scaling relations.

1. Introduction

Silica networks have become important model systems for the study of network structures with fractal geometry and their physical properties. This is to some extent due to the easy way in which they can be fabricated by skilled chemists but, equally important, because they allow for considerable variation in parameters like density, ρ, correlation length, ξ , and intrinsic particle size, a. These systems are therefore well suited for systematic studies of scaling relations between the structural parameters and properties like elasticity (Youngs modulus), sound velocity, thermal conductivity, heat capacity and vibrational density of states. These properties are strongly affected by the morphology, i.e. the connectivity of the silica networks, and it is the purpose of this presentation to review the status of the experimental studies in this field.

Silica networks are made of aggregated clusters of spherical SiO_2 particles of radius, a, which can be made quite monodisperse. The networks result from processes like colloidal aggregation of SiO_2 particles in suspension e.g. the commercially available LUDOX™ [1]; flame hydrolysis of $SiCl_4$ to give products like Cab-O-Sil [2] hydrolysis and polymerization of silicid acid esters [3] leading to gelformation which followed by hypercritical drying give aerogels.

The latter procedure produces homogeneous monolithic samples with densities in the range from ~ 0.05 g/cm³ to ~ 2.0 g/cm³, i.e. maximum densities near that of SiO_2-glass [3,4]. The morphology of the networks produced in the different manners seems to be similar to that of branched polymer networks in which the monomers correspond to the intrinsic silica particles of size, a.

The structure of silica network systems can be studied by scattering experiments using x-rays, neutron and light scattering techniques [5,6,7]. The ideal scattering law is

$$I_q \propto \int \langle \rho_0 \rho_r \rangle e^{i\vec{q}\vec{r}} d\vec{r} \propto q^{-D} \qquad (1)$$

where the latter scaling holds for a fractal system. D is the Hausdorff or fractal dimension [8,9] and $\langle \rho_0 \rho_r \rangle$ is the density correlation function. This simple law is found to represent the scattering from colloidal aggregates over

Editors: D. Richter · A.J. Dianoux · W. Petry · J. Teixeira

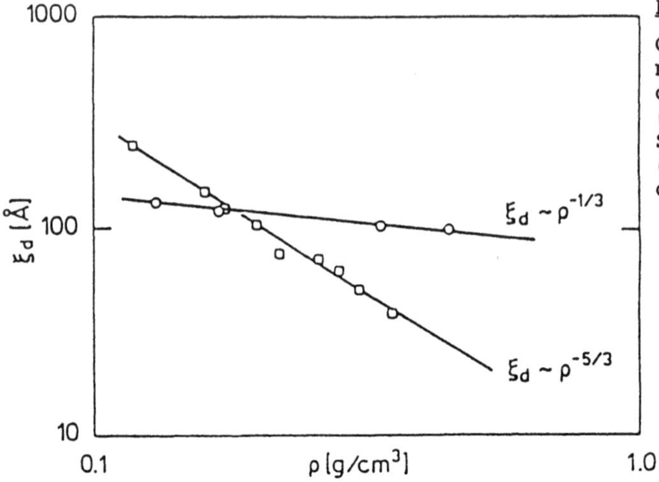

Fig. 1.

Observed relation between macroscopic density and crossover length, for aerogels (□) [13] and compressed Cab-O-Sil (o). Both sets are SANS (small angle neutron scattering) data.

several decades in momentum transfer, q [10]. However, in general the scattering is modified by the effect of the finite correlation length, ξ_d and the particle size, a. To a first approximation the finite particle size enters as a form factor F_q so that

$$I_q \propto (1+S_q)F_q \qquad (2)$$

and the effects of the finite correlation length can be accounted for by the emperical form devised by Freltoft et al. [11]:

$$S_q = C(D-1)\xi_d^D \frac{\Gamma(D-1)\sin[(D-1)\arctan(q\xi_d)]}{[1+q^2\xi_d^2]^{(D-1)/2}(D-1)q\xi_d} \qquad (3)$$

A more detailed form that takes into account the cutoff in the correlation function at the near neighbour particle distance has also been devised by Freltoft et al. [6].

The variation of the correlation length, ξ_d with density has been studied in both Cab-O-Sil [6] and aerogels [7] using small-angle neutron scattering, SANS. For Cab-O-Sil the density variation is achieved by varying the degree of mechanical compression of the same powdered starting material and for aerogels the variation is controlled in the production phase by the concentration of reactants. The results are shown in Fig. 1 for variations in the same density range. In both cases a power-law can be used to represent the data $\xi \propto \rho^{-\zeta}$ but the exponent, ζ turn out to be very different, namely $\sim 1/3$ for Cab-O-Sil and $\sim 5/3$ for aerogels. Courtens et al [7] have pointed out, that for systems that are mutually self-similar one expects

$$\xi_d \propto \rho^{\frac{1}{D-d}} \qquad (4)$$

where d is the dimension of the embedding space. Within the experimental accuracy this is consistent with the observed exponent for aerogels but not for the compressed powders of Cab-O-Sil.

2. Dynamics

2.1. Elasticity

The elastic properties of silica networks can be studied directly in samples of densities above ~ 0.1 g/cm³ by measurements of the bending or the torsional force constants. For both aerogels [12] and compressed powders of Cab-O-Sil [13] one finds power-law behaviour as function of $(\rho - \rho_c)$ where ρ_c corresponds to the critical density for percolation. The exponents, τ, that are derived from the data are $\tau = 2.9$ for Cab-O-Sil, $\tau = 3.8$ for low density aerogels and $\tau = 3.3$ for densified aerogels. Some of this difference is probably caused by the difference in the structural properties that are reflected in very different relations between correlation lengths and densities mentioned above.

Information about the elastic properties can also be derived from measurements of the sound velocity. This has been done for aerogels [7] and consistent values for the exponents have been found.

The experimental values for the exponents have been used to make conclusions about the scalar or tensorial character of network elasticity [14,15], but such deductions may depend on the mutual self-similarity, which as mentioned above is not found to hold for Cab-O-Sil samples with different densities.

2.2. Fractons

The elasticity and sound velocities represent the long wavelength limit of the vibrational excitation of the silica networks. When the wavelengths become comparable to the correlation length, ξ_d, the nature of the excitation changes from phonons to fractons [16] as illustrated schematically in Fig. 2.

This shows the three regions that govern the elastic and thermal properties of such networks. In these regions the dominating excitation are phonons, fractons and phonons in the monomer particles respectively. The phonon regime and the phonon-fractor crossover have been studied extensively by Courtens et al. using Brillouin scattering techniques [17]. We note one important point, namely that the crossover length scale ξ_c, also called the Cauchy-length, which is deduced from the analysis of the dynamical studies data, does not coincide with the length scale ξ_d that is deduced from the structural studies. In fact, ξ_c exceeds

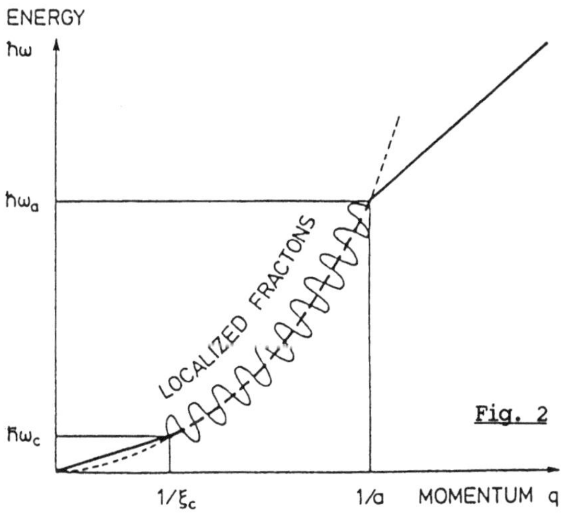

Fig. 2 Schematic representation of the dispersion of vibrational excitations in silica networks.

ξ_d by a factor of apprximately five. This may be important for the analysis of amorphous materials like glasses and polymers where ξ_d is close to the atomic scale but where ξ_c may be large enough to warrant a description of the dynamics using the fracton concept.

In the fracton region the density-of-states is given by

$$Z(\omega) \; \alpha \omega^{d_s-1} \qquad (5)$$

where d_s is the spectral dimension. The density-of-states can be measured directly by incoherent inelastic neutron scattering. This has been done for a sample of compressed Cab-O-Sil, $\rho = 0.2$ g/cm³, by considering the difference between the scattering from a hydroxylated and a dried sample [18]. In this manner the incoherent scattering component can be selected under the assumption that the hydroxyl groups are tightly bound to the surface of the SiO_2-particles. The result is illustrated in Fig. 3. The fracton density of states corresponds to the low energy part of the spectra and it is seen that there is an apparent change of exponent as function of temperature, probably due to anharmonic effects.

The crossover to the phonon region occurs at energies that are smaller than the instrumental resolution but the crossover to the particle region can be seen clearly at $\hbar\omega \sim 2$ meV in the 136 K data. The particle modes are the elastic modes of the SiO_2 particles.

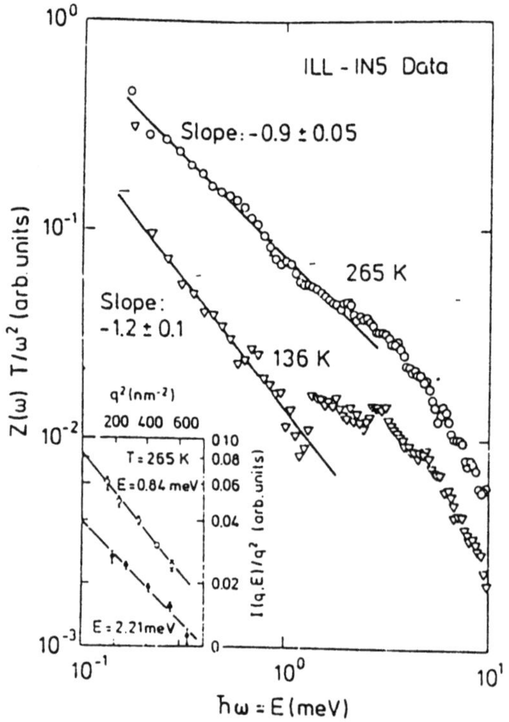

Fig. 3 Density-of-states x T/ω^2 versus energy for compressed Cab-O-Sil with $\rho = 0.2$ g/cm³ at different tempeatures. The slopes below $\hbar\omega = 1$ meV corresponds to $Z(\omega)\alpha\omega^{d_s-1}$. The insert illustrates the extrapolation procedure [18].

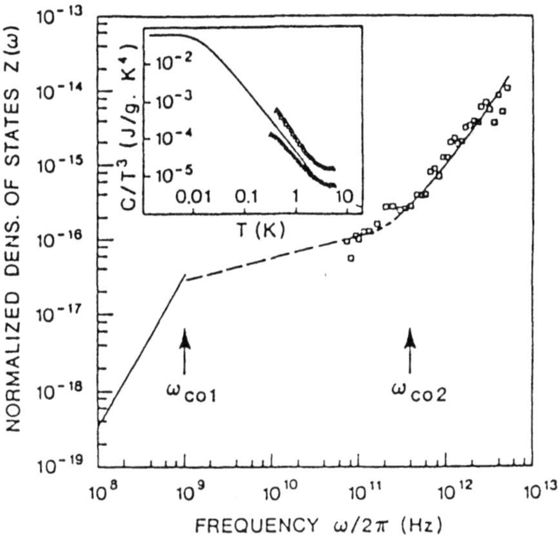

Fig. 4 The normalized density of states $Z(\omega)$ combined from different experiments on aerogels. The absolute scale results from a fit of the high-energy region to an approximate expression for the density-of-states of spheroidal particles (solid line). The line of slope 2 in the phonon region is calculated using the known parameters of this system. It extends up to ω_{co1}. The dashed line in the intermediate region is traced with the slope $d_s-1 = 0.3$ known from Brillouin results and made to coincide with the data points below ω_{co2}. The solid line in the insert shows the low-T specific heat calculated from $Z(\omega)$. It is compared with experimental data on other samples. This figure is cited from [21].

2.3 Thermal properties

The density-of-states is also directly reflected in the heat capacity, C_p. Low-temperature data for aerogels are consistent with the picture of the different regimes that are illustrated in Fig. 2. In particular, the values of C_p are found to be less than the asymptotic value C_D (debye limit at low energy) that is deduced from the elastic properties, even though the crossover temperature could not be reached in the experiment by Calemczuk et al. [19]. The same authors also measured the thermal conductivity, K, of different aerogels and their data also shows evidence of three different regimes. At the lowest temperatures below 0.1 K one expects a phonon regime with $K \propto T^2$, this is followed by a fracton regime $0.1K<T<5K$ where a linear temperature dependence is expected [20]. At still higher temperatures the internal modes in the SiO_2-particles may contribute and the thermal conductivity rises more rapidly. So far there has been no theoretical analysis of the high temperature regime.

3. Summary

The experimental results from the dynamics of silica networks can be summarized in a representation of the density-of-states, $Z(\omega)$. This has been done by Vacher et al. [21] and their results are shown in Fig. 4. They find a gratifying consistency in the absolute values of $Z(\omega)$ in three different frequency domains separated by the crossover values ω_{co1} and ω_{co2}. They also find consistency with the specific heat measured on similar samples.

The qualitative picture and the quantitative agreements between different types of experiments have been further corroborated by the work presented in the papers by Conrad et al. and by Reichenauer et al. that follow the present contribution.

In this brief paper we have reviewed the structural and dynamical properties of silica networks. The scattering experiments have shown to give full information on the structural properties like fractal dimension, D, and the crossover length scales ξ_d and a for a variety of systems. The dynamic properties reveal three frequency regimes separated by the crossover frequencies ω_{co1} and ω_{co2}. Below ω_{co1} homogeneous phonon modes prevail whose life-times are dominated by disorder scattering. In the regime $\omega_{co1}<\omega<\omega_{co2}$ one finds the localized fracton modes with the characteristic spectral dimension d_s and the relatively poor thermal conductivity. Above ω_{co2} the internal modes of the constituent particles make significant contributions to the density-of-states and to the thermal conductivity. It is important to note that the crossover length scale, ξ_c, associated with ω_{co1} does not necessarily coincide with the length scale, ξ_d, derived from structural studies of the same system.

4. Acknowledgements

This paper has benefitted from the experience gained from fruitful collaboration with S.K.Sinha, P.Schofield, H.Blank, and F.W.Poulsen.

5. References

1. LUDOX™ is a trademark from E.I.du Pont de Nemours and Co., Wilmington, Delaware, USA.

2. G.D.Ulrich and J.W.Riehl, J.Colloid. Interface Sci., 87, 257 (1982)

3. S.Henning and L.Svensson, Physica Scripta, 23, 697 (1981)

4. C.J.Brinker, K.D.Keefer, D.W.Schaefer and C.S.Ashley, J.non-Cryst.Solids, 48, 47 (1982).

5. D.W.Schaefer, J.E.Martin, P.Wiltins and D.S.Cannell, Phys.Rev.Lett. 52, 2371 (1984)

6. T.Freltoft, J.K.Kjems and S.K.Sinha, Phys.Rev. B33, 269 (1986)

7. E.Courtens and R.Vacher, Z.Phys.B Condensed Matter 68, 355 (1987)

8. T.Witten and L.M.Sander, Phys.Rev.Lett.47, 1400 (1981)

9. B.B.Mandelbrot, The Fractal Geometry of Nature (Freeman, San Francisco, 1983)

10. J.E.Martin and A.J.Hurd, J.Appl.Cryst., 20, 61 (1987) gives the most recent review.

11. S.K.Sinha, T.Freltoft and J.K.Kjems in Kinetics of Aggregation and Gelation, F.Family and D.P.Landau editors, North Holland, Amsterdam (19843), p.87.

12. T.Woignier, J.Phalippou, R.Sempere and J.Pelous, J.Phys.France 49, 289 (1988)

13. J.Forsman, J.P.Harrison and A.Rutenberg, Can.J.Phys. 65, 767 (1987)

14. S.Alexander, J.Phys.France 45, 1939 (1984)

15. Y.Kantor and I.Webman, Phys.Rev.Lett. 52, 1891 (1984)

16. S.Alexander and R.Orbach, J.Phys.(Paris)Lett. 43, L625 (1982)

17. E.Courtens, J.Pelous, J.Phallipou, R.Vacher, T.Woignier, Phys.rev.Lett. 58, 128 (1987)

18. T.Freltoft, J.K.Kjems and D.Richter, Phys.Rev.Lett. 59, 1212 (1987)

19. R.Calemczuk, A.M.de Goer, B.Salce, R.Maynard and A.Zarembowitch, Europhys.Lett. 3, 1205 (1987)

20. O.Entin-Wohlman and R.Orbach in Time-Dependent Effects in Disordered Materials. Edited by R.Pynn and T. Riste, Plenum, New York (1987), p.243.

21. R.Vacher, T.Woignier, J.Pelous, G.Coddens and E.Courtens, to be published. E.Courtens, R.Vacher "Random Fluctuations and Pattern Growth: Experiments and Models" (ed. H.E.Stanley and N.Ostrowsky) NATO ASI proceedings, Kluwer Academic Publishers, Dordrecht 1988.

Crossover in the Vibrational Density of States of Silica Aerogels

H. Conrad[1], *G. Reichenauer*[2], *and J. Fricke*[2]

[1]Institut für Festkörperforschung, KFA Jülich, D-5170 Jülich, Fed. Rep. of Germany

[2]Physikalisches Institut der Universität Würzburg, Am Hubland, D-8700 Würzburg, Fed. Rep. of Germany

High resolution neutron backscattering has been applied for the determination of a crossover in the vibrational density of states $g(\omega)$ of three base catalyzed aerogels with mass densities $\rho = 0.12$, 0.18 and 0.25 gcm^{-3}. A Debye behaviour has been found for frequencies lower than a ρ-dependent crossover value ω_{co}. Above ω_{co} a power law ω^x with a spectral exponent $x = 1 \pm 0.5$ has been observed. The observed behaviour resembles a phonon-fracton crossover. The effective sound velocities $1/v_e^3 = 1/v_L^3 + 2/v_T^3$ have been determined from the ratio of total elastic to total inelastic one-quantum scattering and found to be in good agreement with ultrasonic data.

1. Motivation

Since the introduction of the concept of fractals /1/, a great deal of work has been devoted to the exploration of phenomena which might be explained in these novel terms /2,3/. Porous substances seem to be the paradigm of fractal solids and therefore X-ray and neutron small angle scattering as a tool for structural investigations have been applied to elucidate the question of fractality. Unfortunately, SAXS or SANS do not necessarily offer a unique interpretation of the results, in the sense that scattering profiles may be e.g. equally well described by pore size distributions or by large pores with fractal boundaries /4/. The same is true for the complementary case of particle aggregates, where a smooth particle size distribution may give the same result as a power law correlation between the centers of equal particles forming a presumed fractal body /5/. In any case a deviation from the asymptotic Q^{-4}-behaviour of the scattered intensity $I(Q)$ is used as a criterion for fractal structures. But one has to bear in mind that a Q^{-p}-law ($p < 4$) should be observed at least over one decade to be significant. In that sense aerogels may be viewed as fractals /6/.

Although obviously structure yields much more direct information, dynamics may help to shed additional light on doubtful results. There is theoretical work on the vibrational properties of fractals predicting a crossover from ordinary propagating excitations (phonons) at low frequencies ω to localized excitations, called fractons, at higher frequencies /7,8/. In the vibrational density of states $g(\omega)$ this is reflected in a crossover from ordinary quadratic Debye behaviour to $g(\omega) \propto \omega^{\bar{d}-1}$ for $\omega > \omega_{co}$ where ω_{co} is the crossover frequency. The quantity $\bar{d}$ was termed fracton dimensionality and conjectured /7/ to assume the universal value $\bar{d} = 4/3$. In Brillouin scattering measurements on neutrally reacted aerogels the crossover frequencies and $\bar{d}$ could be indirectly determined from the dispersion of the observed phonons /9/. Inelastic neutron scattering, on the other hand, is a direct way to measure $g(\omega)$ and thereby checking the above predictions about deviations from a Debye behaviour. It is known that for an incoherently scattering sample (i.e. no interferences between different scatterers), the one-quantum scattering cross section is directly proportional to $g(\omega)$. This factorization is also possible for amorphous solids, which scatter neutrons both incoherently and coherently like aerogels, if certain conditions hold (see below).

Springer Proceedings in Physics, Vol. 37 **Dynamics of Disordered Materials**
Editors: D. Richter · A.J. Dianoux · W. Petry · J. Teixeira

2. Neutron scattering and the density of states $g(\omega)$

The total (i.e. coherent plus incoherent) one-quantum scattering cross section can be written according to BUCHENAU /10/

$$\frac{d^2\sigma_{tot}^{\{1\}}}{d\Omega\, d\omega} = \frac{k_f}{k_i}\frac{\hbar Q^2}{2M_{av}}\frac{g(\omega)}{\omega}\left[\frac{1}{\exp(\hbar\omega/k_BT)-1}(+1)\right]\cdot\frac{1}{2}\int_{-1}^{1} dz\, I^{(o)}(Q+q_\omega z) \qquad (1)$$

where k_i and k_f are the moduli of the incoming and scattered wavevectors; $\vec{Q}=\vec{k}_i-\vec{k}_f$ is the momentum transfer vector of the scattering event (divided by Planck's constant $\hbar$); the energy transfer is defined by $\hbar\omega = (\hbar^2/2m)(k_i^2-k_f^2)$ with m the neutron mass; the expression in brackets is the Bose factor with +1 for energy loss scattering, k_B is Boltzmann's constant. $I^{(o)}(Q)$ is the total elastic scattering, q_ω is an average over all long-wavelength eigenmodes of the amorphous solid at the considered frequency. The integral averages over all propagation directions. M_{av} is the average atomic mass of the aerogel-forming molecule SiO_2. An essential condition for (1) to be fulfilled is that the harmonic approximation holds. This has been verified in the present experiment by determining the temperature dependence of the Debye-Waller factor (see sect. 5).

From (1) follows that in the limiting case $\hbar\omega \ll k_BT$ the intensity of inelastically scattered neutrons is proportional to $g(\omega)/\omega^2$ which is a constant for Debye solids. If, on the other hand, the density of states of a fractal solid beyond ω_{co} varied like $\omega^{\bar{d}-1}$ with $\bar{d} < 3$ as mentioned above, it should be observable in the neutron inelastic regime as decreasing intensity with increasing energy transfer provided the crossover frequencies are close to values accessible by existing spectrometers. If we take the recent Brillouin results cited above /9/, we obtain as an order of magnitude estimate $\hbar\omega$ = ...4...40.µeV. These figures limit the choice to the two spectrometer types with the highest energy resolution obtainable, the backscattering spectrometer or the spin echo spectrometer. We chose the former one, which has an instrumental resolution of 1 µeV and a maximum dynamic range of ±20 µeV.

A remark is suitable at this point concerning the inelastic intensities to be expected in the energy transfer window below ±20 µeV, which should include the Debye range. In this range, $g(\omega)/\omega^2 = (V_{av}/6\pi^2)\cdot(1/v_L^3 + 2/v_T^3)$, which is very small in common solids due to the large sound velocities of the order of several $10^3\ ms^{-1}$. In the silica aerogels investigated the sound velocities are of the order of 100 ms^{-1}. Therefore the inelastic scattering, hardly observable in that range in common solids, is stronger by a factor of about 10^4.

3. Experimental

The neutron spectra have been measured with the improved backscattering spectrometer BSS1 in the new neutron guide hall at the Jülich research reactor FRJ-2. The incoming neutrons have an energy centered at 2.08 meV. For the time being it is covering a range of only ±11 µeV for technical reasons. For the present measurement we chose 5 scattering angles such that we covered a Q-range between 0.5 and 1.9 $Å^{-1}$. Three base-catalyzed silica aerogels have been investigated at temperatures between 12 K and 360 K. The mass densities of the samples are ρ = 0.12; 0.18 and 0.25 gcm^{-3}. The aerogels were baked for 40 hours at 500 °C, in order to reduce the strong incoherent scattering from physisorbed H_2O and chemisorbed OH-groups; under N_2-atmosphere the aerogels then were packed in an indium sealed aluminium cylinder of 3 cm diameter.

4. Data Evaluation and Results

Typical energy scans are presented in Figure 1. The essential feature is the composite form of the spectra consisting of an elastic line superimposed on an inelastic distribution. The shape of this inelastic scattering depends strongly on the mass density of the aerogels. The sample with ρ = 0.12 gcm^{-3} exhibits a

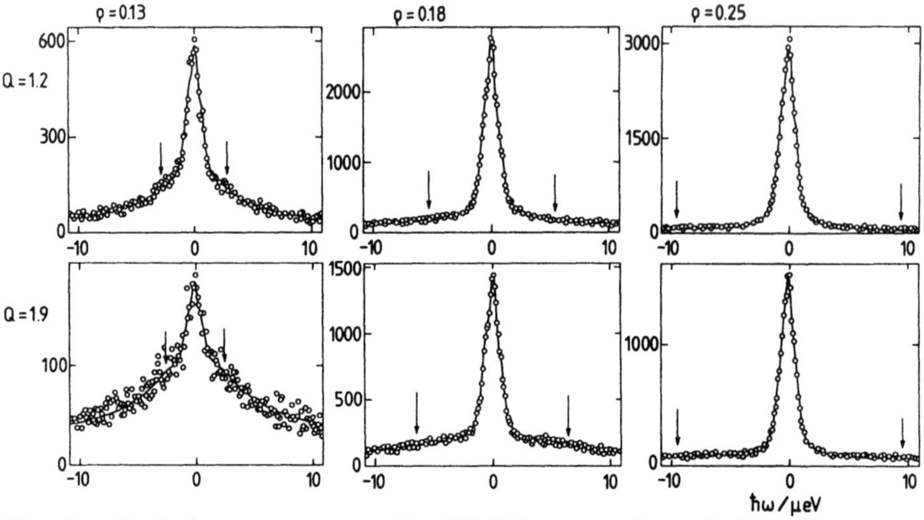

Fig. 1 - Typical energy scans at T = 300 K for two values of the momentum transfer Q for the three investigated aerogel specimens of different densities ρ. The arrows indicate the fitted positions of the crossover energies $h\omega_{co}$. The curves through the data points are the results of the fitting procedure described in the text. Note that these curves in the elastic regime ($\hbar\omega \approx 0$) represent the independently measured instrumental line shapes and are not just a guide to the eye.

pronounced energy dependence suggesting an underlying spectral density of states of the form ω^x with $x < 2$ for $\omega > \omega_{co}$. The sample with $\rho = 0.25\ \mathrm{gcm}^{-3}$, on the contrary, only produced a nearly energy independent contribution over the inelastic energy range we employed during this measurement. This suggests that the latter sample exhibits Debye behaviour over most of the measured range. In other word, ω_{co} is close to or beyond 11 μeV.

In order to evaluate the spectra quantitatively we used a scattering cross section consisting of an elastic line $c_0\delta(\omega)$ and an inelastic contribution, which for $\hbar\omega \ll k_BT$ is proportional to $g(\omega)/\omega^2$, as already mentioned. The symmetric function $g(\omega)$ is constructed such that it is $g_D(\omega) = G_D\omega^2$ for $|\omega| < \omega_{co}$ (Debye or phonon part) and $g_F(\omega) = G_D\omega_{co}{}^2\cdot(\omega/\omega_{co})^x$ for $|\omega| \geq \omega_{co}$ (fracton part). If so defined, $g_D(\omega)$ and $g_F(\omega)$ are continuous at $\omega = \omega_{co}$, which is a reasonable but not necessary choice supported by the data. In particular, there is no indication in the data of a peak near the crossover frequency postulated for reasons of conservation of mode numbers /11/ or as a consequence of a flat dispersion curve /8,9/. Figure 2 shows the scattering cross section based on a density of states function as just described.

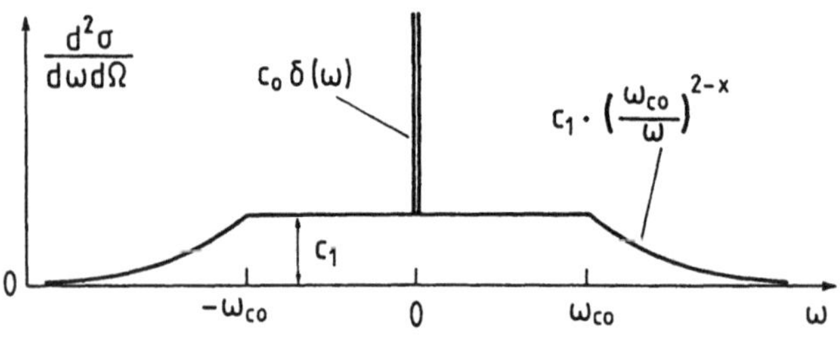

Fig. 2 - Theoretical scattering cross section based on the assumption of a crossover from Debye behaviour to fractal behaviour at $\omega = \omega_{co}$.

Table 1. Cross over frequencies $\nu_{co} = \omega_{co}/2\pi$ and fractal dimension for three silica aerogels (brackets indicate that the value may be not significant due to large statistical errors)

$\rho\,(gcm^{-3})$	0.12	0.18	0.25
ν_{co}(GHz)	0.8 ± 0.2	1.7 ± 0.4	(2.2)
$\bar{d}$	1.9 ± 0.5	2.1 ± 0.6	(2.4)

The scattering cross section has to be folded with the experimental resolution function and fitted to the data points varying the parameters c_o, ω_{co}, x and c_1. Typical fit results are the continuous lines through the data points in Figure 1. If we now identify the empirical exponent x with the fractal notation, $x \equiv \bar{d}-1$, we obtain the spectral dimension $\bar{d}$ given in Table 1 together with the fitted cross over frequencies $\nu_{co} = \omega_{co}/2\pi$.

One can even go one step further and calculate an effective sound velocity v_e defined by $1/v_e^3 = 1/v_L^3 + 2/v_T^3$ from the values of the parameters c_o and c_1; v_e then can be compared with values deduced from ultrasonic data. In this way one will confirm the validity of the approximations leading to (1). In the Debye limit, according to (1), the total scattering cross section is $d^2\sigma/d\Omega d\omega = c_o\delta(\omega) + (Q^2/2M_{av})k_BTG_D\cdot c_o \equiv c_o\delta(\omega) + c_1$. Taking the wellknown relation between the Debye density of states and the effective sound velocity, i.e. $g(\omega) = (V_{av}/6\pi^2 v_e^3)\omega^2 \equiv G_D\omega^2$ with the average atomic volume V_{av}, we can calculate v_e from the ratio of elastic to inelastic scattering

$$v_e^3 = \frac{V_{av}}{6\pi^2}\,\frac{Q^2}{2M_{av}}\,k_BT\,\frac{\pi\Gamma c_o}{c_1} \tag{2}$$

The factor $\pi\Gamma$ enters because of the integration over the instrumental resolution, Γ being the half width at half maximum. The effective sound velocities calculated from the neutron data according to (2) are given in Table 2 together with the corresponding data from ultrasonic measurements.

The sound velocities calculated from the neutron data are mean values of the results obtained from sets of up to 35 spectra per sample taken at 5 different Q-values and up to 7 different temperatures between 12 K and 360 K. The errors quoted are single standard deviations. The agreement with ultrasonic results is reasonably good. Nevertheless we won't stress that quality too much, because the third power of the sound velocity enters the neutron scattering cross section. Therefore larger deviations of the fit parameters c_o and c_1 in (2) within one set of spectra are essentially smoothed.

Table 2. Comparison of effective sound velocities $v_e^{-3} = v_L^{-3} + 2v_T^{-3}$ from neutron scattering data according to (2) and from ultrasonic measurements /12/.

$\rho\,(gcm^{-3})$	$v_e\,(ms^{-1})$ neutrons	$v_e\,(ms^{-1})$ ultrasonic
0.12	92 ± 9	84 ± 5
0.18	154 ± 10	136 ± 8
0.25	210 ± 14	181 ± 12

5. Discussion and Conclusions

The obvious agreement of the sound velocities obtained from ultrasonic measurements and from neutron scattering justifies the application of (1) to both incoherently and coherently scattering aerogels for a density of states analysis. On the other hand, as at a first glance one might be tempted to interpret the spectra alternatively in terms of e.g. rotational diffusion of protons of residual OH-groups, we will discuss and exclude this possibility with several arguments. Firstly, the amount of residual OH-groups from the sample preparation varies proportional to the mass density ρ of the samples. Therefore, the topology of the intensity distribution in the neutron spectra should be equal for all samples. From Figure 1 it is obvious that the intensity profiles differ markedly for the three specimens. Secondly, the intensity variation of both elastic and inelastic (quasielastic) scattering from rotationally diffusing protons obey Q-dependences described by spherical or cylindrical Bessel functions. As can be seen in Fig. 3 there is no agreement at all between the experimental elastic intensities and the theoretical form factors.

Finally, rotational motion of an OH-group is a thermally activated process. This in turn should be observable as a step in the temperature dependence of elastically scattered neutrons. In contrast we found in a log-lin plot of intensity versus temperature a straight line between 12 K and 360 K as shown in Fig. 4, thereby verifying the validity of the harmonic approximation implied in (1).

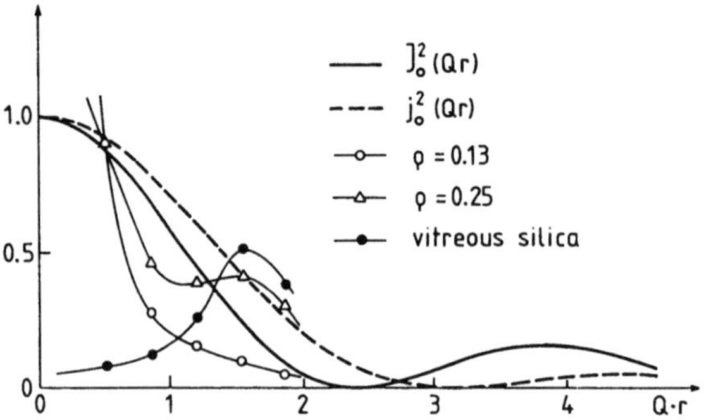

Fig. 3 - Comparison of theoretical form factors for rotational diffusion with experimental elastic intensities. Rotational radius R = 1 Å.

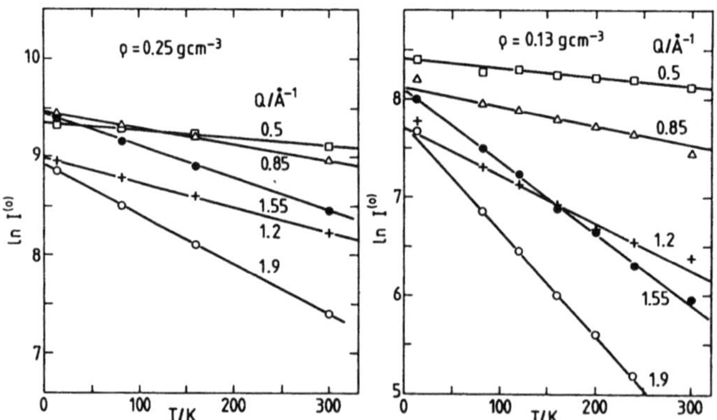

Fig. 4 - Elastic intensity as a function of temperature for two samples (ρ = 0.25 gcm^{-3} and ρ = 0.12 gcm^{-3}) and 5 momentum transfers Q. The figures at the lines are corresponding numerical values of Q in units of A^{-1}. No indication of any discontinuity is observed.

From the slopes of the straight lines in Fig. 4 we have extracted the mean square displacements $\langle u^2 \rangle = -(\Delta \ln I^{(o)}/\Delta T)T/Q^2$, wherefrom we got the root mean square vibrational amplitudes $\langle u^2 \rangle^{1/2} = 1$ Å for $\rho = 0.12$ gcm^{-3} and $\langle u^2 \rangle^{1/2} = 0.7$ Å for $\rho = 0.25$ gcm^{-3}. These are, as far as we know, the largest amplitudes of harmonic vibrations ever determined for a solid.

We have presented the first direct observation of a crossover from ordinary Debye vibrational characteristics to a different spectral behaviour resembling features predicted for fractal structures. Within experimental accuracy no peak was detected in the crossover region as postulated by some authors /8,11/, but rather a monotonous transition as found in a recent numerical calculation /14/. Having in mind that a deviation from an ω^2-dependence alone does not imply fractal connectivity of the underlying structure /13/, a conclusion our investigated aerogels might be fractals were based right on this crossover. On the other hand, our values for the fractal dimension $\bar{d}$ (see Table 1) appears to be far from the conjectured universal value 4/3 , even if the estimated experimental error quoted (single standard deviation) is taken into account. Although the fractal concept is very attracting and possibly applicable to aerogels, the deviation of our $\bar{d}$-values from existing predictions could also mean that we have observed a different type of porosity. Physically this deviation from a Debye behaviour (in the sense of Aharony et al. /11/) is the manifestation of a lack of vibrational modes due to missing material in the porous body. Thus we believe that fractal connectivity is only a special case of porosity. Any type of porosity on an adequate length scale should be the origin of a mode deficiency leading to a spectral density varying as ω^x with $x < 2$ and $\omega > \omega_{co}$.

We would like to add a final remark on the numerical values for ω_{co} and $\bar{d}$ for the most dense sample (compare last column in Table 1). These figures need not to be taken too seriously as the reliability of the fitting procedure breaks down if the crossover reaches the edge of the dynamic range of the spectrometer. Nevertheless, even this case supports our interpretation.

ACKNOWLEDGEMENT

The authors are grateful to U. Buchenau for stimulating this work and for many fruitful discussions. The authors are indebted to Mrs L. Schätzler for her excellent work with the computer codes for data evaluation as well as to B. Huy for technical assistance. One of the authors (HC) wants to express the great pleasure he had in collaborating with R. Schätzler in setting to work the new spectrometer.

REFERENCES

1. B.B. Mandelbrot, Fractals: Form, Chance and Dimension; San Francisco, 1977
2. L. Pietronero and E. Tosatti (eds), Fractals in Physics; Elsevier 1986
3. F. Family and D.P. Landau (eds), Kinetics of Aggregation and Gelation; Elsevier, 1984
4. H.D. Bale and P.W. Schmidt, Phys. Rev. Lett. 53, 596 (1984)
5. S.K. Sinha et al. in /3/
6. R. Vacher, T. Woignier, J. Pelous, E. Courtens; Phys. Rev. B37, 6500 (1988)
7. S. Alexander and R. Orbach, J. Physique Lett. 43, L625 (1982)
8. O. Entin-Wohlmann, S. Alexander, R. Orbach and Kin-Wah Yu; Phys. Rev. B29, 4588 (1984)
9. E. Courtens, J. Pelous, J. Phalippou, R. Vacher and T. Woignier, Phys. Rev. Lett. 58, 128 (1987)
10. U. Buchenau, Z. Phys. B. 58, 181 (1985)
11. A. Aharony, S. Alexander, O. Entin-Wohlmann, R. Orbach; Phys. Rev. B31, 2565 (1985)
12. R. Goswin and G. Reichenauer, Universität Würzburg; unpublished
13. J.A. Krumhansl, Phys. Rev. Lett. 56, 2696 (1986)
14. K. Yakubo and T. Nakayama, Phys. Rev. B36, 8933 (1987)

Low Frequency Vibrations in Silica Aerogels

G. Reichenauer[1], *J. Fricke*[1], *and U. Buchenau*[2]

[1]Physikalisches Institut der Universität Würzburg, Am Hubland, D-8700 Würzburg, Fed. Rep. of Germany
[2]Institut für Festkörperforschung, KFA Jülich, D-5170 Jülich, Fed. Rep. of Germany

Low frequency vibrations in highly porous silica aerogels (mass densities ρ = 120 kg/m³ and 250 kg/m³) have been studied by inelastic neutron scattering for frequencies between 20 GHz and 2 THz. Absolute values for the densities of states were determined by evaluating the total inelastic intensity. The temperature dependence of the inelastic scattering between 5 K and 300 K indicates that the modes observed are essentially harmonic; the wavelengths of the excitations are found to be large compared to atomic distances as the dynamic structure factor is well explained by sound-wave-like motions.

1. Introduction

Recently the dynamic properties of amorphous porous media have attracted growing interest /1-8/. An especially intriguing material in this respect is silica aerogel, which is produced via a sol-gel process with subsequent supercritical drying of the wet gel in an autoclave. In this way porosities of up to 98% are accessible /9/. On the other hand densification down to non-porous silica glass can be achieved by thermal treatment. Thus it is possible to investigate the dynamics of these disordered solids upon systematic change of structural parameters.

Within the last years various techniques were used to investigate the dynamics of porous silica /4-8/. If inelastic neutron scattering is applied, one is generally restricted to evaluate the incoherent inelastic scattering in order to determine the vibrational density of states (further abbreviated by DOS); if the coherent and the incoherent scattering contributions can be separated properly (without changing the structure of the specimen by sample preparation), the incoherent scattering provides an amplitude-weighted DOS /10/; under certain conditions, however, both, coherent *and* incoherent inelastic scattering data can be used simultaneously to calculate the DOS $g(\nu)$ on an absolute scale /11,12/.

2. Experiments

Inelastic neutron scattering was performed at the time-of-flight spectrometer SV5 at the cold source of the reactor FRJ-2/Jülich, F.R.G. Additional data were taken with the triple axis spectrometer SV4/Jülich. The wavelengths of the incident neutrons were 6 Å (3 Å) at the SV5 and 2.4 Å at the SV4 instrument. The frequency resolution achieved was 0.02 THz (SV5) and 0.26 THz (SV4), respectively. The temperature dependence of the spectra was investigated between 5 and 300 K.

The measurements were performed for base-catalyzed silica aerogels of two different mass densities ρ = 120 kg/m³ and 250 kg/m³. In order to reduce the dominating incoherent scattering due to large amounts of adsorbed H_2O and chemically surface-bonded OH- and CH_3-groups, the aerogels were first baked at 500°C for about two days and subsequently sealed in a nitrogen-filled aluminum cylinder (this was necessary to get a sufficient coherent scattering contribution in order to test the applicability of the equation used for the evaluation of our data (1)). The comparison of the average elastic intensity of the aerogels (calculated

Springer Proceedings in Physics, Vol. 37 Dynamics of Disordered Materials
Editors: D. Richter · A.J. Dianoux · W. Petry · J. Teixeira

using a vanadium standard) with the theoretically expected value yields an estimate for the ratio between the incoherent and the coherent scattering of about 2:1. This corresponds to about 0.3 H-atoms per SiO_2-unit. As the residual hydrogen is strongly bonded to oxygen atoms at the internal aerogel surface, its motion is not expected to contribute to frequency modes below the THz-range.

3. Data Evaluation and Results

The *elastic* scattering profile for $Q \gtrsim 1$ Å^{-1} (fig. 1) is found to be equivalent to the scattering from vitreous and fumed silica /13,14/. At small Q's the elastic scattering of the aerogels is strongly enhanced due to small angle scattering.

The *inelastic* scattering data were corrected for cryostat background (using the 5K data set), detector efficiency, tail of the elastic line and multiple scattering contributions.

For the interpretation of the inelastic scattering data, we applied the results for the one-phonon cross section derived recently /12/:

$$\frac{d^2\sigma^{(1)}}{d\Omega\, d\nu} = \frac{k_f}{k_i}\,\frac{h\,Q^2}{2M_{av}}\,f_B\,\frac{g(\nu)}{\nu}\,\frac{1}{2}\int_{-1}^{1} I_0(Q + q_\nu \cdot z)\,dz \qquad (1)$$

with $\underline{Q} = \underline{k_i} - \underline{k_f}$ the scattering vector, f_B the Bose-factor, defined with respect to the detailed balance condition, $g(\nu)$ the normalized DOS, M_{av} the average mass of a sample atom, q_ν the absolute value of the average wave vector corresponding to modes of frequency ν and I_0 the total elastic intensity. The integral just smeares out the well structured elastic scattering by an effective q_ν due to a sound wave like motion of the scatterers. Equation (1) implies that the harmonic approximation holds for the vibrational modes investigated and that q_ν is a well defined quantity for a given frequency. The simple form of the integral in (1) applies for $q_\nu \ll Q$; $q_\nu \geq Q$ results in a slight shift of the structure factor and an additional smoothing of its profile.

The harmonicity of the probed modes were checked by plotting the inelastic intensity, taken for a small frequency range, versus Debye-Waller-factor e^{-2W} times Bose-factor f_B. As shown in fig. 2 the 250 kg/m³ aerogel is dominated by harmonic modes at least up to 200 K. The drop in inelastic intensity at 300 K is correlated to a change of the dynamic structure factor (see fig. 3a inset). Surprisingly, this behaviour is not detected for the lower density sample (fig. 3b).

The validity of the assumed Q-dependence of the inelastic intensity can be checked by comparison of the theoretical and experimental dynamic structure factor (fig. 3). The agreement of the experimental data with a sound wave structure factor is good. In contrast, the dynamic structure factor of the non-porous material

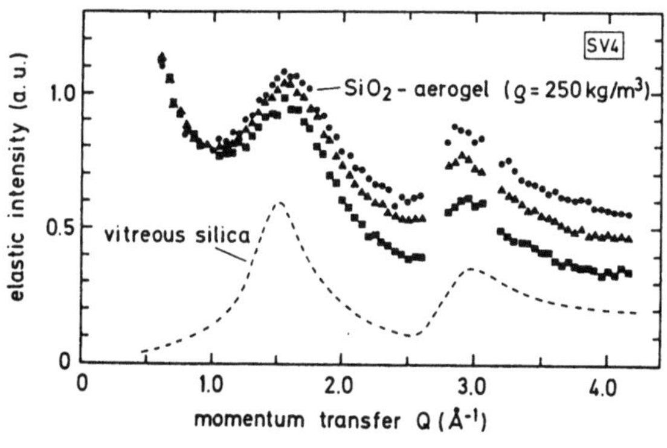

Fig. 1 Elastic scattering intensity (250 kg/m³ -aerogel) versus momentum transfer Q (SV4-data) for ● 5 K, ▲ 100 K and ■ 200 K; the elastic intensity is corrected for multiple scattering

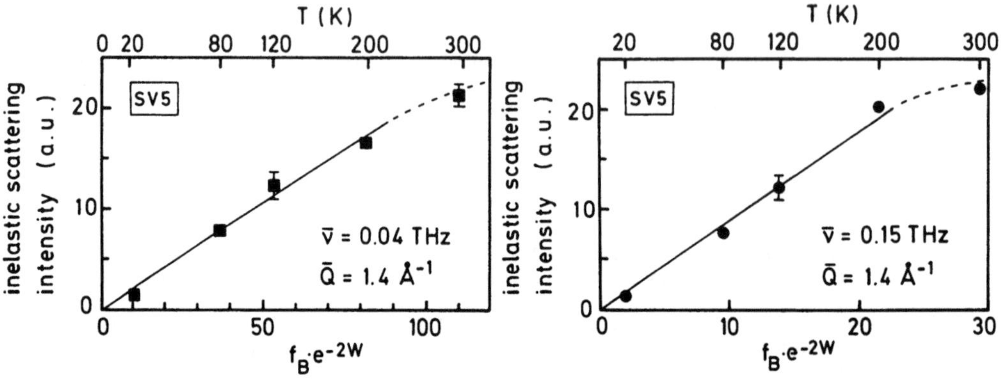

Fig. 2: Inelastic scattering from the 250 kg/m³ -aerogel versus Bose-factor f_B times Debye-Waller-factor $\exp^{-2W}$ for $\overline{Q}$ = 1.4 Å^{-1} and two averaged frequencies $\overline{\nu}$: ■ $0.027 \le \nu \le 0.047$ THz; ● $0.12 \le \nu \le 0.19$ THz. The line is a guide to the eye

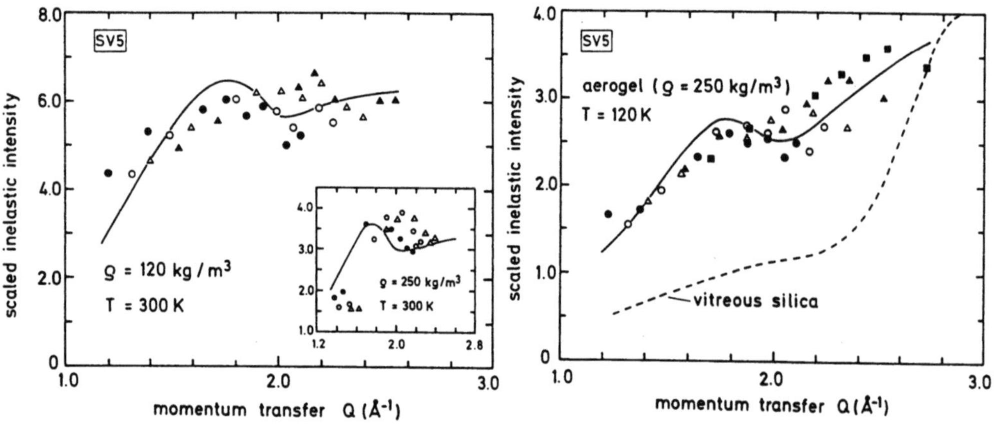

Fig. 3: Experimental and theoretical (——) dynamic structure factor; the theoretical values - and the corresponding experimental data points - for different frequencies are scaled to a common curve: ● 0.09 THz, o 0.26 THz, Δ 0.45 THz, ▲ 0.70 THz, ■ 1.07 THz; a) structure factor for both aerogels at 300 K; b) structure factor for the 250 kg/m³ -aerogel at 120 K and for vitreous silica /13/ (not scaled relative to each other)

is completely different (fig. 3b); its profile can be explained in terms of correlated rocking motions of interconnected SiO_4-tetraheda /13/. We suggest, that these modes cannot be detected in the aerogels, as they are masked by a huge contribution of long wavelength excitations.

As (1) only holds for the one-phonon cross section, whereas the experimental data contain also multi-phonon terms, the data have to be corrected for these contributions, too. If the one-phonon DOS were known for all frequencies, the multi-phonon scattering could be calculated by successive convolution procedures /10/. Thus, the problem of multi-phonon correction could be solved iteratively. In order to simplify this procedure, we approximated $g(\nu)/\nu^2$ (for $h\nu \ll k_BT$) by a function which allows analytical convolutions and describes the known parts of the DOS reasonably well:

(i) In the low frequency limit the DOS should be equal to the Debye-DOS, which can be calculated from ultrasonic data (table 1); the Debye approximation should hold up to wavelengths λ ($\lambda = v/\nu$, where v is the polarization averaged sound velocity), which are in the order of the size of the largest inhomogeneities within the specimen; the corresponding frequencies -often called crossover frequencies ν_{co} /2/- can be estimated to be in the GHz for the aerogels investigated /6,16/.
(ii) The DOS in the 20 GHz to 2 THz range should be similar to the one derived by a zeroth order evaluation of the experimental data according to (1) just neglecting the multi-phonon contributions.

These conditions can be sufficiently fullfilled if $g(\nu)/\nu^2$ is approximated by the sum of a sharp Lorentzian and a broad Gaussian function, both centered at the origin (fig. 4).

The experimental inelastic intensity and the corresponding multi-phonon corrections calculated by use of the analytic function are given in fig. 5. As the multi-phonon terms caused by the DOS below the frequency resolution of our instrument give an integral contribution to the intensity within the frequency range covered by the measurement, we even can extract some information on the DOS for frequencies below the resolution limit. If we define ν_{co} by $g(\nu_{co})/\nu_{co}^2 = 0.5\, g_{Debye}(\nu)/\nu^2$, best consistency between the function used to calculate the multi-phonon contributions and the final DOS was achieved for the values given in table 1.

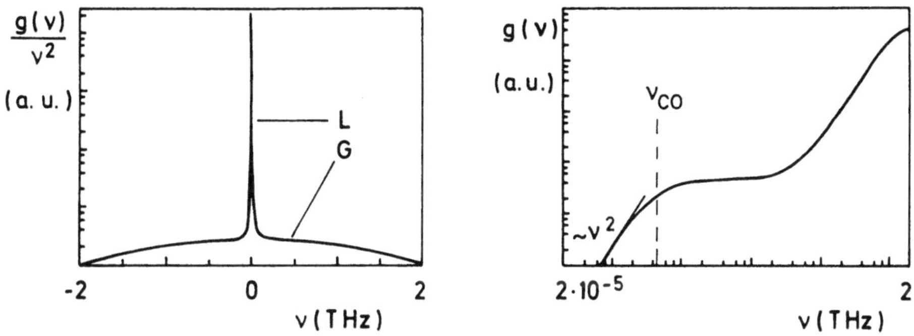

Fig. 4: a) Frequency dependence of the analytical function $g(\nu)/\nu^2$, with L = extremly narrow Lorentzian and G = broad Gaussian function; b) corresponding log-log-plot of the DOS versus frequency with ν_{co} being the cross-over frequency

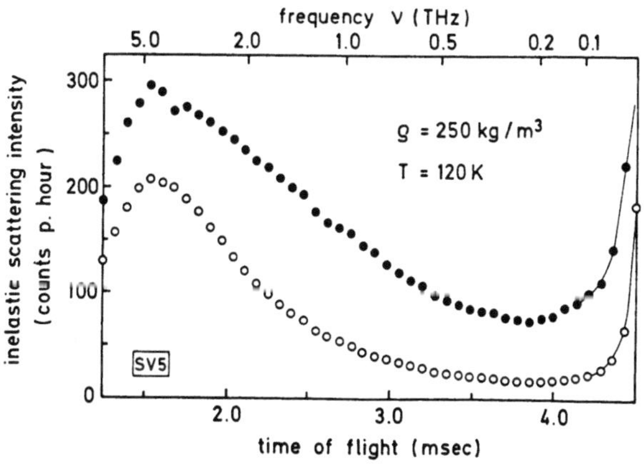

Fig. 5: • : Inelastic scattering in time-of-flight representation, corrected for cryostat-background, multiple scattering, tail of the elastic line and detector efficiency; o : multiphonon-scattering determined via selfconsistent calculation of the DOS

Table 1. Crossover frequencies and average sound velocities v for aerogels of different densities; the sound velocities v are calculated from ultrasonic data; * priv. comm. R. Vacher

ρ /kgm^{-3}/	ν_{co} /GHz/	ν_{co} /GHz/*	v /ms^{-1}/
120	0.48	0.25	120 ± 10
250	3.0	1.4	260 ± 15

The final DOS for the two aerogels investigated are shown in fig. 6. Even at the highest frequencies, they exceed the DOS of bulk silica by a factor of 6. This shows, that the elastic connectivity of these samples must be lower than that of vitreous silica even in the nm- to Å- range. On the low frequency side, the lines with slope 2 in the double logarithmic plots (fig. 6) indicate the Debye region. Though the sound-wave-DOS for the two specimens differ by a factor of 20, both DOS agree within experimental error at higher frequencies; this is understandable if we assume both samples to be built from similar 1 to 10 nm porous building-blocks /15/.

Very recently, Conrad et al. /16/ have performed measurements with strongly improved resolution using the backscattering technique, which allows direct access to the crossover region. Moreover, experiments within the frequency range 2 to 20 GHz are under way; thus we hope to present soon a complete picture of the DOS in silica aerogels, which can be considered as model materials for highly porous disordered solids.

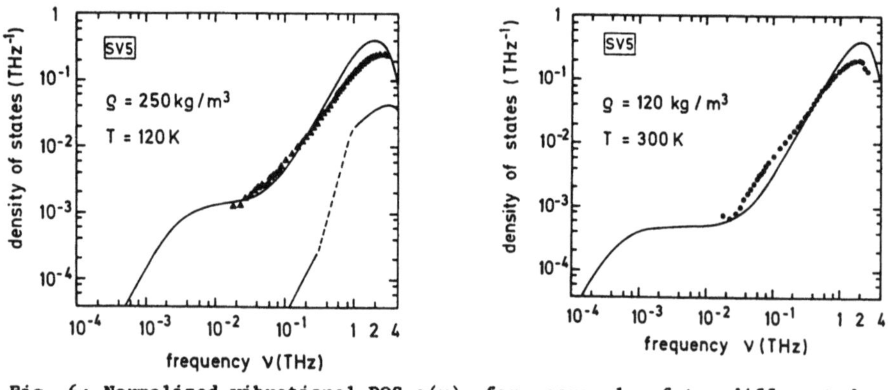

Fig. 6: Normalized vibrational DOS $g(\nu)$ for aerogels of two different densities. a) $g(\nu)$ for ρ = 250 kg/m^3 and T = 120 K; the curve in the right corner represents the DOS for vitreous silica /13/. b) DOS for ρ = 120 kg/m^3 and T = 300 K. The full curves in both plots give the DOS approximated by ν^2 times sum of a Lorentzian and a Gaussian function with parameters adjusted for each case; for low frequencies the functions reproduce the Debye-behaviour $g(\nu) \sim \nu^2$

Acknowledgement : We are grateful to René Vacher / University Montpellier for providing information on the crossover frequencies. We also want to thank Michael Monkenbusch / KFA Jülich for performing complementary measurements at the SV4-spectrometer and Hans Grimm / KFA Jülich for stimulating discussions.

References

1. S. Alexander, R. Orbach: J. Phys. (Paris) Lett. 43, L625 (1982)
2. R. Orbach: Science 231, 814 (1986)
3. I. Webman, G.S. Grest: Phys. Rev.B 31, 1689 (1985)
4. J.M. Grace, A.C. Anderson: Phys. Rev.B 33, 7186 (1986)
5. A. Boukenter, B. Champagnon, E. Duval, J. Dumas, J.F. Quinson, J. Serughetti: Phys. Rev. Lett. 57, 2391 (1986)
6. E. Courtens, J. Pelous, J. Phallipou, R. Vacher, T. Woignier: Phys. Rev. Lett. 58, 128 (1987)
7. T. Freltoft, J. Kjems, D. Richter: Phys. Rev. Lett. 59, 1212 (1987)
8. R. Calemczuk, A.M. De Goer, B. Salce, R. Maynard, A. Zarembowitch: Europhys. Lett. 3, 1205 (1987)
9. S.S. Kistler: Nature 127, 741 (1931)
10. W. Marshall, S.W. Lovesey: Theory of Thermal Neutron Scattering (Clarendon Press, Oxford, 1971)
11. J.M. Carpenter, C.A. Pelizzari: Phys. Rev.B 12, 2391 and 2397 (1975)
12. U. Buchenau: Z. Phys. B 58, 181 (1985)
13. U. Buchenau, M. Prager, N. Nücker, A.J. Dianoux, N. Ahmad, W.A. Phillips: Phys. Rev. B 34, 5665 (1986)
14. D.Richter, L. Passell: Phys. Rev. B 26, 4078 (1982)
15. J. Fricke, G. Reichenauer: J. Non-Cryst. Solids 95 & 96, 1135 (1987)
16. H. Conrad, J. Fricke, G. Reichenauer: High-Resolution Neutron Spectroscopy of the Crossover in the Vibrational Density of States of Silica Aerogels, 2nd International Symposium on Aerogels, Montpellier/France, Sept. 1988

Low Frequency Light Scattering in Lithium Borate Glasses; a Cross-Over from Phonons to Fractons?

L. Börjesson

Department of Physics, Chalmers University of Technology,
S-412 96 Göteborg, Sweden

1. INTRODUCTION

Low frequency vibrational excitations in disordered solids have received considerable theoretical attention during recent years [1-6]. Some of the anomalies found experimentally, e.g. the anomalous low temperature properties of glasses and the thermal conductivity plateau in disordered materials, have been interpreted as being due to a cross-over in the vibrational spectrum from translational invariance (phonons) to highly localized acoustic excitations on a fractal network (fractons) [2,7,8] at a characteristic length-scale ξ. The cross-over from phonons to fractons has successfully been demonstrated recently by direct spectroscopic techniques for materials of known fractal structures [9,10]. Furthermore, for some materials of so far unknown fractal structure, namely the superionic AgI doped borate glasses, the behaviour of the low frequency Raman spectrum [11] as well as the low temperature specific heat [12] are also in consistence with a cross-over from phonons to fractons. From these studies a fractal correlation length of about 30 Å were deduced which is thought to be related to a subnetwork of AgI within the glass, though the fractal structure has yet to be determined.

The aim of the present study is to rationalize the observations made for the AgI doped borate system by chosing a system which is very similar, however, with a different medium range ordering of the dopant salt, namely the LiCl-doped borate glasses $(LiCl)_x$-$(Li_2O$-$2B_2O_3)_{1-x}$. The LiCl doped borate glasses have many properties in common with the AgI analogues e.g. a huge enhancement of the ionic conductivity with increasing salt concentration [13] and a similar local arrangement of the host boron-oxygen network [14]. In both systems the dopant salt is introduced into voids or interstices of the host matrix [13,14]. The medium range order is, however, significantly different in the AgI and the LiCl doped glasses. In the AgI doped glasses a new type of medium range order of a characteristic length ~ 8 Å builds up when AgI is introduced to the glass, while no such ordering effect is observed in the case of LiCl doping [15,16].

By correlating Raman results of AgI and LiCl doped borate glasses it may therefore be possible to make conclusions on the structural nature of the localization of acoustic vibrations reported for the case of the AgI glasses. The observed low frequency Raman spectra will be interpreted both in terms of a phonon-fracton cross-over model and in the frame-work of the Martin-Brenig theory [17,18] which is relevant for low lying vibrational modes corresponding mainly to acoustic phonons.

2. EXPERIMENTAL

Three samples of different LiCl content, x = 0, 0.5, 0.7 were used for the Raman experiments. The preparation method of the glasses is the same as that described previously [19]. The equipment for the Raman experiment includes an Argon-ion Laser (Spectra Physics), operated at 5145 Å with a power of 500 mW, a double monochromator (SPEX 1403) with the slits set to give a resolution of ~0.8 cm^{-1}, a cooled photo-multiplier (RCA 31034-76) and photon counting equipment. A 90° degrees scattering angle was employed. Raman spectra in $||\perp$ and $||||$ polarizations were recorded in the range 2-300 cm^{-1} for each sample at room temperature.

Springer Proceedings in Physics, Vol. 37 **Dynamics of Disordered Materials**
Editors: D. Richter · A.J. Dianoux · W. Petry · J. Teixeira

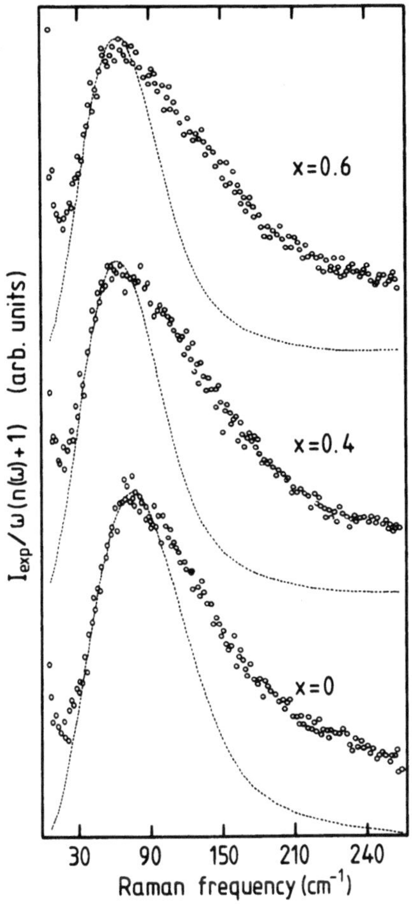

Fig.1 Normalized low frequency Raman spectra , corresponding to $C(\omega)$, for $(LiCl)_x$-$(Li_2O\text{-}2B_2O_3)_{1-x}$ glasses (circles). Dotted lines represents fits to eq. (6).

3. RESULTS

In fig. 1 the normalised $||\perp$ Raman spectra of the x=0, x=0.5 and x=0.7 glasses are shown. For all three samples a broad peak is observed around ~65 cm^{-1}. The peak maximum is observed to shift slightly to lower frequencies as the LiCl content increases. The intensity of the peak is approximately unchanged for all three samples. The findings are in sharp contrast to the reported case of AgI doped borate glasses for which the intensity was observed to increase enormously and the peak position to shift considerably towards lower frequencies with increasing concentration of the dopant salt [20].

4. DISCUSSION

It has been shown that the Raman scattering intensity from a disordered material is related to the vibrational density of states $g(\omega)$ [21]. The observed intensity $I(\omega)$ is however modulated by the light-vibration coupling coefficient $C(\omega)$ according to

$$R(\omega) = I(\omega)\omega/(n(\omega)+1) = C(\omega)g(\omega) \qquad (1)$$

where $R(\omega)$ is the normalized Raman intensity, $n(\omega)+1$ is the Bose-Einstein population factor for Stoke scattering which has to be replaced by $n(\omega)$ for anti-Stokes scattering ($n(\omega)=[\exp(h\omega/kT)-1]^{-1}$).

For a structure which is fractal at intermediate length scales between the particle size a and the correlation length ξ there exist at least three different vibrational regimes [1,4,5]. For the low frequencies which correspond to wavelengths $\lambda>\xi$ the material behaves like a homogeneous system. Then the vibrational waves are only weakly localized acoustic phonons. For higher frequencies λ approaches ξ and the vibrational spectrum obeys a cross-over to a new region, the fracton region. The fracton region, which is characterized by strong localization and a different dispersion relation, persists until λ approaches a. At still higher frequencies the vibrations are predominantly bulk and surface modes of the particles. The various vibrational regions will give rise to separated regions in the vibrational density of states $g(\omega)$ of different frequency dependences. In the acoustic range $g(\omega) \sim\omega^2$, while for the fractons the relation $g(\omega) \sim\omega^{\tilde{d}-1}$, where $\tilde{d}$ is the spectral dimension, is expected to hold [1]. In the Raman spectra $g(\omega)$ will however be modulated by the coupling constant. The Raman intensity $R(\omega)$ is then expected to be $\sim\omega^4$ in the phonon region [17,18]. For the fractons two theories of contradictory expectations have been suggested. On one hand Boukenter et al [22] derived

$$R(\omega) \sim\omega^{(\tilde{d}/D)(2d_{\phi}+D)-1}, \qquad (2)$$

under the assumption that the coupling constant is scale independent. In eq. (2) d_{ϕ} is the "super-localization" exponent and D is the Haussdorf dimension. Fontana et al [11] simplified this expression to $R(\omega) \sim\omega^{2\tilde{d}/d+\tilde{d}-1}$, where d is the Euclidean dimensionality, for a percolating system consisting of both finite and infinite clusters (D=d=3) and for $d_{\phi}=1$. On the other hand, Tsujumi et al. [20] argue for a scale dependent elasto-optic coupling constant and obtained

$$R(\omega) \sim \omega^{-2(\tilde{d}-1)} \qquad (3)$$

in their simplest expression (for $d_{\phi}=(2-\tilde{d})D/\tilde{d}$). The two expressions for $R(\omega)$ differ significantly. For a system exhibiting both phonon and fracton regimes the cross-over frequency, ω_{CO}, is dependent of the fractal characteristic length ξ according to

$$\omega_{CO} = (a/\xi)\,\omega_D, \qquad (4)$$

where ω_D is the Debye frequency [7,11,12].

In order to examine the frequency dependence of the Raman intensity we plot $R(\omega)$ vs ω in a log-log representation in fig. 2 for the present glasses. The figure shows similar behaviour of the three glasses: a linear region at low frequencies followed by a cross-over at about 75 cm^{-1} to another linear region of smaller slope. The cross-over frequency is found to change with LiCl content (see table 1). The effect is however small, almost negligible, in contrast to the findings for the AgI doped borate glasses, in which the cross-over frequency is highly dependent on AgI concentration [11]. In the low frequency region the slopes of the curves in fig. 2 decrease with increasing LiCl concentration. For the undoped glass (x=0) the low frequency region exhibit a $\omega^{3.2}$ behaviour, while for the glass with highest LiCl concentration the corresponding frequency dependency is $\omega^{2.8}$. The behaviour indicate a progressive deviation of $R(\omega)$ from the expected Debye behaviour (ω^4) with increasing LiCl concentration.

Assuming that the high frequency linear region in fig. (2) represents fractons, $\tilde{d}$ is obtained from the slope of the curves using either eq. (2) or eq. (3) depending on the model used. By applying the approach by Boukenter et al. [22] (eq. (2)) $\tilde{d}$ increases from 1.27 for the undoped glass to 1.47 for the highest doped glass, while application of eq. (3) instead causes $\tilde{d}$ to decrease from 1.28 to

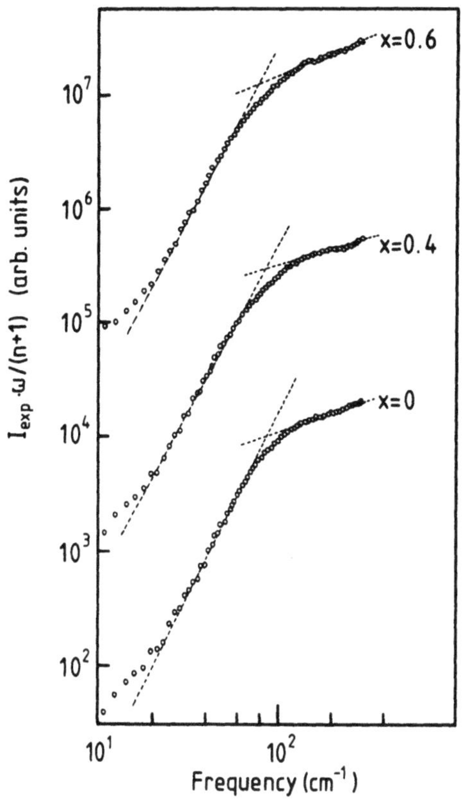

Fig.2 Log-log plot of the reduced Raman spectra, corresponding to $C(\omega)g(\omega)$, for $(LiCl)_x$-$(Li_2O$-$2B_2O_3)_{1-x}$ glasses. Dashed lines are linear fits.

1.11 with increasing LiCl content. The various values obtained are near the value expected for a three dimensional percolating network $\tilde{d}$=4/3 [1]. Under the assumption that the cross-over frequencies between the two linear regions in fig. (2) correspond to phonon-fracton cross-overs, the fractal characteristic lengths ξ can be evaluated from eq. (4) using reported sound velocities [19,23] to calculate the Debye frequencies and a taken as the mean nearest neighbour atomic distance (B-O) in the glass (1.4 Å). The values obtained for ξ are approximately equal, ~6.6 Å, for the three differently doped glasses (see table 1). The value of ξ is very small and it is hard to conceptually think of a material which is structurally self-similar only between 1.4 Å and 6.6 Å. Thus, it is concluded that the LiCl doped borate glasses can not be considered to have a fractal structure despite the observation of a cross-over between two regions of different frequency dependences of the Raman intensity, which is characteristic of a cross-over from phonons to fractons, and the realistic values obtained for $\tilde{d}$.

Since the above analysis failed to give evidence of fractons in the present glasses a model for light scattering from homogeneous glasses is applied to the data. The Martin and Brenig model [17,18] attribute the low frequency Raman scattering of glasses to the scattering of Debye like acoustic phonons. This model considers both electrical and mechanical disorder which affects the photoelastic constants and the strain tensor, respectively. The spatially fluctuating quantities are assumed to exhibit Gaussian correlations with the same structural correlation range (SCR) 2σ. The SCR, which represents the extent of the short range ordering in which a phonon can propagate without any significant damping, is directly related to the position of the low frequency "Boson peak" of glasses according to

Table 1. Properties of $(LiCl)_x$-$(Li_2O$-$B_2O_3)_{1-x}$ glasses. See text for explanation.

x	ω_{max} (cm^{-1})	v_l * (m/s)	v_t * (m/s)	2σ [ω_{max}] (Å)	2σ [fit] (Å)	ω_D (cm^{-1})	ω_{co} (cm^{-1})	ξ (Å)
0	77	6840	3860	5.3	5.5	419	86	6.8
0.4	68	6390	3620	5.6	6.2	375	79	6.6
0.6	66	6130	3470	5.6	6.0	354	75	6.6

* from ref. 19 and 23

$$2\sigma = \frac{v_t}{\omega_{max}} \qquad (5)$$

where v_t is the transverse sound velocity and ω_{max} is the frequency at maximum intensity of the peak. The complete frequency dependence of the Raman coupling constant $C_{||\perp}$ in $||\perp$ scattering geometry is given by

$$C_{||\perp}(\omega) = A\omega^2 \left[3\left(\frac{v_l}{v_t}\right)^5 \exp\left[-\frac{\omega^2\sigma^2}{v_t^2}\right] + 2\exp\left[-\frac{\omega^2\sigma^2}{v_l^2}\right] \right]. \qquad (6)$$

The structural correlation length 2σ can then be obtained either from ω_{max} by using eq. (5) or from a fit of the low frequency Raman spectrum to eq. (6) assuming a ω^2 dependence for $g(\omega)$. In fig. 1 the solid lines represent fits of the low frequency Raman spectra to eq. (6). The discrepancy between the fit and the experimental data at frequencies higher than ω_{max} is expected since eq. (6) is only valid for $2\sigma\omega/v_t<1$, i.e. at the low frequency side of the peak. The values of the correlation lengths obtained from the values of ω_{max} and the fits are presented in table 1. They are found to be approximately equal. Table 1 shows that the correlation length is more or less independent of LiCl content. The value of 5.5 Å for the x=0 glass agrees with that reported by Lorösch et al. [23]. Note also the close similarities of the values of the SCR and the characteristic length obtained from the analysis within the framework of a phonon-fracton cross-over. This may just be accidental because the LiCl doped glasses do not obey the criteria for a fractal structure and therefore the analysis is questionable. Still the coincidence of the values obtained using the two different models is noteworthy.

Introduction of AgI into the borate glasses causes a large increase of the low frequency Raman intensity and a considerable shift of the Boson peak to lower frequencies which indicate an increasing medium range correlation length [11,20]. In contrast, the present study shows that for introduction of LiCl in the borate glasses the low frequency Raman spectum is more or less unaffected. We therefore conclude on the basis of the light scattering results that there must be a significant structural difference in the medium range order between LiCl and AgI doped borate glasses. Such a difference has recently been observed by neutron diffraction studies of the doped borate glasses. An intense sharp diffraction peak at low Q ($Q \approx 0.8$ Å^{-1}) was observed for borate glasses of large AgI concentrations, while no such peak was present in the diffraction pattern of the undoped silver borate glass[15]. The peak corresponds to a real space characteristic length of about 8 Å. In contrast, for the lithium borate glasses no such ordering effects in the medium range structure was observed when LiCl was introduced into the glass. A structural hypothesis is then that in case of LiCl doped glasses LiCl goes into interstices of the borate network in a random manner and is consequently homogeneously distributed in the glass structure. Therefore, introduction of LiCl into the host glass does not give any additional low frequency light scattering. The situation is different for the case of AgI doping. Here introduction of AgI creates medium range ordering effects which give rise to additional low frequency light scattering as well as to the diffraction peak observed at small Q with neutron diffraction. It has then been suggested that the salt tend to build up small clusters within the borate network possibly with a fractal structure [11,12]. However, the value of the characteristic lengths reported for the AgI doped borate glasses from the phonon-fracton analysis (30 Å) [11,12] differ considerably from that reported for the neutron diffraction (8 Å) [15].

We conclude that despite the similarities between LiCl and AgI doped glasses, in the local strucure as well as in the conductivity properties, important differences have been observed in the low frequency Raman spectra. Neither the intensity nor the position of the low frequency peak is much affected by introduction of LiCl in contrast to the reported case of AgI doped glasses [11,20]. The differences of the low frequency light scattering between borate glasses subject to LiCl and AgI doping is attributed to different medium range ordering properties of the dopant salts within the host glass. The LiCl seems to be randomly distributed in interstices of the boron-oxygen network, while AgI tends to create a distinct medium range order probably in form of microclusters of AgI. The observations are in support of recent neutron diffraction results [15,16]. Despite that the reduced Raman intensities for the LiCl doped glasses are well described by two frequency regions of different power laws separated by a cross-over frequency it is not appropriate to interpret the data within the phonon-fracton framework since it gives an unacceptable low value of about 6.6 Å for the fractal characteristic length. However, about the same value (6 Å) is obtained for the structural correlation length using the Martin-Brenig model independent of LiCl concentration.

References

1) S. Alexander and R. Orbach, J. Phys. (Paris), **43**, L-625, (1982).
2) B. Derrida, R. Orbach and K.-W. Wu, Phys. Rev. **B 29**, 6645, (1984).
3) Y. Gefen, A. Aharony and S. Alexander, Phys. Rev. Lett., **50**, 77, (1983).
4) A. Aharony, O. Entin-Wohlman, S. Alexander and R. Orbach, Phil. Mag. **B 56**, 949, (1987).
5) A. Aharony, S. Alexander, O. Entin-Wohlman and R. Orbach, Phys. Rev. Lett. **58**, 132, (1987).
6) G. Deutscher, R. Maynard and O. Parodi, Europhys. Lett. **6**, 49, (1988).
7) S. Alexander, C. Laermans, R. Orbach and H.M. Rosenberg, Phys. Rev. **B 29**, 6645, (1984); ibid 28,4615,1983.
8) J.E. Graebner, B. Golding and L.C. Allen, Phys. Rev. **B 34**, 5696, (1986); ibid **34**, 5788, (1986).
9) E. Courtens, R. Vacher, J. Pelous and T. Woignier, Europhys. Lett. **6**, 245, (1988).
10) Y. Tsujimi, E. Courtens, J. Pelous and R. Vacher, Phys. Rev. Lett. **60**, 2757, (1988).
11) A. Fontana, F. Rocca and M.P. Fontana, Phys. Rev. Lett. **58**, 503, (1987).
12) A. Avogadro, S. Aldrovandi and F. Borsa, Phys. Rev. **B 33**, 5637, (1986).
13) D.P. Button, R.P. Tandon, H.L. Tuller, and D.R. Uhlmann, Solid State Ionics **5**, 655, (1981)
14) W. Soppe, J. Kleerebezem and H.W. den Hartog, J. Non-Cryst. Solids **93**, 142, (1987)
15) L. Börjesson, L.M. Torell, U. Dahlborg and W.S. Howells, Phys. Rev. **B** (in press).
16) L. Börjesson, L.M. Torell and W.S. Howells, Phil.Mag. **B** (in press).
17) A. Martin and W. Brenig, Phys. Stat. Solidi **B 64**, 163, (1974).
18) R.J. Nemanich, Phys. Rev. **B 16**, 1655, (1977).
19) L. Börjesson and L.M. Torell, Solid State Ionics, **25**, 85, (1987).
20) G. Carini, M. Cutroni, A.Fontana, G. Mariotto, and F. Rocca, Phys.Rev. **B 29**, 3567, (1984).
21) R. Shuker and R.W. Gammon, Phys. Rev. Lett. **25**, 222, (1970).
22) A. Boukenter, B. Champagnon, E. Duval, J. Dumas, J.F. Quinson, and J. Serughetti, Phys. Rev. Lett. **57**, 2391, (1986)
23) J. Lorösch, M. Couzi, J. Pelous, R. Vacher, and A. Levasseur, J. Non-Cryst. Solids **69**, 1, (1984)

Index of Contributors

FSC
www.fsc.org
MIX
Papier aus verantwortungsvollen Quellen
Paper from responsible sources
FSC® C105338

If you have any concerns about our products,
you can contact us on
ProductSafety@springernature.com

In case Publisher is established outside the EU,
the EU authorized representative is:
Springer Nature Customer Service Center GmbH
Europaplatz 3, 69115 Heidelberg, Germany

Printed by Libri Plureos GmbH
in Hamburg, Germany